EXPERIMENTAL PHYSICAL CHEMISTRY

EXPERIMENTAL PHYSICAL CHEMISTRY

A Laboratory Textbook

Arthur M. Halpern
Northeastern University

James H. Reeves
University of North Carolina at Wilmington

Scott, Foresman/Little, Brown College Division
SCOTT, FORESMAN and COMPANY
Glenview, Illinois   Boston   London

Library of Congress Cataloging-in-Publication Data

Halpern, Arthur M.
  Experimental physical chemistry.

  Includes index.
  1. Chemistry, Physical and theoretical—Laboratory
manuals. I. Reeves, James H. II. Title.
QD457.H27   1988   541.3'07'8   87–35760
ISBN 0–673–39786–6

1 2 3 4 5 6 7 8 9 10 — SEM — 94 93 92 91 90 89 88

Printed in the United States of America

PREFACE

Many students feel that physical chemistry is among the most difficult and challenging courses in science and engineering. Faculty members, on the other hand, often say it is one of the more rewarding courses to teach. Unifying physical concepts with mathematical rigor, which is the essence of physical chemistry, has a beauty that often escapes students, perhaps because they are so briefly exposed to the subject. The physical chemistry laboratory, we believe, has a crucial function—placing the often abstract concepts encountered in lectures in an experimental context. The laboratory not only introduces students to the methods enabling them to do rigorous measurement, record keeping, and report writing; it also emphasizes the *experimental* origin of scientific knowledge and development. Topics such as the Joule-Thomson coefficient, the second virial coefficient, the standard heat of formation, the standard cell potential, and an absolute rate constant take on new meaning when students measure them in the laboratory. We believe that such direct observation and experimentation help students begin to appreciate (and even enjoy) how compelling the mathematical description of chemistry is, and thus how physical chemistry is the key to scientific understanding and prediction.

We have felt the need for a direct, up-to-date laboratory textbook for physical chemistry, presenting the fundamental topics encountered in the lecture course and giving sufficient guidance to students as they perform the experiments in the laboratory. We do expect students to read before the laboratory session the theoretical background and procedural guidelines provided with each experiment, but we have observed that many students benefit from easy access to the book in the laboratory. A number of the experiments have duplicate data sheets; use of this material is, of course, optional. Although we realize that these devices are controversial, we believe that many students need this guidance in learning how to organize their recorded data and observations, especially because they are simultaneously learning and following new experimental techniques.

We intend this book to provide a few experiments, both basic and advanced, to be carried out with modern apparatus that still does not obscure the fundamental experimental methodology. A highly accurate result attained with a state-of-the-art technique will mean little to a student who must treat the experimental apparatus as a black box. We also recognize that students need to use the ubiquitous laboratory microcomputer in direct acquisition, management, and analysis of data. Accordingly, several experiments can be done with a dedicated microcomputer (as optional equipment); this device essentially replaces the strip-chart recorder, eliminating much of the unnecessary drudgery in data reduction. We also recommend that commercially available, general-purpose software be used for rigorous and meaningful analysis of data.

The constraints on developing and presenting these experiments are considerable: the experiments must *work*; they must provide quantitatively acceptable results, and do so in finite time; topics must be fundamental and experimental methods should be as simple as possible. Many of these experiments are derived from articles that appeared in the primary scientific literature; a few are based on doctoral dissertations. The finite time constraint is particularly treacherous because many of the experiments deal with measurements of systems that are presumed to be *at equilibrium*. Equilibrium is much easier to conceptualize than to realize. If some of the experiments on phase equilibrium come close to failing, it is usually because measurements are made on systems that are not sufficiently close to equilibrium.

Awareness of safety and liability issues has expanded in recent years in many areas of society. Accordingly, we include brief, specific safety precautions for each experiment. Surely, prudence and personal responsibility will lead an experimentalist to follow procedures that are commensurate with contemporary safety standards. Although it is the laboratory instructor's responsibility to enforce proper safety procedures and rules, we hope that a sense of personal and professional responsibility can be inculcated at the same time.

The experiments in this book have been performed and evaluated during the past several years in the junior-level physical chemistry laboratory courses at Northeastern University and the University of North Carolina at Wilmington. In preparing these experiments for publication, we have benefited from the experience and response of the students who have, themselves, acted as objects of an experiment. Much of the ability to fuse this experience into reasonably "converged" procedures stems from the many end-of-term post mortems conducted with the help of the graduate teaching assistants, and we acknowledge with pleasure these helpful discussions with: Dr. M. Nosowitz, Dr. S. L. Frye, Dr. M. W. Legenza, and Messrs. C. J. Ruggles, A. Taaghol, and S. Gozashti (NU). Preparing this book would have been much more onerous, and perhaps impossible, without encouragement from our colleagues. We are grateful to Professors J. L. Roebber, R. N. Wiener, P. W. LeQuesne, and Mr. B. J. Lemire (NU) and Professor C. R. Ward (UNCW), and especially to Dr. J. Wronka (NU) whose enthusiasm and guidance were helpful during the formative stages of the project.

The manuscript was improved by the constructive advice and suggestions of the referees and we wish to acknowledge their help: Leslie Forster, University of Arizona; Stephen Gregory, College of Notre Dame of Maryland; Norman Hackerman, Rice University; James Kaufman, Curry College; Charles Marzzacco, Rhode Island College; Joseph Noggle, University of Delaware; John Schug, Virginia Tech; and Edmund White, Southern Illinois University.

ARTHUR M. HALPERN
Boston, Mass.

JAMES H. REEVES
Wilmington, N.C.

CONTENTS

EXPERIMENTAL
PHYSICAL
CHEMISTRY

□ PART ONE

Error Analysis in the Physical Chemistry Laboratory

TREATMENT OF EXPERIMENTAL DATA

Knowledge that is gained by experimentation is derived through a series of observations, measurements, and analyses. The quality of this knowledge (i.e., its reliability) depends not only on the quality of the measurements but also on the rigor and objectivity with which the data are analyzed. Central to the critical analysis of experimental data is a thorough understanding of the sources and magnitudes of the errors associated with it. This section discusses some fundamental considerations about errors: their origin, their impact on data analysis, and several methods of evaluating and reporting them.

ERRORS

We all make mistakes, and sometimes the equipment we use makes them for us. There are *always* limitations in discovering the "ultimate reality" about a system that we wish to characterize. These limitations cause discrepancies between our experimental result and the "true value" of the quantity of interest. Indeed, the fact that the true value is defined in statistics as the mean of the sample population composed of an *infinite* number of measurements implies that any result obtained from a finite set of data may be in error to some extent.

It is helpful to separate the sources of error into two categories, *systematic* and *random*. A systematic error arises from a bias that is placed on the measurement either by the instrument itself or by a consistently improper method of reading the instrument. For example, a balance not properly zeroed (tared) will provide mass readings that are erroneously high (or low) despite the care with which the balance is read. The same is true for an uncalibrated thermometer or pressure gauge. Thus, by virtue of the misuse of the measuring instrument, all measurements are thrown off in the same direction; they are biased.

Random errors are different. They arise from *intrinsic* limitations in the sensitivity of the instrument and in the ability of the user to interpret the instrument's output. Suppose that a series of mass measurements are made on an object with a very sensitive digital balance and the object is carefully removed and replaced between measurements in such a way that *no* actual change in mass takes place.

The mass indicated by the balance is likely to be slightly different each time the measurement is made. These fluctuations in the reading would be caused by the inherent inability of the electronic and mechanical components of the balance to function absolutely reproducibly. If one used an analog balance requiring the interpretation of a dial or scale (for the last one or two significant figures), additional uncertainty would be introduced because of the random errors associated with converting the dial reading to a number. This process, which is known as an analog to digital conversion, is automatically performed by the digital balance with an integrated circuit known as an A/D converter. Although the digitization process can produce random errors in either case, the errors are likely to be greater when the conversion is done "by eye" on the analog device.

It is possible to observe random errors when, for example, the output of a noisy electronic signal is displayed on a meter or oscilloscope. The pattern observed in random fluctuations of this type (i.e., its distribution of values) can be characterized in a definite mathematical way.

Systematic and random errors can be related to the terms *accuracy* and *precision*. An experimental measurement has high precision if the random errors (fluctuations) are small. Many significant figures are justified. A measurement is accurate if there are small systematic errors. In this case there is no intrinsic bias to the measurement, and its value approaches the true or accepted one. It is important to realize that there is no necessary relationship between accuracy and precision, since an experiment can have small random errors and still give inaccurate results due to large systematic errors. In the same way, the results obtained from an experiment plagued by large random errors may still be accurate in the sense that the "true" or accepted value lies within the limits of error reported.

Two other types of errors should also be mentioned. The first is sometimes referred to as a "blunder." This characterization speaks for itself. Whether this error originates with the experimenter or with the equipment, a blunder is a mistake. It is often evidenced by the fact that a particular data point "sticks out like a sore thumb." It is considered acceptable to throw this data point out, and there are rigorous statistical tests that can be applied to justify this decision. One of these is discussed below. Unlike random errors, a blunder can, in principle, be avoided through more careful attention to detail. Systematic errors can also be eliminated through proper instrument calibration and more careful experimental practices.

The second type of error, which might be referred to as a "model error," is less obvious and potentially more serious than a blunder. In this case the assumption that the system under study will behave in the predicted fashion proves to be incorrect. Suppose, for example, that a student were to measure the pressure-volume behavior of some real gas at constant temperature in an effort to demonstrate Boyle's law. If the experiment were done with proper care, deviations from the expected behavior (a hyperbola following the equation $PV = $ constant) would be observed for the high-pressure measurements. The problem in this case is not poor data but the ideal gas *model,* which is not valid under these conditions. Model errors can be difficult to detect because it is often easier to mistrust your data than to question the validity of the theory that is supposed to be demonstrated by them. It must be remembered, however, that most of the significant new theories in sciences have come from individuals who believed in their data and challenged the conventional wisdom on which an incorrect theory was based. Data are the facts on which science is built, and theories that do not conform to the facts *must* be modified or rejected.

ESTIMATING UNCERTAINTIES

Mean and Standard Deviation

If a measurement of some property p of a system is repeated several times, a statistical treatment of the data can provide an indication of the reliability of the result. The *mean* or average value of p on which N independent measurements have been made is defined as

$$<p> = p_1 + p_2 + p_3 + \cdots + p_N = \frac{1}{N} \sum_{i=1}^{N} p_i, \qquad (1.1)$$

where p_i is the value of the ith determination. If the errors associated with the measurements are completely random in nature, the data are said to follow a *Gaussian* distribution and $<p>$ represents the best estimate of the true value of the property that can be obtained from the data. The basis for this assertion, which is known as the principle of maximum likelihood, is discussed in the appendix to this section.

It is possible to consider how each of the measurements p_i compares with the mean value by computing its deviation d_i, where

$$d_i = p_i - <p>. \qquad (1.2)$$

If this is done for each value in the set p_i, the mean of the deviation, $<d>$, can be computed:

$$<d> = \frac{1}{N} \sum_{i=1}^{N} d_i = \frac{1}{N} \sum_{i=1}^{N} (p_i - <p>) = <p> - <p> = 0. \qquad (1.3)$$

That the mean deviation of any data set will be zero follows from the fact that there is as much chance that a measurement will be higher than the mean by an arbitrary amount as there is that it will be lower by the same amount. Thus the mean deviation provides no meaningful statistical information.

Important statistical information *can* be obtained from the sum of the *squares* of the deviations. The sample *variance*, s^2, which is defined as

$$s^2 = \frac{1}{N-1} \sum_{i=1}^{N} d_i^2 = \frac{1}{N-1} \sum_{i=1}^{N} (p_i - <p>)^2, \qquad (1.4)$$

provides an estimate of extent of dispersion or scatter of the data about the mean. The factor $(N-1)$ is used in the denominator since in a set of N measurements, there are $N-1$ *independent* pieces of information besides the mean. If the parenthetical expression on the right-hand side of equation (1.4) is expanded, the sample variance is shown to be related to the difference between the mean square of the p_i and the square of the mean:

$$s^2 = \frac{N}{N-1} (<p^2> - <p>^2). \qquad (1.5)$$

It is a fundamental characteristic of multiple measurements of experimental data that the mean square value ($<p^2>$) is different from the square of the mean ($<p>^2$).

The *standard deviation* is obtained by taking the square root of the variance:

$$s = \left[\frac{1}{N-1} \sum_{i=1}^{N} (p_i - <p>)^2 \right]^{1/2} \qquad (1.6)$$

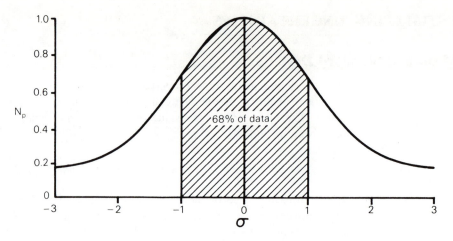

FIGURE 1.1   Random distribution of p_i values. The ordinate represents the relative frequency of observing a given value of p, and the abscissa shows the values of p relative to the mean, $<p>$, in standard deviation units.

Root mean square (rms) quantities similar to s are frequently encountered in science and engineering. Two examples are rms speed in the kinetic theory of gases and rms voltage in electronics. The standard deviation is particularly useful in statistical analysis because it provides a quantitative measure of the dispersion of the experimental data about the mean. Thus, if N is large and the data are randomly distributed, 68 percent of the data fall within one standard deviation unit of $<p>$ (cf. Figure 1.1).

EXAMPLE 1:   Six independent measurements of the barometric pressure were taken during a laboratory period. The values obtained were 762.2 mm, 761.8 mm, 762.0 mm, 761.5 mm, 762.2 mm, and 760.0 mm. Calculate the average value and standard deviation of these data.

Solution:

$$<P> = \frac{762.2 + 761.8 + 762.0 + 761.5 + 762.2 + 760.0}{6} = 761.6$$

$$d_1 = 0.6 \quad d_2 = 0.2 \quad d_3 = 0.4 \quad d_4 = -0.1 \quad d_5 = 0.6 \quad d_6 = -1.6$$

$$s = \left\{ \frac{(0.6)^2 + (0.2)^2 + (0.4)^2 + (-0.1)^2 + (0.6)^2 + (-1.6)^2}{5} \right\}^{1/2}$$

$$= 0.84 \qquad \qquad \square$$

Standard Errors and Confidence Limits

If two series of measurements are made on the same system, the average value of p determined from the first series will (probably) be different from the value obtained from the second. If a large number of these series were performed, the respective mean values would be symmetrically distributed about the "true value," and the standard deviation of the distribution, known as the *standard error of the mean*, $s_{<p>}$, would be given by the relationship

$$s_{<p>} = \frac{s}{(N-1)^{1/2}}. \tag{1.7}$$

TABLE 1.1   Partial Table of Critical *t* Factors

Degrees of Freedom	*t* (90%)	*t* (95%)	*t* (99%)
1	6.31	12.7	63.7
2	2.92	4.30	9.92
3	2.35	3.18	5.84
4	2.13	2.78	4.60
5	2.01	2.57	4.03
6	1.94	2.45	3.71
8	1.86	2.31	3.36
10	1.81	2.23	3.17
15	1.75	2.13	2.95
20	1.72	2.09	2.85
30	1.70	2.04	2.75
∞	1.64	1.96	2.58

Once $s_{<p>}$ is determined, a confidence limit λ can be obtained. This quantity defines the range on both sides of $<p>$ within which the "true value" can be expected to be found with a given degree of "confidence." The range is the product of $s_{<p>}$ with t, a factor whose value depends on N and the degree of confidence desired. A brief t table is provided (Table 1.1). More complete tables are available in elementary statistics books such as Downie and Starry.[1]

EXAMPLE 2:   Calculate the standard error of the mean and the 95 percent confidence limit for the data presented in Example 1.

Solution:

$$s_{<P>} = \frac{0.84}{\sqrt{5}} = 0.376 \qquad \lambda \text{ (95 percent)} \qquad (0.376)(2.57) = 0.96 \qquad \square$$

Thus, the value of the barometric pressure could be reported to be 761.6 ± 0.96 (95 percent confidence).

Discarding Data

In the set of data used in Examples 1 and 2, it is apparent that the sixth pressure reading is substantially smaller than the others. One method for deciding whether this measurement can be justifiably discarded is to evaluate the mean pressure without this suspect point and then determine if this point deviates from the mean by more than four times the *average deviation* of the other points.[2] The average deviation is defined as

$$d_{av} = \frac{1}{N} \sum_{i=1}^{N} |d_i|. \tag{1.8}$$

Two simple additional rules are: (1) do not discard more than one value in five, and (2) do not discard two data values if they are the same. In the present case, the $<P> = 761.9$, and the average deviation of the set of five data points is

$$d_{av} = \frac{1}{5} (0.16 + 0.14 + 0.06 + 0.44 + 0.26) = 0.21.$$

The last point deviates from the new mean by more than $4(0.21) = 0.84$, suggesting that it is in fact reasonable to discard it. The new values of s, $s_{<p>}$, and λ (95 percent) are 0.30, 0.15, and 0.41, respectively (the t factor in this case is 2.78). Thus, it would be appropriate to report the pressure to be 761.9 ± 0.41.

ESTIMATING ERRORS: INDIVIDUAL JUDGMENT

Although statistical analysis provides methods for determining the limits of reliability of experimental data, it is not always possible to repeat measurements a sufficient number of times to generate credible statistical information. Often the uncertainties associated with measurements must be estimated by the experimenter. Factors such as the specifications and capabilities of the equipment, fluctuations in experimental conditions, the nature and purity of the substances, and the experimenter's skill must be taken into account. Assigning realistic uncertainties to experimental data is a difficult and often frustrating task. Nevertheless, accurate error estimates are essential to the professional presentation of experimental results. A result has little scientific credibility if a reasonable estimate of its error limits is not given.

In the physical chemistry laboratory, it is often possible to compare the uncertainties predicted from statistical considerations with those generated through "educated guesses." Moreover, since the accepted value of the quantity sought in the experiment is usually available in the literature, the *accuracy* of the results can be evaluated. In this sense the physical chemistry laboratory provides an opportunity to develop the skills required to evaluate the quality and reliability of experimental data.

A first approach in estimating the uncertainties in measurements is to assess the limitations of the equipment used. Often the manufacturer provides specifications or guidelines. For instance, the percent error to be expected when a volume is measured using volumetric glassware is frequently etched on the glassware itself. In other cases, the manufacturer provides literature or documentation containing information about the accuracy and reproducibility that can be obtained with the equipment. In the absence of such information, it is often assumed that the limit on the precision is the smallest increment that can be accurately read or estimated by the experimenter. For example, a thermometer graduated in degrees can typically provide temperature estimates to the nearest 0.2 to 0.5 degree, and the volume reading of liquid contained in a 50-mL burette can usually be estimated to the nearest 0.02 mL.

Error estimates could be improved if the results obtained from previous experiments using the same equipment were available to provide some insight into its reliability. This is readily accomplished by calibrating the instrument first (i.e., by carrying out experiments on known systems before the system of interest is studied).

PROPAGATION OF ERRORS

Once reliable estimates of the uncertainties associated with each of the individual measurements have been made, their combined effect on the quantity of interest must be assessed. This procedure, known as the propagation of errors, is best studied by starting with some simple examples.

Suppose the perimeter P of a rectangle is to be determined from the measurement of the length of the sides a and b. Since $P = 2(a + b)$, the uncertainty in P, ϵ_P, which is present by virtue of uncertainties in the measurements a and b (ϵ_a and ϵ_b), can be estimated from

$$P + \epsilon_P = 2(a + \epsilon_a + b + \epsilon_b)$$

by subtracting $P = 2(a + b)$ from both sides:

$$\epsilon_P = 2(\epsilon_a + \epsilon_b). \tag{1.9}$$

This calculation represents a "worst case" analysis since it is assumed that the errors in both measurements will affect the determination of P in the same way (ϵ_a and ϵ_b have the same sign). If these uncertainties are the results of random errors, however, it is equally as likely that ϵ_a and ϵ_b will cancel one another (be of different sign) as it is that they will be additive. An approach that compensates for this possibility and thus provides a more realistic measure of the actual uncertainty is to square equation (1.9)

$$\epsilon_P^2 = 4(\epsilon_a^2 + \epsilon_b^2 + 2\epsilon_a\epsilon_b),$$

and drop the cross term ($2\epsilon_a\epsilon_b$). The justification for ignoring this term is the fact that as the product of two independent uncertainties, its value may be positive or negative for any given measurement, and thus its net effect over a series of measurements is negligible. The uncertainty in P thus becomes

$$\epsilon_P = 2(\epsilon_a^2 + \epsilon_b^2)^{1/2}. \tag{1.10}$$

This example illustrates the general principle that the uncertainty in a result derived from the addition or subtraction of measured variables depends on the *absolute* uncertainties in the measurements. The situation is different when the calculation involves multiplication or division. Consider, for example, the area (A) of the rectangle ($A = ab$). Here the uncertainty ϵ_A can be calculated from

$$A + \epsilon_A = (a + \epsilon_a)(b + \epsilon_b) = ab + a\epsilon_b + b\epsilon_a + \epsilon_a\epsilon_b.$$

By ignoring the last term (as the product of two independent uncertainties), dividing by $A = ab$, and squaring the result (see above), we get

$$\frac{\epsilon_A}{A} = \left[\left(\frac{\epsilon_a}{a} \right)^2 + \left(\frac{\epsilon_b}{b} \right)^2 \right]^{1/2} \tag{1.11}$$

The uncertainty calculation in this case involves the relative or fractional errors (ϵ_a/a and ϵ_b/b) in each of the measurements.

These examples demonstrate specific relationships that apply to error analysis when results are derived by addition and subtraction *only,* or by multiplication and division *only.* It is possible to establish a general method of evaluating these relationships. Suppose that the quantity Q is a function of a series of n measurables, x_i: $Q = Q[x_1, x_2, x_3, \ldots, x_n]$, and that each measurement differs from its true value, μ_i, by an amount θ_i [$\theta_i = x_i - \mu_i$]. In order to determine the effect that each θ_i has on the result, we ask the question "by how much does the value of Q change when x_i changes (i.e., $\Delta Q/\Delta x_i$)?" The error in Q ($\theta_Q = Q - \mu$) resulting from θ_i is then calculated from the expression

$$\theta_{Q,i} = \left(\frac{\Delta Q}{\Delta x_i} \right) \theta_i,$$

or, in the limit where both ΔQ and Δx_i are small,

$$\theta_{Q,i} = \left(\frac{\partial Q}{\partial x_i}\right)\theta_i.$$

To determine the combined effect of all the errors [θ_1 through θ_n] on the value of θ_Q, the individual contributions are added:

$$\theta_Q = \sum_{i=1}^{n} \theta_{Q,i} = \left(\frac{\partial Q}{\partial x_1}\right)\theta_1 + \left(\frac{\partial Q}{\partial x_2}\right)\theta_2 + \cdots + \left(\frac{\partial Q}{\partial x_n}\right)\theta_n. \tag{1.12}$$

Use of equation (1.12) is restricted to the case in which the θ_i's, the *actual error* in each of the measurements, are known. θ_i cannot be obtained from a finite series of measurements since μ_i, the *true value* of the measurable x_i, is the average of an *infinite* number of measurements. Thus, some other measure of the uncertainty associated with each of the observations must be used. If N measurements are made on Q, equation (1.12) can be recast to evaluate the deviation in each calculation of Q that results from deviations in measurements from which it was derived. For the jth determination of Q,

$$d_{Q(j)} = \left(\frac{\partial Q}{\partial x_1}\right)d_1(j) + \left(\frac{\partial Q}{\partial x_2}\right)d_2(j) + \cdots + \left(\frac{\partial Q}{\partial x_n}\right)d_n(j), \tag{1.13}$$

where $d_1(j) = x_1(j) - <x_1>$, for example. From the definition of sample variance [cf. equation (1.4)]

$$s_Q^2 = \left[\frac{1}{N-1}\right] \sum_{j=1}^{N} d_{Q(j)}^2.$$

Substitution of equation (1.13) into this result gives

$$\begin{aligned}
s_Q^2 &= \left[\frac{1}{N-1}\right] \sum_{j=1}^{N} \left[\left(\frac{\partial Q}{\partial x_1}\right)d_1(j) + \left(\frac{\partial Q}{\partial x_2}\right)d_2(j) + \cdots + \left(\frac{\partial Q}{\partial x_n}\right)d_n(j)\right]^2 \\
&= \left[\frac{1}{N-1}\right]\left[\left(\frac{\partial Q}{\partial x_1}\right)\sum_{j=1}^{N} d_1(j) + \left(\frac{\partial Q}{\partial x_2}\right)\sum_{j=1}^{N} d_2(j) + \cdots \right. \\
&\quad \left. + \left(\frac{\partial Q}{\partial x_n}\right)\sum_{j=1}^{N} d_n(j)\right]^2 \\
s_Q^2 &= \left[\left(\frac{\partial Q}{\partial x_1}\right)s_1 + \left(\frac{\partial Q}{\partial x_2}\right)s_2 + \cdots + \left(\frac{\partial Q}{\partial x_n}\right)s_n\right]^2.
\end{aligned} \tag{1.14}$$

The squared sum (in brackets) in equation (1.14) is composed of two sets of terms, the sum of squares, $(\partial Q/\partial x_i)^2 s_i^2$, and the sum of cross terms, $(\partial Q/\partial x_i)(\partial Q/\partial x_k)s_i s_k$. Since the cross terms may be positive or negative, their sum will be significantly smaller than the sum of squared terms, and their effect on s_Q^2 can be ignored. Thus, s_Q^2 can be obtained from

$$s_Q^2 = \left(\frac{\partial Q}{\partial x_1}\right)^2 s_1^2 + \left(\frac{\partial Q}{\partial x_2}\right)^2 s_2^2 + \cdots + \left(\frac{\partial Q}{\partial x_n}\right)^2 s_n^2. \tag{1.15}$$

Although equation (1.15) was derived assuming that N observations were carried out on *each* measurable x_i, it has also been shown to be valid when the number of observations made on each of the x_i's is different. Thus equation (1.15) has wide applicability and can be expressed in the more general form

$$\epsilon_Q^2 = \left(\frac{\partial Q}{\partial x_1}\right)^2 \epsilon_1^2 + \left(\frac{\partial Q}{\partial x_2}\right)^2 \epsilon_2^2 + \cdots + \left(\frac{\partial Q}{\partial x_n}\right)^2 \epsilon_n^2, \qquad (1.16)$$

where ϵ_i is the best estimate of the uncertainty in x_i.

The application of equation (1.16) to calculations that require only addition and subtraction produces an expression equivalent in form to equation (1.10), while multiplication and division can be treated using equations similar in form to equation (1.11). More complicated expressions must be treated on an individual basis by the direct application of (1.16). The following examples are designed to clarify these points.

EXAMPLE 3:   Calculate the molecular weight of an unknown gas if 5.05 g of the gas is contained in a 1.4-L vessel at 765 mm pressure and 22°C. The mass measurement was carried out on a triple-beam balance that may be read to the nearest 0.01 g, the volume has a precision of 0.1 L, the pressure was measured using a transducer with an absolute accuracy of 1 torr, and the temperature was read to the nearest degree.

Solution: The molecular weight (MW) may be calculated from the relationship

$$\text{MW} = \frac{m}{n} = \frac{mRT}{PV} = \frac{(5.05)(0.08206)(295)}{(1.4)(765/760)} = 86.75 \text{ g mol}^{-1} \qquad \square$$

In order to determine the net uncertainty in this result using propagation of errors, equation (1.16) can be utilized.

$$\epsilon_{MW} = \left[\left(\frac{RT}{PV}\right)^2 \epsilon_m^2 + \left(\frac{mR}{PV}\right)^2 \epsilon_T^2 + \left(\frac{-mRT}{VP^2}\right)^2 \epsilon_P^2 + \left(\frac{-mRT}{PV^2}\right)^2 \epsilon_v^2\right]^{1/2}.$$

Dividing by MW $= mRT/PV$, a result similar to equation (1.11) is obtained:

$$\frac{\epsilon_{MW}}{\text{MW}} = \left[\left(\frac{\epsilon_m}{m}\right)^2 + \left(\frac{\epsilon_T}{T}\right)^2 + \left(\frac{\epsilon_P}{P}\right)^2 + \left(\frac{\epsilon_V}{V}\right)^2\right]^{1/2}.$$

Thus, the relative error in the molecular weight is seen to be the square root of the sum of the squares of the relative errors in m, T, P, and V. R does not appear in this equation since it is a constant whose value is known very accurately. Thus, it is always possible to carry enough significant figures in R so its uncertainty in a particular calculation is insignificant compared with those of the other parameters. In this example, the relative uncertainty in the molecular weight is given by

$$\frac{\epsilon_{MW}}{\text{MW}} = \left[\left(\frac{0.01}{5.05}\right)^2 + \left(\frac{1}{295}\right)^2 + \left(\frac{1}{765}\right)^2 + \left(\frac{0.1}{1.4}\right)^2\right]^{1/2} = 0.0715,$$

and $R = 0.08206$ contributes a relative uncertainty of $(0.00001/0.08216) = 0.00012$, a value that is negligible compared with 0.0715. The value of the molecular weight that should be reported from these data is 87 ± 6.2. Note that the primary source of error in this experiment is the volume measurement since its relative uncertainty far exceeds those of the other variables. Thus, it would be fruitless to perform the mass measurement on a more sensitive balance or to use a more accurate thermometer unless the method of determining the volume of the container were significantly improved.

EXAMPLE 4:   Suppose a parameter of interest W was related to the measurable parameters x and y and the constant a by the equation

$$W(x,y) = \frac{xy}{x + a}.$$

Calculate W and its uncertainty for the case $x = 0.55 \pm 0.03$, $y = 20 \pm 1$, and $a = 0.05 \pm 0.01$.

Solution: By direct calculation, $W = 18.33$. Equation (1.16) may be applied to determine the uncertainty of this number.

$$\epsilon_w^2 = \left[\frac{y}{x + a} - \frac{xy}{(x + a)^2}\right]^2 \epsilon_x^2 + \left[\frac{x}{x + a}\right]^2 \epsilon_y^2 + \left[\frac{xy}{(x + a)^2}\right]^2 \epsilon_a^2$$

$$\frac{\epsilon_w^2}{W^2} = \left[\frac{a\epsilon_x}{x(x + a)}\right]^2 + \left(\frac{\epsilon_y}{y}\right)^2 + \left(\frac{\epsilon_a}{x + a}\right)^2 = 0.0028 \qquad \square$$

Thus, $W = 18 \pm 1$.

EXAMPLE 5:   In 1844, J. L. Poiseuille showed that the flow rate, v, of a liquid through a straight cylindrical tube of radius, r, and length, ℓ is given by the relationship

$$v = \frac{\pi r^4 P}{8\eta\ell}$$

(provided that the flow is laminar), where P is the pressure difference between the ends of the tube and η is the viscosity of the liquid. Using this relationship, calculate the viscosity of a liquid (along with its uncertainty) if the liquid flows through a tube of length 10 ± 1 cm and radius 1 ± 0.2 mm at a rate of 170 ± 20 cm^3 min^{-1} while under a pressure difference of 400 ± 10 Pa.

Solution: The calculation must be done with all variables in the same units. Converting to cgs units, $P = 4000 \pm 100$ dynes cm^{-2}, $r = 0.1 \pm 0.02$ cm, $v = 2.83$ cm^3 s^{-1}. Solving for η:

$$\eta = \frac{\pi(0.1)^4(4000)}{8(2.83)(10)} = 5.55 \times 10^{-3} \text{ g cm}^{-1}\text{ s}^{-1} = 0.555 \text{ cP}.$$

From equation (1.16),

$$\left(\frac{\epsilon_\eta}{\eta}\right)^2 = \left(\frac{4\epsilon_r}{r}\right)^2 + \left(\frac{\epsilon_P}{P}\right)^2 + \left(\frac{\epsilon_v}{v}\right)^2 + \left(\frac{\epsilon_l}{l}\right)^2$$

$$= \left[4\left(\frac{0.2}{1}\right)\right]^2 + \left(\frac{10}{400}\right)^2 + \left(\frac{20}{170}\right)^2 + \left(\frac{1}{10}\right)^2$$

$$= 0.664$$

$$\frac{\epsilon_\eta}{\eta} = 0.815 \qquad \square$$

Thus, $\eta = (5.5 \pm 4.5) \times 10^{-3}$ poise. Note that the presence of the r^4 term in the Poiseuille equation causes the uncertainty in r to dominate the overall uncertainty calculation. Thus, although the radius was known to a precision of 20 percent, it contributed an 80 percent uncertainty to the viscosity.

REPORTING OF DATA/SIGNIFICANT FIGURES

The method of reporting data illustrated in these examples is to be preferred in most situations. This method gives a clear indication of the reliability of the information reported. However, if scientific data are presented without error limits, care must be taken to report these values to the correct number of significant figures. One very poor (and unprofessional) habit practiced by many beginners is to report a result using all the figures that appear on the calculator display or computer output. Only the digits *known with certainty* should appear in the number that is reported. Consider, for example, a situation in which the length of a rod measured with a meter stick is found to be 0.53 m. This value indicates that the length is between 0.525 and 0.535 m. Citing a value of 0.53 m is equivalent to specifying the length to be 0.53 ± 0.005 m. Moreover, it would be inadvisable to report this measurement to be 530 mm, since that suggests that the precision is ±0.5 mm (0.0005 m). This ambiguity can be avoided if data of this sort are reported using standard scientific notation. Thus, 5.3×10^{-1} m is *precisely* the same result as 5.3×10^2 mm. Note that this method of indicating precision suffers from the limitation that it generally exaggerates the uncertainty in the measurement. If the rule stated above were to be strictly followed, the result of Example 3 would have to be reported as 2×10^1, indicating an uncertainty of ±5, rather than ±1.1, as indicated by the propagation of errors.

When two or more numbers are combined, the precision of the result is limited to that of the *least* precise number used in the calculation. Again, a distinction must be made between addition/subtraction and multiplication/division. When data are added or subtracted, the number of significant figures in the result is determined by considering the *absolute* precision of each of the data points. Consider the calculation

	Absolute Precision
100.7	±0.05
0.61	±0.005
+226	±0.5
327	±0.5

Here the number of significant figures in the calculation is limited by the data point with the lowest absolute precision, 226.

EXAMPLE 6:   Use the propagation of errors technique to determine the predicted uncertainty in this result.

Solution: Applying equation (1.16) to this result,

$$\epsilon = [(0.5)^2 + (0.05)^2 + (0.005)^2]^{1/2} = 0.50 \qquad \square$$

Note that, as predicted, the largest absolute uncertainty dominates this calculation.

If the same three numbers were multiplied together, their *relative* or fractional precisions would be the deciding factor.

	Relative Precision	
100.7	0.05/100.7 =	0.00050
0.61	0.005/0.61 =	0.0082
×226	0.5/226 =	0.0022
1.4×10^4	500/14,000 =	0.036

In this case the relative precision of the result (3.6 percent) is larger than those of any of the data. However, if an additional significant figure were added to the result to give $1.40 \times 10^4(\pm 50)$, the implication would be that the product is more precisely known than is the number 0.61 from which it is calculated.

The rule of thumb in multiplication/division is to limit the result to the same number of significant figures present in the factor with the fewest significant figures. As with all rules of thumb, this must be applied with some caution. For example, consider the product of

$$0.99 \times 10.2 = 10.1.$$

If we followed our ''rule'' and reported the result to two significant figures (since 0.99 has only two), the relative uncertainty in the result, 10 ± 0.5, would be 5 percent. By contrast, the least precise number, 0.99, has a relative uncertainty of approximately 0.5 percent (0.005/0.99). Thus, the extra significant figure in the result is justified in this case.

Mathematical operations involving significant figures should be performed using data rounded off to one or two more figures than the number justified in the answer. In this way, rounding errors can be avoided. If the data in Example 3 had been rounded to the nearest whole number *before* the addition was performed, the result would have been

$$101 + 1 + 226 = 328,$$

which is one unit higher than the value without such rounding.

CURVE FITTING BY THE METHOD OF LEAST SQUARES

Often the data obtained in a series of measurements must be mathematically manipulated to provide a desired result. For example, consider a gas that is thought to follow the equation of state

$$Z = \frac{PV}{RT} = B_0 + B_1 P, \tag{1.17}$$

where B_0 and B_1 are constants and the other symbols have their usual meaning. Measurements of Z, called the compressibility factor, are made at each of a series of pressures in an effort to determine B_0 and B_1. Since two values are to be determined, a *minimum* of two measurements must be carried out. In order to improve the reliability of the results, however, it is desirable to acquire several sets of (P, Z) data.

Suppose our experiment provided the data shown in Table 1.2.

TABLE 1.2   Compressibility Factors of a Gas at Four Pressures

P (atm)	Z
1	0.9950
10	0.9675
25	0.9489
50	0.9063

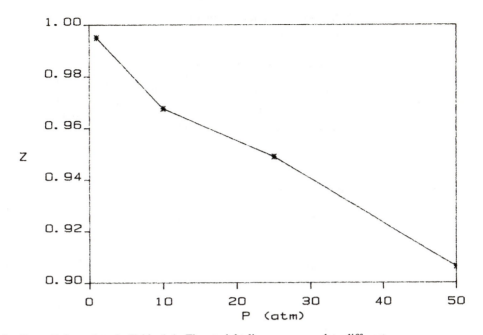

FIGURE 1.2 *Z* vs. *P* from data in Table 1.1. The straight lines accommodate different pairs of data points.

Since equation (1.17) is linear, with a slope of B_1 and y intercept of B_0, the values of these constants can be obtained from the straight line that "best" represents the data. The question is, "how is the best line to be determined?" As illustrated in Figure 1.2, quite different results are obtained for B_0 and B_1 if a straight line is drawn through different pairs of data points. These differences arise because each determination of Z is subject to random errors and will have some uncertainty associated with it. We seek the line that comes as close to each of the data points as possible.

The least squares technique has been shown to provide the best estimate of the value of a quantity p that can be made from a series of measurements p_i (cf. the appendix to this section). This approach can also be used to determine the slope (m) and y intercept (b) that define the best straight line that can be drawn through a series of x,y data. The measured value of some quantity (y_i) will differ from the value predicted by a linear equation ($y = mx + b$) by an amount r_i (called the residual):

$$r_i = y_i - y = y_i - mx_i - b. \qquad (1.18)$$

The least squares approach identifies the best straight line as the one for which R, the sum of the *squares* of the residuals, has the smallest possible value:

$$R = \sum_{i=1}^{N} r_i^2 = \sum_{i=1}^{N} (y_i - mx_i - b)^2. \qquad (1.19)$$

The values of m and b that provide the minimum value for R (and hence the best straight line) will be those which satisfy the equations

$$\left[\frac{\partial R}{\partial m}\right]_b = 2 \sum_{i=1}^{N} (y_i - mx_i - b)(-x_i) = 0, \qquad (1.20)$$

$$\left[\frac{\partial R}{\partial b}\right]_m = 2 \sum_{i=1}^{N} (y_i - mx_i - b)(-1) = 0. \qquad (1.21)$$

Using the following definitions

$$S_x = \sum_{i=1}^{N} x_i \quad S_{x2} = \sum_{i=1}^{N} x_i^2 \quad S_{xy} = \sum_{i=}^{N} x_i y_i \quad S_y = \sum_{i=1}^{N} y_i \quad S_{y2} = \sum_{i=1}^{N} y_i^2,$$

the solutions to these equations (called *normal equations*) are

$$m = \frac{1}{D}(NS_{xy} - S_x S_y) \quad \text{and} \quad b = \frac{1}{D}(S_{x2}S_y - S_x S_{xy}), \quad (1.22)$$

where $D = NS_{x2} - (S_x)^2$. Since the calculations required involve summations, microcomputers and calculators are well suited to the task of determining least squares parameters.

EXAMPLE 7: Calculate the slope and intercept of the best fit straight line that can be generated from the data given in Table 1.2.

Solution: For these data, $S_x = 86$, $S_y = 3.8177$, $S_{x2} = 3226$, $S_{y2} = 3.6479$, and $S_{xy} = 79.7075$. Thus,

$$B_1 = m = \frac{4(79.7075) - (86)(3.8177)}{4(3226) - (86)^2} = -1.723 \times 10^{-3}$$

$$B_0 = b = \frac{(3226)(3.8177) - (86)(79.7075)}{4(3226) - (86)^2} = 0.9915 \qquad \square$$

A display of this line plotted with the data points is shown in Figure 1.3.

Linear regression is a powerful tool that takes the guesswork out of obtaining best fit information. However, like all mathematical tools, it must be used with caution. The technique makes no distinction between data that are truly linear and data that may not be linear but can have a straight line drawn through them. There are statistical quantities that can be calculated to test the goodness of fit of the

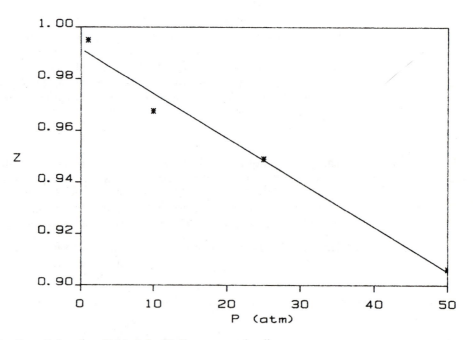

FIGURE 1.3 *Z* vs. *P* data from Table 1.2 with linear regression line.

TABLE 1.3   Data of Table 1.2 with Two Additional Z (P) Points.

P (atm)	Z
1	0.995
10	0.9675
25	0.9489
50	0.9063
100	0.8433
200	0.749

best line. Among the most frequently applied is the correlation coefficient r, which is defined as

$$r = \frac{NS_{xy} - S_x S_y}{[(NS_{x^2} - S_x^2)(NS_{y^2} - S_y^2]^{1/2}}. \qquad (1.23)$$

For a perfectly correlated line, $r = 1$ (if $m > 0$) or -1 (if $m < 0$). If there is no correlation (i.e., if the data were circular or completely random) or if the slope is exactly 0, r will be 0. Using the data from Example 7, $r = -0.991$.

While r can give some indication of goodness of fit, there is no substitute for *graphing the data* and *looking at the result*. Since the errors are assumed to be random, the y_i values should be scattered about the best fit straight line in a random pattern. (Such a pattern is observed in the data from Table 1.2) The residuals should likewise be either positive or negative without any pattern. If the data show a pattern of curvature, it is possible that the results do not conform to the linear model proposed.

Consider the Z vs. P results given in Table 1.3, which include additional measurements obtained at pressures of 100 and 200 atm. Using the least squares procedure, B_0 and B_1 for these data are found to be 0.979 and -1.21×10^{-3}, respectively. These results are significantly different from the values determined in Example 7 (where only the first four points were considered) because the last two measurements were made at pressures where equation (1.17) does not apply (i.e., the Z vs. P data are no longer linear). The failure of the linear model in this case is not indicated by the correlation coefficient (-0.990), since it is virtually identical to that for the linear data of Example 7 (-0.991). When the higher-pressure points are included in the graph (Figure 1.4), however, it is apparent that the points are better represented by a curve.

Errors in Least Squares Analysis

In linear regression analysis, it is assumed that all the error in the measurement lies in the y_i data. If the uncertainties (ϵ_{y_i}) associated with these data are known, equation (1.16) can be used to relate these to the expected uncertainties in m and b. Thus,

$$\epsilon_m = \left[\sum_{i=1}^{N} \left(\frac{\partial m}{\partial y_i} \right)^2 (\epsilon_{y_i})^2 \right]^{1/2} = \left[\frac{1}{D^2} \sum_{i=1}^{N} (Nx_i - S_x)^2 (\epsilon_{y_i})^2 \right]^{1/2} \qquad (1.24)$$

If the ϵ_{y_i} values are the same, equation (1.24) reduces to

$$\epsilon_m = \left(\frac{N}{D} \right)^{1/2} (\epsilon_y), \qquad (1.25)$$

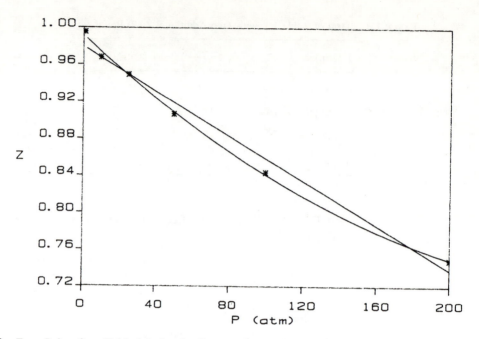

FIGURE 1.4   *Z* vs. *P* data from Table 1.3 showing linear and second-order fits.

where ϵ_y is the value of each of the ϵ_{y_i}. For the intercept

$$\epsilon_b = \frac{1}{D^2} \left[\sum_{i=1}^{N} (S_{x^2} - S_x x_1)^2 (\epsilon_{y_i})^2 \right]^{1/2} \tag{1.26}$$

This relationship reduces to

$$\epsilon_b = \left(\frac{S_{x^2}}{D} \right)^{1/2} (\epsilon_y), \tag{1.27}$$

if all the ϵ_{y_i} are the same.

Next consider a situation in which the uncertainties in y_1 are not known. In this case statistics may be used to estimate these uncertainties from the residuals, a reasonable approach if there are no systematic errors associated with the data. The procedure involves generating the 95 percent confidence limits by first determining the standard deviation of the *residuals* (s_r) according to the relationship

$$s_r = \left(\frac{1}{N-2} \sum_{i=1}^{N} r_i^2 \right)^{1/2}, \tag{1.28}$$

and then multiplying this by the appropriate t factor. The uncertainties are then generated from the equations

$$\epsilon_m = \left(\frac{N}{D} \right)^{1/2} (t s_r) \quad \text{and} \quad \epsilon_b = \left(\frac{S_{y^2}}{D} \right)^{1/2} (t s_r). \tag{1.29}$$

In Example 7,

$$s_r = \left(\frac{7.424 \times 10^{-5}}{2} \right)^{1/2} = 6.092 \times 10^{-3}.$$

Thus,

$$\epsilon_m = \left(\frac{4}{5508} \right)^{1/2} (4.30)(6.092 \times 10^{-3}) = 7.1 \times 10^{-4},$$

and

$$\epsilon_b = \left(\frac{3226}{5508}\right)^{1/2} (4.30)(6.092 \times 10^{-3}) = 0.02.$$

Weighted Linear Regression

Linear regression of the type discussed above relies on a number of critical assumptions. Among the most important is that all the uncertainty in an x,y pair lies with the y variable and that the uncertainty in each of the y values is the same. If the form of the equation required to provide a linear relationship produces an independent variable that contains a significant uncertainty, the set of conditions that produced the minimum in the residual of the line may not be the set that actually minimizes the uncertainty in the data. In this case a "nonlinear" least squares technique such as simplex is recommended.

When measurements of the dependent variable incur different uncertainties, a *weighted* least squares technique can be used. In this approach, each residual r_i is divided by a factor proportional to its uncertainty ϵ_{y_i}. Then R, the sum of the squares of the residuals, becomes

$$R = \sum_{i=1}^{N} \frac{r_i^2}{\epsilon_{y_i}^2} = \sum_{i=1}^{N} \frac{1}{\epsilon_{y_i}^2} (y_i - mx_i - b)^2. \tag{1.30}$$

Using the same minimization technique described above, and redefining the summation symbols:

$$S_1' = \sum_{i=1}^{N} \frac{1}{\epsilon_{y_i}^2} \qquad S_x' = \sum_{i=1}^{N} \frac{x_i}{\epsilon_{y_i}^2} \qquad S_y' = \sum_{i=1}^{N} \frac{y_i}{\epsilon_{y_i}^2},$$

$$S_{xy}' = \sum_{i=1}^{N} \frac{x_i y_i}{\epsilon_{y_i}^2} \qquad S_{x2}' = \sum_{i=1}^{N} \frac{x_i^2}{\epsilon_{y_i}^2} \qquad D' = S_1' S_{x2}' - (S_x')^2,$$

the slope and intercept for weighted least squares become

$$m = \frac{1}{D'} (S_1' S_{xy}' - S_x' S_y') \qquad b = \frac{1}{D'} (S_{x2}' S_y' - S_x' S_{xy}'). \tag{1.31}$$

Weighted least squares has utility in a number of applications in the physical chemistry laboratory. One common situation arises when the linear equation requires the log of a measured quantity to be the dependent variable. Consider, for example, the measurement of vapor pressure. Most pressure-sensing devices will provide a constant error in the measurement of P. If the dependent variable to be plotted for linear regression is $\ln P$, however, its uncertainty is not constant but decreases as P increases. Thus,

$$\epsilon_{\ln P} = \frac{\partial \ln P}{\partial P} \epsilon_P = \frac{\epsilon_P}{P}.$$

An appropriate weighting factor in this case would be $\epsilon_{y_i} = 1/P$. Weighted linear regression can also be applied to cases in which the presence of one or two disparate data points tends to skew the best fit line. One such approach is discussed by Mortimer.[3] The line is fit with all the data using nonweighted least squares, and the residual r_i for each point is determined. Then the line is refit using the weighted regression technique with weighting factors equal to the residuals ($\epsilon_{y_i} = r_i$). In this way the effect of the disparate data is minimized.

Error Analysis in Weighted Regression

Assuming that the uncertainties come from the same population as the data, the following relationships apply to weighted regression analyses:

$$\epsilon_m = \left[\sum_{i=1}^{N} \left(\frac{\partial m}{\partial y_i} \right)^2 (\epsilon_{yi})^2 \right]^{1/2} = \left[\sum_{i=1}^{N} \frac{1}{D'^2} \left(\frac{S_1' x_i - S_x'}{\epsilon_{yi}^2} \right)^2 (\epsilon_{yi})^2 \right]^{1/2}$$

$$= \frac{1}{D'} (S_1'^2 S_{x_2}' - S_1' S_x'^2)^{1/2}, \tag{1.32}$$

$$\epsilon_b = \left[\sum_{i=1}^{N} \left(\frac{\partial b}{\partial y_i} \right)^2 (\epsilon_{yi})^2 \right]^{1/2} = \frac{1}{D'} \left[S_1' S_{x_2}'^2 - S_{x_2}' (S_x')^2 \right]^{1/2} \tag{1.33}$$

Error Analysis in Nonlinear Regression

Equation (1.16) and the discussion accompanying error analysis in linear regression provide the guidelines for estimating uncertainties for quantities derived from experimental data using nonlinear regression. The uncertainties ϵ_{yi} can be estimated either from experimental considerations or from confidence limits generated from statistical analysis. In the case of fitted data, the confidence interval is the product of the standard deviation of the residuals s_r and the t factor.

Notes

1. N. M. Downie and A. R. Starry, ''Descriptive and Inferential Statistics,'' Harper & Row (New York), 1977.
2. H. W. Salzberg, J. I. Morrow, S. R. Cohen, and M. E. Green, ''Physical Chemistry Laboratory,'' pp. 18–19, Macmillan (New York), 1978.
3. R. G. Mortimer, ''Mathematics for Physical Chemistry,'' p. 303, Macmillan (New York), 1981.

Further Readings

P. R. Bevington, ''Data Reduction and Analysis for the Physical Sciences,'' McGraw-Hill (New York), 1969.

J. N. Noggle, ''Physical Chemistry,'' pp. 902–908, Little, Brown (Boston), 1985.

H. D. Young, ''Statistical Treatment of Experimental Data,'' McGraw-Hill (New York), 1962.

□ Appendix

Average Value and the Principle of Least Squares

As mentioned previously, the basis of the assertion that the average value is the best estimate of the true value of a result that can be obtained from the available data is the *principle of maximum likelihood*. This principle states that the set of measurements of p that have been made is the *most probable* set that could be obtained. The probability distribution function, p_i is directly related to the probability of obtaining a measurement, p_i:

$$p_i = \frac{1}{\sigma\sqrt{2\pi}} e^{-(p_i - \mu)^2/2\sigma^2}, \qquad (1.34)$$

where μ is the "true value" of the property and the "population variance" σ^2 measures the dispersion of the (infinite) sample population. The probability of obtaining all N data points is the product of the individual probabilities. If all the measurements of p are from the same sample population (i.e., have the same variance),

$$p_{tot} = \prod_{i}^{N} p_i = \left[\frac{1}{\sigma\sqrt{2\pi}}\right]^{N} \exp\left(-\sum_{i=1}^{N} \frac{(p_i - \mu)^2}{2\sigma^2}\right). \qquad (1.35)$$

The "most probable" set of measurements are the set for which p_{tot} is a maximum, and this will occur when the exponential term has its *minimum* value (hence the term "least squares"):

$$\frac{d}{dp}\left[\frac{1}{N}\sum_{i=1}^{N}\frac{(p_i - \mu)^2}{2\sigma^2}\right] = 0,$$

$$\frac{1}{2\sigma^2 N}\sum_{i=1}^{N}\left[\frac{d}{dp}(p_i - \mu)^2\right] = 0,$$

$$\frac{1}{\sigma^2}\sum_{i=1}^{N}(p_i - \mu) = 0, \qquad (1.36)$$

$$\frac{1}{N}\sum_{i=1}^{N}P_i = <p> = \mu.$$

If the measurements come from different sample populations, the maximum probability occurs when

$$\frac{d}{dp}\sum_{i=1}^{N}\frac{(p_i - \mu)^2}{2\sigma_i^2} = 0.$$

Evaluation of this expression gives

$$\mu = \frac{\displaystyle\sum_{i=1}^{N}(x_i/\sigma_i^2)}{\displaystyle\sum_{i=1}^{N}(1/\sigma_i^2)}. \qquad (1.37)$$

The right-hand side of equation (1.37) corresponds to the weighted average of the data, where each weighting factor $w = 1/\sigma^2$.

Thus, the assumption that the principle of maximum likelihood holds for a set of measurements leads to the result that the μ, the true value of p, is the average of the measurements. Since our (finite) set of measurements may not be the "most probable" set, $<p>$ is said to be the best estimate of μ that can be made from the available data.

PART TWO

Measurement and Control: Temperature, Pressure, and Electrical Measurements

□ Temperature

Temperature is one of the most fundamental and universally required measurements made in the scientific laboratory. Absolute temperature is a direct measure of the average kinetic energy of the molecules that comprise the system. This fact, which follows directly from kinetic molecular theory, requires that the translational, or kinetic, energy of every atom or molecule be zero at 0 K (zero-point vibrational energy, however, persists at 0 K). As the temperature is raised above 0 K, both the average kinetic energy of the molecules in a system and the distribution of energies about the average are increased. Because the kinetic energy of reacting species plays a key role in most chemical processes, variations in temperature that occur during the course of an experiment can have a significant effect on the results. Thus, temperature often becomes a variable that must be monitored and/or controlled if the data collected during an experiment are to be meaningful.

In principle, any property that depends on temperature could be used to measure it. In practice, the volume of a fluid at fixed pressure is often the basis for determining temperature. Indeed, the fact that the volume of an ideal gas is directly proportional to temperature at constant pressure (Charles's law) forms the basis for the ideal gas temperature scale, which is defined by the relationship

$$T = (273.16) \lim_{P \to 0} \left(\frac{V}{V_{tr}} \right) \quad \text{(constant } P, n\text{)}. \quad (2.1)$$

In equation (2.1), V is the volume of the gas at the measured temperature, T, V_{tr} is the volume of the gas at the triple-point temperature of water (273.16 K), and the $P \to 0$ requirement signifies the fact that all gases exhibit ideal behavior in the limit of zero pressure. Note that according to this scale, $T = 0$ is the temperature at which the gas volume is zero, a result that follows directly from the definition of an ideal gas. (In the case of imperfect, or real gases, liquefaction will take place before 0 K is reached and thus Charles's law is no longer valid.) It can be shown that the ideal gas temperature scale is identical to the thermodynamic scale, which is the fundamental temperature scale of science.[1]

TABLE 2.1   IPTS-68 Reference Temperatures[2]

Equilibrium Point	K	°C
Triple Point of Hydrogen	13.81	−259.34
Boiling Point of Hydrogen	20.28	−252.87
Boiling Point of Neon	27.102	−246.048
Triple Point of Oxygen	54.361	−218.789
Boiling Point of Oxygen	90.188	−182.962
Triple Point of Water	273.16	0.01
Boiling Point of Water	373.15	100
Freezing Point of Zinc	692.73	419.58
Freezing Point of Silver	1235.08	961.93
Freezing Point of Gold	1337.58	1064.43

An exact, if tedious, approach to measuring temperature can be inferred from equation (2.1). The volume of a fixed amount of gas is measured at some arbitrary constant pressure, first with the gas in thermal equilibrium with the body of interest, and then with the gas in thermal equilibrium with water at its triple point. Next, the pressure is lowered and the procedure is repeated. After a series of volume ratios has been obtained at successively lower pressures, the ratios are plotted against pressure, and the data are extrapolated to $P = 0$. The result is multiplied by 273.16 to provide the temperature of the body of interest. While this method has the advantage of providing a well-defined value of the temperature, it is impractical for routine measurements. A better approach is to establish standards by measuring the thermodynamic temperatures of a series of readily observable and reproducible physical phenomena (using the ideal gas thermometer or some other absolute temperature-measuring device) and to calibrate the temperature-sensing devices with them. The International Practical Temperature Scale (IPTS) establishes standards with which such calibrations can be performed (see Table 2.1).

Properties that have proved practical for the construction of temperature-sensing devices include the volume of a fluid, electrical resistance of a metal or semiconductor, the voltage produced by a thermocouple, and the voltage or current produced by an integrated circuit temperature transducer.

THE MERCURY-IN-GLASS THERMOMETER

The mercury-in-glass thermometer represents temperature as the height of a column of mercury in a glass capillary. Because the coefficient of thermal expansion of mercury is greater than that of glass, an increase in temperature causes the level of mercury to rise in the capillary. The thermometer is calibrated by measuring the height of the mercury at two well-defined temperatures on the thermodynamic scale. Temperatures between the two calibration points are determined by linear interpolation. Thus, if the calibration points used are the $H_2O(s) - H_2O(l)$ and $H_2O(l) - H_2O(g)$ equilibria at 1 atm, each of a series of one hundred equally spaced marks inscribed between the two points on the capillary would correspond to one "degree" Celsius (or Kelvin). This presupposes that the capillary bore is uniform and that the difference between the coefficients of thermal expansion for mercury and glass is constant over the temperature range. The latter assumption has been shown to be correct to within 0.5 percent between −30 and +150°C.[3]

Mercury-in-glass thermometers are convenient, reliable, easy to read, and reasonably inexpensive. For these reasons, they have traditionally been the instrument of choice for measuring temperature in the chemistry laboratory. However, they also have some limitations that must be considered when highly accurate temperature information is required. One important source of error is the fact that unless all the mercury contained in the thermometer is immersed in the medium, a temperature gradient will exist between the mercury bulb and the emergent stem. The effect this gradient will have on the accuracy of the temperature measurement will be a function of the size of the gradient. Many thermometers indicate a recommended depth of immersion and are calibrated to compensate for this gradient effect if they are immersed to this depth and if the exposed stem is at 20°C. Even when this is done, however, small emergent stem errors remain. Additional limitations of the mercury-in-glass thermometer include the fact that this type of thermometer is fragile, may have a slightly nonuniform bore size, and is relatively slow in response to changes in temperature.

A problem common to all thermometers that rely on fluid volumes to indicate temperature information is that the information cannot be easily transmitted to an external reading or storage device. Thus they are not suited for applications that require continuous temperature monitoring, temperature control, or remote temperature sensing. Although temperature controllers designed to be used with mercury-in-glass thermometers are available, these devices are complicated, fragile, and expensive. In applications of this type, the sensor of choice is a *temperature transducer* that converts thermal energy, i.e., temperature information, to another form of energy, such as voltage, which can be readily transmitted and interfaced to a reading device.

THE THERMOCOUPLE

If wires constructed from two *different* metals (*A* and *B*) are joined at both ends and one of the ends is at a different temperature from the other, a continuous current will flow in the wires. This condition is illustrated in Figure 2.1a. If the circuit is now broken at one end, the net open circuit voltage, E_{TC}, is a function of the metal combination and the temperature difference ($T_2 - T_1$). For small temperature differences the relationship between temperature and voltage is linear:

$$E_{TC} = \alpha \, (T_2 - T_1),$$

where α is called the *Seebeck coefficient,* in honor of the scientist who first discovered the effect (1821).

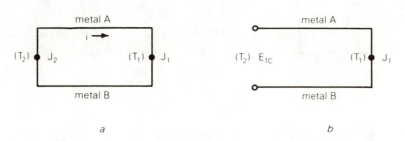

FIGURE 2.1   (a) Current flow through wires of two different metals joined at both ends; each end is at a different temperature. (b) Potential developed across the two wires if one junction is broken.

Any combination of different metals will exhibit the thermocouple effect. In fact, when operated in reverse, i.e., when a current is passed through a junction of two dissimilar metals, the metal pair will become hot or cold, depending on the direction of current flow. This phenomenon, known as the *Peltier effect,* is the basis of thermoelectric heating (or cooling) devices.

Thermocouples in common use fall into two categories. The *noble metal thermocouples,* which are the most stable (and expensive), are combinations of platinum/rhodium and platinum. An example from this group is the type S thermocouple, which consists of a platinum–10 percent rhodium positive element and a pure platinum negative element. The *base metal thermocouples* have no specified chemical composition because a wide variety of compositions within a given type produce the same voltage vs. temperature curve (within the error limits inherent to the wires themselves). In fact, constantan, which is the negative element on type E (nickel–10 percent chromium/constantan), type J (iron/constantan), and type T (copper/constantan) thermocouples, is not an alloy of specific composition but a generic name for a whole series of copper-nickel alloys.

The universal nature of the thermocouple voltage complicates the procedure for measuring it. Any connection to the open leads of Figure 2.1b designed to measure E_{TC} will create at least one additional thermocouple junction, since the terminals of the measuring device are constructed from a metal different from at least one of the thermocouple wires. Referring to Figure 2.2a, we see that in addition to the voltage produced at $J1$, we have the possibility of producing voltages at $J2$ and $J3$.

The situation is improved with the arrangement pictured in Figure 2.2b, where the creation of another A/B junction ($J4$) assures that identical Cu/A junctions occur at each of the terminals. This practice is followed in Experiment 3 (Joule-Thomson coefficient) in which a pair of Cu/constantan thermocouples is used to determine the temperature difference of a gas as it passes from high to low pressure through a porous plug. The thermocouples are connected in series such that both *copper* leads are connected to the voltage-reading device which often contains binding posts made of copper as the input connectors. The copper-to-copper arrangement further reduces extraneous junction voltages. In Figure 2.2b, as long as the voltages at $J2$ and $J3$ are in opposition and the junctions are at the same temperature, their net effect on the measured voltage is zero. Thus, the measured voltage, V, will be related to the temperature (in Celsius) by the expression

$$V = \alpha_{A/B}(T_1 - T_{ice}) = \alpha_{A/B}T_1.$$

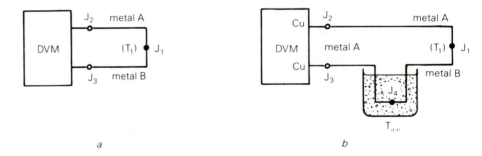

a   *b*

FIGURE 2.2   (a) Single-ended thermocouple with junction at T_1 connected to a readout device; $J2$ and $J3$ are the interconnection junctions. (b) Dual-junction thermocouple. One junction is ice bath referenced; the same thermocouple metal makes the junction to the readout device.

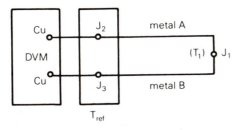

FIGURE 2.3   Single-junction thermocouple interfaced to a readout device by a compensating circuit.

Because the ice point has been selected by the National Bureau of Standards (NBS) as the fundamental reference point for their extensively tabulated thermocouple voltages, the NBS tables can be used to convert the voltage generated by the circuit of Figure 2.2b directly to T_1.

If temperature acquisition is put under microprocessor control, the need for an ice bath can be eliminated (ice-point compensation), and this represents a considerable increase in convenience. As a consequence of an empirical law of thermocouple behavior known as the "law of intermediate metals," the circuit shown in Figure 2.3 gives a result equivalent to that of Figure 2.2b.

Thus, with $J2$ and $J3$ at T_{ref}, the measured voltage becomes

$$V = \alpha_{A/B} (T_1 - T_{ref}).$$

If T_{ref} is determined with a device such as a thermistor, which produces a resistance directly related to its absolute temperature, T_1 can be calculated.

The principal advantage of thermocouple temperature sensors of the type shown in Figure 2.3 is the range of temperatures that can be measured. According to NBS specifications, a type S thermocouple can be used to monitor temperatures between -50 and $+1768°C$. Additional advantages include low cost, durability, rapid response, and reproducibility. Their primary disadvantage is the extremely low level of the voltages they produce. The Seebeck constant, α, ranges from 6 to 60 $\mu V/°C$. Thus the measuring device must be capable of resolution of 0.6 to 6 μV if a precision of $\pm 0.1°C$ is to be achieved. Low-level signals of this type are vulnerable to noise pickup in the conducting leads, and elaborate noise rejection techniques are often required. In any case, a high-sensitivity voltmeter is needed for direct thermocouple readout. Alternatively, an *instrumentation amplifier,* a highly accurate, low-distortion device, may be used to raise the thermocouple output level to the millivolt or volt range for subsequent voltmeter readout.

RESISTANCE THERMOMETERS

Materials that change resistance in response to a change in temperature include metals and semiconductors. The resistance of metals increases as their temperature is raised. This phenomenon has been used to create *resistance temperature detector* (RTD) devices from nickel, tungsten and copper, and platinum. The platinum-based RTD (PRTD) is by far the most widely used and stable of these devices. It is the interpolation standard between $-182.96°C$ (the oxygen reference temperature point) and approximately $+630°C$. The resistance change of a typical PRTD is on the order of 0.1 to 0.4 ohms per °C, and the device exhibits a resistance of

25 to 100 ohms (Ω) at 0°C. The response is reasonably linear over a limited temperature range. The chief advantages of a PRTD are its stability, reliability, and nearly linear response. The primary disadvantages include the relatively high cost, slow response time, the possibility of errors due to the resistance of the lead wires, and self-heating errors. Three- and four-wire RTD thermometers provide compensation for leads resistances, significantly reducing the error introduced by this effect. The problem of self-heating stems from the fact that the measuring device used with the PRTD must provide a current in order to carry out the resistance measurement. The electrical energy associated with the current (i^2R) is converted to heat within the RTD, changing its temperature. The size of the error introduced by this effect depends on the thermal conductivity of the medium and the magnitude of the current flowing through the device. A typical value is 0.5°C mW^{-1} in free air. Thus a current of 1 mA passed through a 100-Ω RTD produces a temperature rise of

$$(0.5 \ °C \ mW^{-1})(1 \times 10^{-3} \ A)^2(100\Omega)(1000 \ mW \ W^{-1}) = 0.05 \ °C \ s^{-1}.$$

Self-heating effects are much less serious when the probe is immersed in media such as liquids or solids that have much higher thermal conductivities than does still air. With proper attention to detail, RTD devices can achieve temperature precision of about ±0.001°.

The resistance of a semiconductor decreases with an increase in temperature. The effect, which is often substantial, provides the basis for the *thermistor* thermometer. Thermistors are particularly well suited for the detection of small temperature differences. They are also inexpensive, durable, small, and capable of rapid temperature response. Their main drawback is the fact that their response is extremely nonlinear. Frequently, combination thermistors are available that demonstrate a linear response over the range of temperatures most commonly measured in the laboratory, e.g., −50 to 100°C. Because small changes in temperature induce large resistance changes in these devices, errors due to lead resistance are not likely to be significant. However, self-heating errors can be considerable, especially when the device has been constructed with a small thermal mass for rapid temperature response. Thermistors are also less rugged than either RTD or thermocouple thermometers and must be handled with care.

SEMICONDUCTOR TEMPERATURE TRANSDUCERS

The semiconductor temperature transducer is an integrated circuit that provides an output in the form of either a voltage or a current that is linear with temperature. Because they are semiconductors, these devices share all the disadvantages of thermistors (except, of course, nonlinearity). For laboratory measurements in the range of −50 to 125°C, however, they represent an excellent, inexpensive choice.

Notes

1. I. N. Levine, "Physical Chemistry," 2nd ed., pp. 13–14, McGraw-Hill (New York), 1983.
2. "Temperature Measurement Handbook and Encyclopedia," p. T7, Omega Engineering Co. (Stamford, Conn.), 1987.
3. H. W. Salzberg, J. I. Morrow, S. R. Cohen, and M. E. Green, "Physical Chemistry Laboratory," p. 53, Macmillan (New York), 1978.

Further Readings

See reference 1.
R. P. Benedict, "Fundamentals of Temperature, Pressure, and Flow Measurements," Wiley (New York), 1969.
Dashcon-1 Manual, Metrobyte Corp., Taunton, Mass.

□ Pressure

Pressure is one of the fundamental state variables in thermodynamics. It is ubiquitous in topics in physical chemistry, appearing as both a dependent and an independent variable; it would be impossible to discuss the topic of gases or of phase equilibrium without invoking the concept of pressure. Accordingly, the measurement of pressure is indispensable in experimental physical chemistry. In this section, the definition and units of pressure are reviewed, and various methods of pressure measurement are discussed.

DEFINITION AND UNITS

Pressure is defined as force per unit area. A dimensional analysis of pressure using SI (System International) units, i.e., meters, kilograms, seconds—mks, shows that

$$P = \frac{F}{A} = \frac{\text{kg m s}^{-2}}{\text{m}^2} = \text{kg m}^{-1}\text{ s}^{-2}.$$

The unit force, 1 kg m s^{-2}, is known as the *newton* (N), and hence pressure can be expressed as N m^{-2}. This pressure unit is further defined as 1 *pascal* (Pa), named after the 17th-century French scientist and writer, Blaise Pascal. Thus, the dimensions of pressure are

$$P \text{ (SI)} = 1 \text{ Pa} = 1 \text{ N m}^{-2} = 1 \text{ kg m}^{-1}\text{ s}^{-2}.$$

A pressure of one pascal is very small in the context of human experience. A 150-lb person standing on one foot exerts a pressure of about 33,000 Pa. The pressure exerted by the atmosphere (at sea level) is equivalent to about 10^5 Pa. For this reason, a larger unit, typically kPa (10^3 Pa) is frequently used.

In centimeter-gram-seconds units (cgs), pressure is expressed as

$$P = \frac{F}{A} = \frac{\text{g cm s}^{-2}}{\text{cm}^2} = \text{g cm}^{-1}\text{ s}^{-2},$$

and since the unit force is called the *dyne*, the cgs unit pressure is 1 dyne cm^{-2}. The relationship between the cgs pressure and the SI unit is

$$1 \text{ dyne cm}^{-2} = 1 \text{ (g/s}^2 - \text{cm) } (10^{-3} \text{ kg/g})(10^2 \text{ cm/m}) = 10^{-1} \text{ kg s}^{-2}\text{ m}^{-1},$$

$$1 \text{ dyne cm}^{-2} = 10^{-1} \text{ Pa}.$$

Another unit of pressure, the *bar,* is conveniently defined as

$$1 \text{ bar} = 10^5 \text{ Pa} = 10^6 \text{ dynes cm}^{-2}.$$

The bar is now accepted as the *standard pressure* in thermodynamics, and is denoted as $p°$.

For historical reasons, certain trivial, or non-SI, pressure units have been developed, and some are still in use. Among the more popular are the atmosphere (atm), mmHg (or torr), and pounds per square inch (psi). The torr is a defined unit and arises historically from practical considerations, i.e., the way in which pressure was commonly measured. To illustrate this, we consider a vertical cylinder of mercury 1 cm high having a radius of r cm. The pressure exerted by this mass of mercury, m is

$$P = \frac{F}{A} = \frac{mg}{A} = \frac{dVg}{A},$$

where d is the density of mercury, V and A are volume and cross-sectional area of the column, respectively, and g is the acceleration due to gravity. At 20°C, this pressure is

$$1 \text{ mmHg} = \frac{d\pi r^2 hg}{\pi r^2} = (13.6 \text{ g cm}^{-3})(10^{-1} \text{ cm})(980.7 \text{ g cm s}^{-2})$$

or

$$1 \text{ torr} \sim 1333 \text{ dynes cm}^{-2} = 133.3 \text{ Pa}.$$

Notice that the value of the pressure is independent of the cross-sectional area of the column.

The standard *atmosphere* (atm) is defined as the pressure exerted by a column of Hg (exactly) 760 mm (76 cm) high at 0°C. Thus, 1 atm is 760 mmHg, or 760 torr. The atmosphere is a phenomenological unit because this height of mercury is close to the pressure exerted by the atmosphere (at sea level) on a typical day. In SI units, 1 atm is (760 torr)(13.3 Pa torr^{-1}), or 101 kPa. By international agreement, the atmosphere is defined as

$$1 \text{ atm} = 101.325 \text{ kPa} = 1.01325 \text{ bar}.$$

Finally, we consider another trivial, or nonstandard, unit, *pounds per square inch* (psi). This pressure unit is based on English standard measurements, and although widely used in industrial or engineering applications, it is becoming rapidly replaced by the SI unit. By going through unit conversions, we see that 1 atm is related to the psi as follows:

$$1 \text{ atm} = (1.01325 \times 10^5 \text{ kg m}^{-1} \text{ s}^{-2})(2.205 \text{ lb kg}^{-1})(2.540 \times 10^{-2} \text{ m in.}^{-1})/$$

$$(3.86158 \times 10^2 \text{ in. s}^{-2})$$

$$= 14.70 \text{ lb in.}^{-2} \text{ (psi)}.$$

The last factor used in the conversion is the standard acceleration of gravity, which allows pounds (a unit of force) to be used instead of mass (which, in English units, is the slug).

PRESSURE MEASUREMENT

We will describe several common methods of pressure measurement. This is not an inclusive list of such techniques, but it represents the typical approaches used in laboratories (and in the experiments in this book).

Before specific devices are discussed, we should distinguish between two different approaches to pressure measurement. In the first, the pressure of the system is measured in comparison with a reference pressure, usually the ambient (i.e., the surrounding) atmosphere. The pressure of the system measured in this way is called the *gauge pressure*. If, for example, the pressure is determined in psi units, the symbol *psig* is used.

In the second approach, the pressure of the system is measured relative to an ideal vacuum (0 pressure); it is thus called the *absolute pressure*. In psi units, this is represented as *psia*. If pressure is measured relative to a precisely defined standard pressure (e.g., 14.7 psi, 100 torr, etc.), the symbol *psis* is often used.

PRESSURE-MEASURING DEVICES

A device that converts an impulse (a form of energy) to a response (another form of energy) is called a *transducer*. An example of a transducer is a thermocouple; it is a junction of two dissimilar metals that converts a temperature (manifestation of kinetic energy) to a voltage (a form of potential energy). One can generalize the construction and design of a pressure transducer into two parts: (1) an *elastic element* (or sensor) that moves in response to the application of a pressure, and (2) a *readout mechanism* that converts the displacement of the sensor to a measurable result. The readout can be achieved mechanically (e.g., a needle and gauge plate, a height of liquid) or electrically (in which a voltage is produced). In the latter case, the electrical signal can be produced in an active or passive manner. A piezoelectric device directly converts a pressure to a voltage (the principle behind some phonograph pickups). A passive electrical transducer registers changes in capacitance or impedance; this type of device will be described below.

Open-tube Manometer

A diagram of an open-tube manometer is shown in Figure 2.4. The U-tube is filled with a liquid, frequently mercury, although any convenient liquid can be used. This device measures the difference between the pressure of the system, P_X, and the pressure of the environment to which the open tube is exposed, P_{ref}. In a manometer, the liquid acts as both the sensor and the readout device.

The pressure difference may be determined from the density of the liquid, d_L, and the (vertical) displacement in millimeters between the meniscuses, h. To ex-

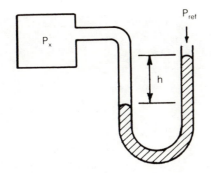

FIGURE 2.4   Open-end (U-tube) manometer showing reference and measured pressures, P_{ref} and P_X, respectively.

press the pressure in torr (mmHg), the ratio of the liquid density to that of mercury, d_{Hg}, is required:

$$P_X - P_{ref} = \frac{hd_{Hg}}{d_L} \quad \text{(torr).} \tag{2.2}$$

In equation (2.2), d_{Hg} is the value at 0°C (13.5950 g cm^{-3}) and d_L is the density of the liquid at the temperature at which the measurement is carried out.

Note that the open-tube device can function as a *null manometer*. Thus if P_{ref} is adjusted so that the levels of the liquid are of equal height, then $P_X = P_{ref}$. In this application, a liquid having a lower density than mercury is often used because it affords higher sensitivity. One common liquid is di-*t*-butylphthalate. It is relatively unreactive and has a density of 1.04 g cm^{-3}. The pressure obtained from an open-tube manometer is the gauge pressure; it is a relative measurement. It can be converted to an absolute pressure only if the value of P_{ref} is known.

The open-tube manometer, shown in Figure 2.4, is used in the *isoteniscope*, a convenient device for measuring the vapor pressure of a liquid (or solid). This apparatus is used (and described in more detail) in Experiment 11.

Closed-tube Manometer

A diagram of this type of manometer is shown in Figure 2.5. A tube is filled with a liquid, e.g., Hg, and the open end is temporarily stoppered. Next, the tube is inverted, and the stopper is removed with the open end immersed in a reservoir of the liquid. This, basically, is a Torricelli barometer, the principle by which many laboratory barometers are designed. The (vertical) height of the surface of liquid in the tube relative to that in the reservoir (in millimeters) is equal to the pressure exerted at the surface of the pool of liquid. The pressure above the liquid in the sealed tube is, strictly speaking, not zero, but the finite vapor pressure of the liquid at the particular temperature. In the case of mercury at room temperature, this is about 2 micrometers (μm), which corresponds to 2×10^{-3} torr and is usually negligible.

Bourdon Gauge

A *Bourdon gauge* (1853) is another mechanical pressure transducer. It is illustrated in Figure 2.6. The sensor is a C-shaped metal tube (brass or stainless steel) having an elliptical cross section. One end of this tube is sealed and the other is connected to the system whose pressure is to be measured. Any pressure difference between the inside and outside of this tube will cause it to distort slightly as indicated in the diagram. (It is reminiscent of the coiled-up paper party favor

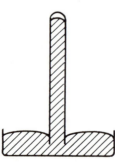

FIGURE 2.5   Closed-end manometer; P_X is ambient pressure.

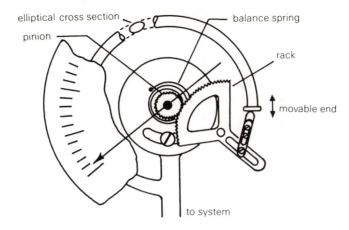
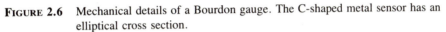

FIGURE 2.6 Mechanical details of a Bourdon gauge. The C-shaped metal sensor has an elliptical cross section.

which, when blown into, extends out to some length.) This small displacement is mechanically amplified by a rack and pinion assembly and converted into pivotal motion. A pointer mounted on a pivot gear indicates the pressure (difference) in conjunction with a calibrated dial.

Most Bourdon gauges are designed to read zero when there is no pressure difference across the elliptical tube. Since the external pressure is equal to that of the atmosphere, this type of device reads "gauge pressure." Some gauges are calibrated to read pressures less than 1 atm, and others will read down to 1 atm below ambient pressure (a vacuum gauge).

Bourdon gauges are usually used with high-pressure gas cylinders and reducing valves. They generally lack the precision needed for quantitative studies, although some gauges are constructed to have acceptable reproducibility and sensitivity (ca. 0.25 percent full-scale accuracy).

Disc Manometer

This is a mechanical transducer that consists of an evacuated, sealed cylindrical metal chamber, usually constructed of stainless steel. The sensor is one of the surfaces that is contoured concentrically; as the pressure exerted on the cylinder changes, the flexible surface moves slightly. This displacement is mechanically amplified and results in the rotation of a pointer that indicates the pressure on a calibrated dial. Because of its design, the disc gauge reads *absolute* pressure. The disc and the mechanical assembly are mounted in a vacuum-tight chamber that is connected to the system whose pressure is to be determined. The aneroid (free of liquid) barometer is an example of a disc gauge.

One of the disadvantages of the disc gauge is that its mechanical components (usually including the dial) are exposed to the gas being studied. Thus corrosive or reactive gases cannot be used. In addition, it is difficult (and usually undesirable) to heat the disc gauge, and this limitation confines its application to ambient temperatures.

Electrical Pressure Transducers

A piezoelectric device acts as an active transducer because it produces an electrical signal as a consequence of an applied pressure. The sensor is a particular type of crystal (e.g., quartz, Rochelle salt) whose unit cell lacks a center of symmetry.

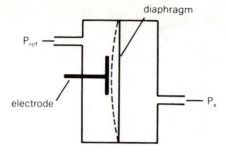

FIGURE 2.7   Schematic diagram of a capacitance manometer sensing head. The movement of the diaphragm relative to the fixed electrode produces changes in the capacitance between these two elements.

These materials produce a surface potential when they are pressurized in specific directions.

Passive transducers are more commonly used. These devices convert changes in pressure to differences in electrical capacitance. A schematic diagram of a *capacitance manometer* is shown in Figure 2.7. The sensor is a thin, flexible diaphragm constructed of an unreactive material (such as Inconel or stainless steel) that serves as one of a pair of electrodes. One side of the diaphragm is exposed to the gas whose pressure is to be measured and the other faces a chamber that is either permanently evacuated (for absolute measurements) or contains a gas at an arbitrary reference pressure (for relative measurements). The application of a pressure (force/area) to the diaphragm causes it to deflect, resulting in a change in the capacitance between the sensor and another, fixed, electrode. This change in capacitance is "conditioned" by associated electronic circuitry and converted to a dc output voltage. The output signal can be displayed on a voltmeter, or fed (through an A/D converter) to a digital computer for further data processing or process control. The capacitance manometer is used in Experiments 4 and 19.

Further Readings

R. P. Benedict, "Fundamentals of Temperature, Pressure, and Flow Measurements," pp. 187–232, Wiley (New York), 1969.

A. Berman, "Total Pressure Measurement in Vacuum Technology," Academic Press (Orlando, Fla.), 1985.

□ Electrical Measurements

A substantial number of experimental measurements produce an electrical signal in the form of a voltage, current, resistance, or capacitance that must be monitored and converted into a number. Because most of the signals that are sampled in the physical chemistry laboratory are from direct-current (dc) sources, the discussion will be confined here to problems associated with their interpretation.

The process used to monitor the signal can often be a source of significant error. An important concept embodied in the Heisenberg uncertainty principle is that the very *process* of measurement perturbs the system and thus affects the

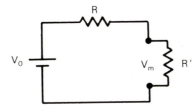

FIGURE 2.8   Circuit showing resistance R in series with a voltage source, V_0.

value of the property measured. A pertinent example is illustrated in Figure 2.8. The magnitude of the open-circuit, or zero-load, voltage, V_0, produced from a voltage source having a series resistance, R, is to be determined. According to Ohm's law ($V = iR$), a measuring device with an effective input resistance (impedance), R', that is connected to the voltage source will cause a current

$$i = \frac{V_0}{R + R'},$$ (2.3)

to flow. The voltage that will be displayed on the measuring device is

$$V_m = V_0 - iR = \frac{V_0 R'}{R + R'}.$$ (2.4)

Thus the relative error ϵ_V introduced in the measurement of V_0 by the measuring device is

$$\epsilon_V = \frac{V_0 - V_m}{V_0} = \frac{R}{R + R'}.$$ (2.5)

It follows from equation (2.5) that the effects of measurement errors of this type, referred to as "loading errors," can be minimized by having a voltage source of *low output impedance, R,* and a measuring device of *high input impedance, R'.*

EXAMPLE 1:   What voltage would be indicated by a measuring device with an input resistance of 1000 Ω if it is used to detect a 1.2-V signal produced by a source with an output resistance of 100 Ω?

Solution: According to equation (2.4),

$$V_m = \frac{(1000)(1.2 \text{ V})}{1100} = 1.09 \text{ V} \qquad \square$$

Thus a source voltage of 1.2 V would be measured as 1.09 V. This represents an error of -9.2 percent.

EXAMPLE 2:   What is the minimum input resistance required for a measuring device so that an error of less than 0.1 percent is introduced by the measurement of a voltage from a source with an output resistance of 50 Ω?

Solution: From equation (2.5):

$$\epsilon_V = 0.001 = \frac{50}{50 + R'}$$

and

$$R' = 49,950 \ \Omega = 49.95 \text{ k}\Omega \qquad \square$$

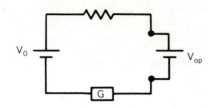

FIGURE 2.9   Same circuit as in Figure 2.8, but in series with a galvanometer and an opposing voltage source.

An alternative method for measuring the output voltage V_0, which is free from the type of errors associated with the circuit of Figure 2.8, is illustrated in Figure 2.9. The source voltage V_0 is connected to an *opposing* voltage V_{op}, in such a way that V_{op} could be varied until a net voltage of zero is obtained $[(V_{op} + V_0) = 0]$. Since the net current flow in this case would also be zero (Ohm's law), the null condition, $V_{op} = -V_0$, may be determined as the value of V_{op} for which no current flow is detected on the galvanometer G, a very sensitive current-measuring device. This approach has the distinct advantage that the voltage developed by the source can be measured under essentially zero load (i.e., reversible) conditions. This technique is the basis of design of the potentiometer (see Experiment 9).

A simplified schematic of a potentiometer is shown in Figure 2.10. A precision variable resistor, denoted by *ABC*, acts as a voltage divider. The current produced by V_{op} is

$$i_{op} = \frac{V_{op}}{R_{AC}}.$$

When the key is depressed, an opposing current is produced by V_0:

$$i_0 = \frac{V_0}{R_{AB}}.$$

The direction of current flow, i.e., the sign of $i_0 + i_{op}$, will be indicated by the deflection of the galvanometer needle. When the net current is zero (no deflection of the needle), the *null condition* exists and V_0 can be determined from the relationship

$$\frac{V_{op}}{R_{AC}} + \frac{V_0}{R_{AB}} = 0,$$

or

$$V_0 = -V_{op}\frac{R_{AB}}{R_{AC}}.$$

Note that the precision resistor *ABC* consists of a fixed resistor and a linear slide wire. The ratio R_{AB}/R_{AC} is thus determined by the position of B along the slide wire. In practice, this position is read directly as a "voltage." The reading is calibrated by connecting a source of accurately known voltage at X (see Figure 2.10), positioning the dial to read the specified value, and adjusting the trimming resistor, T, so that no current flows. This standardizing, or reference, source is also highly precise; its voltage is often known to six significant figures.

Potentiometers are capable of great precision (and accuracy) and are the instrument of choice when very accurate results are required (i.e., potentials accurate

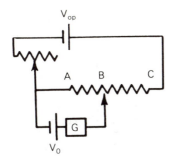

FIGURE 2.10   Schematic diagram of a simple potentiometer.

to ± 0.01 mV). However, these devices are inconvenient for routine measurements. With the introduction of integrated circuits called operational amplifiers, measuring devices such as digital panel meters, digital multimeters, and A/D converters are available with extremely high input resistances, e.g., 10^6 to 10^{14} Ω, with 10^9 Ω being typical. With this type of equipment, source voltages can be measured under virtually zero-load conditions. Additional, significant advantages of these electronic devices are that they provide rapid (virtually instantaneous) readout and that they can be readily coupled to computers for direct on-line data acquisition and storage for subsequent analysis or process control.

Many measurement devices are designed to measure voltages. Signals in the form of current or resistance can also be measured with these devices if the appropriate signal-conditioning circuitry is provided. For example, a signal from a current source can be detected as the voltage drop that occurs across a precision resistor through which the current flows ($i = V/R$). The same loading considerations discussed previously would require that R be as small as possible. If the source current is also small, however, the resulting voltage may be so low that significant detection problems are encountered. In this case, a more sophisticated circuit involving an operational amplifier configured as a "current follower" is often utilized.

A simple technique for determining resistance is to measure the voltage drop that occurs when a known constant current flows through the resistive device. This approach is often used with two-wire RTD thermometers that display resistances proportional to temperature (see pp. 25–26).

A more accurate method for measuring resistance is to include the unknown resistance as one arm of a composite resistance network known as a *Wheatstone bridge*. Such a network is illustrated in Figure 2.11. Since the voltage V_S must drop through R_1 and R_2 on the left, and R_S and R_X on the right, V_A and V_B will be

$$V_A = \frac{V_S R_1}{R_1 + R_2} \quad \text{and} \quad V_B = \frac{V_S R_S}{R_S + R_X}.$$

Like the potentiometer, the Wheatstone bridge can be configured as a nulling device. In this case, the resistance of R_s is varied until the voltage across points *AB* is zero. Because of its sensitivity, a galvanometer is typically used to indicate the null point, at which

$$\frac{R_1}{R_2} = \frac{R_S}{R_X},$$

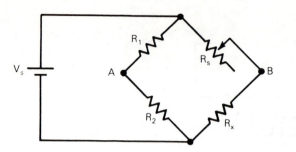

Figure 2.11   Simplified diagram of a Wheatstone bridge for measuring resistance.

or

$$R_X = R_S\frac{R_2}{R_1}.$$

Because the null measurement is made under reversible (i.e., nearly equilibrium) conditions, this constitutes a very sensitive (i.e., precise *and* accurate) technique for measuring resistance. This again illustrates the advantage of a null-measuring device: zero can be measured with great accuracy and precision. However, as in the case of the potentiometer, it is not convenient for continuous or automated detection of resistance, as would be required for on-line temperature detection with an RTD thermometer. In applications of this type, the voltage across *AB* can be monitored, and the resistance R_X can be determined from the relationship (with $R_1 = R_2$)

$$R_X = \frac{R_S(V_S - 2V_{AB})}{2V_{AB} + V_S}.$$

Further Readings

I. Brophy, "Basic Electronics for Scientists," 2nd ed., McGraw-Hill (New York), 1972.

A. J. Diefenderfer, "Principles of Electronic Instrumentation," Saunders (Philadelphia), 1972.

H. V. Malmstadt, C. G. Enke, and S. R. Crouch, "Electronic Analog Measurements and Transducers," Benjamin (Menlo Park, Calif.), 1973.

P. B. Zbar and J. G. Sloop, "Electricity-Electronics Fundamentals," 2nd ed., Gregg Div. McGraw-Hill (New York), 1977.

□ PART THREE

Computers in the Physical Chemistry Laboratory

Computers have become an indispensable part of modern chemistry laboratories. The many dramatic improvements in scientific instrumentation that have taken place in the last 20 years can be traced, in large part, to the introduction of computers and integrated circuits (such as operational amplifiers) to such equipment. Many of these electronic devices were originally developed for, and applied to, signal processing and analysis. Indeed, many modern instruments contain microprocessors similar to the central processing units (CPUs) found in microcomputers. These devices control the use and operation of the instrument and often provide data-manipulation capabilities such as Fourier transform, numerical integration and differentiation, curve fitting, calibration, or standardization. In the physical chemistry laboratory, computers can be a valuable resource in two ways: direct, on-line data acquisition, and data manipulation or analysis.

COMPUTER-ASSISTED DATA ACQUISITION

The errors that are inherent in obtaining data from analog devices, such as meters, strip chart recorders, thermometers, and manometers, can be significantly reduced when the data are acquired directly with the aid of a computer. In any analog measurement, a conversion must be carried out in order to transform the signal generated by the measuring device into a *number*. The signal is usually produced as an electronic level or response—voltage, current, resistance, impedance, etc.—and is often represented as a displacement of the needle of a meter or a recorder pen. One advantage to the direct data acquisition by computer is that this analog-to-digital (A/D) conversion can usually be achieved with little or no loss in accuracy or intrinsic precision. By contrast, when this conversion is carried out by the experimenter (i.e., by reading a meter), the error introduced in the process comes from an uncertainty or imperfection in interpretation or judgment and is often the major contribution to the uncertainty of the measurement. Moreover, even if the random errors associated with reading the device are insignificant compared with others encountered in the experiment, the possibility remains that systematic, gross errors, even blunders, can occur in the reading process. Although data acquired by a computer are not entirely immune to such problems, they are far less likely to occur.

In order to assess the effect of on-line data acquisition on the accuracy of the data collected, it is important to consider both the way the A/D converter works and the accuracy limitations of the device. A complete discussion of the mechanics of A/D operation may be found elsewhere. In this section, the factors that affect A/D accuracy will be reviewed. An A/D converter transforms an analog signal (in the form of a voltage) into binary information by effectively "setting" digital switches, called *bits* (binary digits), to either ON or OFF positions; in this case, the digit 0 corresponds to OFF and 1 denotes ON. The *absolute resolution* of the device, which is equal to ±1/2 the least significant bit (LSB), represents the *minimum* uncertainty that is contributed to the measurement by the device. The absolute resolution is determined by the number of bits available to the A/D converter and by its full-scale input range. For example, an 8-bit A/D converter will represent the input signal as a binary number between 0 (all bits OFF, i.e., binary 0000 0000) and 255 (all bits ON, binary 1111 1111). Stated another way, this device has a relative resolution of 1/256 or 0.39 percent. If the full-scale input range of the source were 0 to 0.50 V, the absolute resolution would then be (0.50 V)(0.0039), or 0.0020 V.

EXAMPLE 1:   Determine the relative and absolute resolution of a 12-bit A/D converter that is designed to accept a full-scale input signal between 0 and 10 V.

Solution: In this case, the signal is converted to a number between 0 and 4095 (i.e., $2^{12} - 1$). Thus the relative resolution is 1 part in 4096, or 0.024 percent, and the *absolute* resolution is (0.00024)(10 V), or 0.0024 V.   □

In addition to limitations imposed by their intrinsic resolution, A/D devices often suffer from other sources of error. Frequently, the net effect of the individual errors is lumped together in the specification known as "overall" or "absolute" accuracy. Typically, this specification would be reported as ±X percent of *range* or ±X percent of *reading,* ±Y bits.

EXAMPLE 2:   Suppose a voltage reading of 3.502 V was determined using a 12-bit A/D converter with an overall accuracy reported to be 0.1 percent of the reading ± 1 LSB. The device is configured to measure a full-scale input of ±10 V. What is the uncertainty in the result due to the use of the A/D device?

Solution: The uncertainty of the A/D device has two parts. The first is an absolute uncertainty (±1 LSB) common to all measurements performed with the device. The error contribution from this source is (1/4096)(20 V), or 0.0049 V. (Note that the full-scale input range is −10 V to +10 V, or 20 V.) The A/D conversion is also subject to an uncertainty of 0.1 percent of the reading, or an additional (±0.001)(3.502 V), or ±0.0035 V. Thus the total uncertainty of the measurement that is contributed by the A/D conversion is 0.0049 V + 0.0035 V, or ±0.0084 V.   □

In the physical chemistry laboratory, the experiments that benefit most from on-line data acquisition are those which require the analysis of a continuously varying signal, since, in these instances, a large number of data points are available to represent the results accurately. For example, to generate a cooling curve, the temperature of a liquid sample must be monitored frequently as the material cools and solidifies. Traditionally, this would be accomplished by transmitting the output of the temperature-sensing device to an analog recorder which represents the temperature as a pen deflection. The pen traces this output on a piece of chart paper that is moving at a selected rate. The data are said to be acquired in "real time.'' The resulting trace inscribed on the chart paper provides the *only* record (sometimes called hardcopy) of the experimental data. Moreover, specific temperature-time information can be recovered only by measuring the position of a point on the trace. This measurement, which essentially constitutes an A/D conversion, may contribute some uncertainty to the temperature value.

The situation is considerably improved if the signal is instead sent to a discrete A/D converter acting as an interface between the output of the temperature sensor and a computer. In this case, the data are automatically (and systematically) converted to digital form and *stored* in the computer memory. Not only is this A/D conversion likely to be much more accurate, but also the resulting data are available for further processing and analysis.

DATA ANALYSIS BY COMPUTER

The most significant impact of the microcomputer in the physical chemistry laboratory is probably on the method of data manipulation and analysis. The tremendous versatility and speed of a computer can make it an invaluable tool that eliminates much of the drudgery of repetitive calculations and also provides new learning possibilities. Many processes in data analysis would literally be impossible without computer assistance. However, like any powerful tool, the computer must be used properly if it is to serve the experimentalist to the full extent of its capabilities. The essential requirement for the efficient and full use of the microcomputer is the *intelligent* application of *proper* software (the computer's instructions).

Writing custom, or dedicated, software for most laboratory applications in physical chemistry tends to be counterproductive. It is generally better policy to make use of existing commercial software that is designed to carry out a number of useful operations on many different types of data. These software "packages" are debugged, tested on many different applications, and are versatile. Most commonly, the basis for this type of utility is a "spread sheet," which provides labeled rows and columns in which data can be entered, organized, and stored. The software also furnishes a wide variety of operations that can be performed on all or part of the stored data. One of the most important requirements of the software is that it provide the user with the ability to carry out required and varied operations on data produced by a wide variety of experiments. For example, linear (or first-order) regression analysis is frequently performed on data acquired in physical chemistry experiments. A program that carries out this procedure has wide application. Other forms of data analysis that are very conveniently carried out with some software packages are polynomial curve fitting, differentiation, and numerical integration.

It must always be remembered, however, that the laboratory computer is only a tool that, one hopes, does exactly what it is told. It cannot transform poor data into good information. Hence the acronym GIGO (garbage in—garbage out) applies well to the analysis of the results of physical chemistry experiments.

□ PART FOUR

Thermodynamics of Gases

□ Experiment 1

Thermodynamics of a Gas Phase Reaction: Reversible Dissociation of N_2O_4

Objective

To determine the equilibrium constant and standard thermodynamic constants for the gas phase reaction $N_2O_4 \rightleftharpoons 2NO_2$.

Introduction

We consider in this experiment the dissociation/dimerization reaction of N_2O_4 and NO_2. The reaction is

$$N_2O_4 \ (g) \rightleftharpoons 2NO_2 \ (g). \qquad (4.1.1)$$

The N—N bond in N_2O_4 is sufficiently weak so that near and above room temperature, appreciable thermal dissociation takes place, producing a measurable equilibrium concentration of NO_2. It is possible to determine experimentally both the monomer (NO_2) and dimer (N_2O_4) compositions, thus allowing the equilibrium constant, K_P, to be measured:

$$K_P = \frac{P_{NO_2}^2}{P_{N_2O_4}}, \qquad (4.1.2)$$

where P_{NO_2} and $P_{N_2O_4}$ are the respective partial pressures of the monomer and dimer (in atmospheres). Because the total pressure of the system never exceeds 1 atm, we can assume the validity of the ideal gas law. Hence, the expression in (4.1.2) represents a true equilibrium constant because the partial pressures are equal to the respective activities.

By determining K_P at different temperatures, the enthalpy and entropy of the dissociation of N_2O_4 can be obtained. The thermodynamic treatment underlying this approach follows.

The standard free energy change associated with a process is related to the equilibrium constant K_P, expressed in (4.1.2), by the relation $\Delta G° = -RT \ln K_P$, or

$$\ln K_P = \frac{-\Delta G^\circ}{RT}. \tag{4.1.3}$$

Here the superscript $^\circ$ indicates the standard state of 1 atm. We wish to express the temperature dependence of $\ln K_p$. From the definition of the Gibbs free energy function for an isothermal process under standard conditions,

$$\Delta G^\circ = \Delta H^\circ - T \Delta S^\circ, \tag{4.1.4}$$

we can express the temperature dependence of (4.1.3) as

$$\ln K_P = \frac{-\Delta H^\circ}{RT} + \frac{\Delta S^\circ}{R}. \tag{4.1.5}$$

Now, differentiating (4.1.5) with respect to T (at constant pressure), and recalling that $(\partial \Delta H/\partial T)_P = \Delta C_P$, and $(\partial \Delta S^\circ/\partial T)_P = \Delta C_P/T$, we have

$$\left(\frac{\partial \ln K_P}{\partial T}\right)_P = \frac{1}{R}\left[\left(\frac{1}{T}\right)(-\Delta C_P) - \frac{(-\Delta H^\circ)}{T^2} + \frac{C_P}{T}\right].$$

After combining terms, we arrive at the useful result

$$\left(\frac{\partial \ln K_P}{\partial T}\right)_P = \frac{\Delta H^\circ}{RT^2}. \tag{4.1.6}$$

Equation (4.1.6) is known as the van't Hoff equation (1885). If the temperature dependence of ΔH° is sufficiently small (as is the case in this experiment), equation (4.1.6) can be integrated to give the result

$$\ln K_P = \frac{-\Delta H^\circ}{RT} + \text{constant}, \tag{4.1.7}$$

where the constant in 4.1.7 is a boundary condition of the integration. Equation (4.1.7) indicates that a plot of $\ln K_P$ vs. $1/T$ should be linear and have a slope equal to $-\Delta H^\circ/R$. Once ΔH° is found, the standard entropy change for the process can be computed from the Gibbs equation, (4.1.4), by interpolating the $\ln K_P$ data to obtain a value of ΔG° at 25°C. The correction for nonideal behavior is very small.[1]

Various methodologies can be employed to measure the equilibrium gas phase composition of the N_2O_4/NO_2 system. One approach might exploit the fact that while N_2O_4 has an even number of electrons, NO_2 has an odd number. The consequence of this is that, as illustrated by the Lewis structures below, the dimer has all electrons paired, while the monomer has one unpaired electron (Figure 4.1.1). The NO_2 monomer is called a *radical* and is paramagnetic;[2] moreover, as is often the case with species containing unpaired electrons, it absorbs light in the visible region of the spectrum. Thus NO_2 has a reddish-brown color, while N_2O_4 is colorless. The colored appearance of the N_2O_4/NO_2 system is due to the absorption of the monomer (from the near ultraviolet throughout much of the visible); N_2O_4 begins to absorb below 400 nm, in the ultraviolet. It should be noted here that absorption of visible light by NO_2 initiates the production of $NO + O$. The formation of highly reactive O atoms by this process is particularly relevant to photochemically generated air pollution.[3] It is one source of troublesome ozone, O_3, which is a major irritant produced in photochemical smog (sunlight acting on air pollutants such as automobile exhaust).

The above discussion suggests that the composition of the system could be determined *spectrometrically* by measuring the amount of light absorbed by a mixture of N_2O_4 and NO_2 at a wavelength where only the latter absorbs appreciably. This is, in fact, a possible approach in determining K_P for this system.[4]

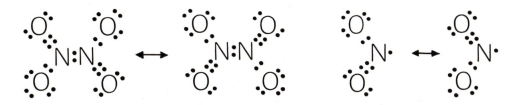

FIGURE 4.1.1   Lewis structures for N_2O_4 and NO_2. Each molecule is represented by two resonance structures.

Another more straightforward method for determining the system composition is to use a *gravimetric* technique in combination with the ideal gas law. In this approach, the *total mass, m_{tot},* of the N_2O_4/NO_2 system is measured under the condition that the *total pressure, P_{tot},* and temperature, T, are known. Experimentally, a bulb (volume, V) is filled with N_2O_4/NO_2 so that the pressure is slightly above 1 atm. The bulb is immersed in a constant temperature bath, and, after equilibration, the bulb is vented, allowing its total pressure (P_{tot}) to become equal to the ambient atmospheric pressure. The mass of the bulb contents (m_{tot}) is then determined gravimetrically.

From the ideal gas law, the number of moles of monomer, n_M, and dimer, n_D, are $P_M V/RT$ and $P_D V/RT$, respectively. P_M and P_D are the partial pressures. The total mass of this system is then

$$m_{tot} = \frac{M_M P_M V}{RT} + \frac{M_D P_D V}{RT} = (P_M + 2P_D)\frac{M_M V}{RT}, \qquad (4.1.8)$$

where M_M and M_D are the molecular weights of the monomer and dimer. This expression can be rearranged to give an equation explicit in the equilibrium partial pressures:

$$P_M + 2P_D = \frac{RT m_{tot}}{V M_M} \equiv a, \qquad (4.1.9)$$

where a is obtained from measurable quantities. Using Dalton's law, we can express the other measurable, the total equilibrium pressure:

$$P_M + P_D = P_{tot}. \qquad (4.1.10)$$

Combining (4.1.9) and (4.1.10), P_D is obtained as

$$P_D = \frac{RT m_{tot}}{V M_M} - P_{tot} = a - P_{tot}, \qquad (4.1.11)$$

and, likewise, P_M is seen to be

$$P_M = 2P_{tot} - \frac{RT m_{tot}}{V M_M} = 2P_{tot} - a. \qquad (4.1.12)$$

Now the equilibrium constant can be expressed

$$K_P = \frac{(2P_{tot} - a)^2}{a - P_{tot}}. \qquad (4.1.13)$$

A related quantity called the *degree of dissociation*, α, can be determined from the above information. α is the ratio of dissociated N_2O_4 molecules to the *total* number of dimer units possible in the system. Since *two* molecules of NO_2 are produced for each N_2O_4 molecule that dissociates, $n_M/2$ represents the mole number of dimers that have dissociated. The total number of dimers possible in the system is then given by the sum of intact dimers plus one-half the number of monomers. α can also be expressed in terms of the equilibrium monomer and dimer partial pressures:

$$\alpha = \frac{n_M/2}{n_M/2 + n_D} = \frac{P_M/2}{P_M/2 + P_D} = \frac{P_M}{P_M + 2P_D}. \qquad (4.1.14)$$

In terms of the measurables, a and P_{tot}, this expression becomes, using (4.1.11) and (4.1.12):

$$\alpha = \frac{2P_{\text{tot}}}{a} - 1. \qquad (4.1.15)$$

Experimental Method

This experiment relies on a *gas* gravimetric technique in which the total mass of an equilibrium mixture of NO_2 and N_2O_4 contained in a glass reaction bulb (ca. 250 mL) is measured. The bulb weighs much more than its gaseous contents; thus, high gravimetric sensitivity *and* reproducibility are required (ca. 0.1 mg). Moreover, the differences in weights of gas between the various temperature "runs" is very small. Thus, this experiment is *very* sensitive to a variety of weighing errors.

You will be given two bulbs. One is the sample vessel, which will be filled with N_2O_4/NO_2, and the other is a sealed ballast bulb having approximately the same volume. The purpose of the ballast bulb is to counteract any changes in the *buoyancy force* that would affect the measurement of the weight of the sample bulb. The buoyancy correction is an application of Archimedes's principle: the weight loss of a solid object immersed in a fluid is equal to the weight of fluid displaced by the object. In other words, the (upward) buoyancy force on the sample bulb is $F_b = M_a g$, where M_a is the mass of air displaced by the particular sample bulb and g is the gravitational constant.

Because the weight of gas in the sample bulb is determined as the difference in two relatively large masses (empty and filled bulb), any change in the air density (arising from changes in temperature and/or atmospheric pressure) that causes a change in buoyancy of the bulb may introduce a relatively large error. By measuring the mass of the ballast just before (or after) weighing the sample bulb, a correction for changes in the bulb's buoyancy can be made.

Another approach, which requires knowledge of the displacement volume of the sample bulb, is to measure the air temperature and pressure in the immediate vicinity of the balance used before (or after) the sample bulb is weighed. The changes in the displacement air mass (determined from the ideal gas law) are then applied as corrections to the measured sample bulb weights. For example, if there is an increase in the air temperature between two measurements corresponding to a decrease of 2 mg of displaced air, 2 mg is added to the observed weight of the sample bulb.

Safety Precautions

□ Always wear safety goggles during this experiment.
□ This experiment requires the use of N_2O_4. This is a corrosive and irritating substance. Although you will not directly come in contact with this chemical in the experiment, you *must* avoid venting this material into the laboratory.
□ The gas-handling manifold, N_2O_4 supply, and equilibration baths must be set up in well-ventilated fume hoods.

- When manipulating the stopcocks, use two hands; never force a stopcock that is hard to turn.
- When handling (and especially walking with) the filled sample bulb, exercise particular care not to drop or break it.
- If a filled bulb breaks, consult your instructor. The laboratory might have to be evacuated until the fumes dissipate.
- This experiment may involve the use of liquid nitrogen, the use of which may be hazardous.
- If you add liquid nitrogen to the trap(s), do so slowly to avoid back splashing. If liquid nitrogen comes in contact with the skin, frostbite may result. If this occurs, consult your instructor immediately.
- Never allow the vacuum pump to pull air through the liquid nitrogen–filled trap(s). This may cause the accumulation of liquid oxygen in the trap, and this can be hazardous.
- The Dewar flask containing the liquid nitrogen must never be tightly closed; an explosion may occur if N_2 boil-off cannot escape.
- Liquid nitrogen must always be used in a well-ventilated room.

Procedure

1. *The success of this experiment largely depends on making these mass measurements as accurately as possible.* If the volume of the sample bulb has already been measured, record this value on your data sheet or in your notebook. If the volume has not yet been determined, it will be measured after the experiment has been completed. Read and record the barometric pressure, as well as the ambient temperature near the balance.

2. The sample bulb will be evacuated and filled using a gas-handling vacuum rack. A schematic diagram of the rack is shown in Figure 4.1.2. A mechanical rotary pump is connected to the manifold of the rack through a cold trap that is filled with a cryogenic medium (either liquid nitrogen or dry ice/acetone or dry ice/isopropyl alcohol). A second trap should be used as a backup. The purpose of these traps is to condense materials that would damage the pump if they were to come into contact with the oil and mechanical parts. If liquid nitrogen is used, it is very important to avoid pumping large volumes of air through the trap because oxygen ($T_b = 91$ K) will condense at the trap temperature (77 K). If the collected liquid oxygen were allowed to warm up (if, for example, the liquid nitrogen coolant were to evaporate) the large pressure buildup could cause the trap to explode. The cryogenic medium is contained in a Dewar flask that, being well insulated, is able to hold this liquid for extended periods of time. During the filling of the bulbs, the level of cryogenic liquid in the Dewar flask should be checked and fluid replaced if necessary. The trap is connected to the manifold through a stopcock. The manifold has several stations (each coupled via a stopcock) to which the sample bulbs are attached through ball joints or O-rings. A stainless-steel pressure/vacuum–reading Bourdon-type gauge and a tank of N_2O_4 are also attached to the manifold.

Before the sample bulbs are evacuated and filled, the instructor will review the design and use of the vacuum rack. The rack will be pumped down, the traps filled, and the manifold stopcock (*M*) opened (see Figure 4.1.2). Attach the bulbs to the manifold stations using the joint clamps. A *minimal* amount of high-vacuum silicone grease should be applied to the ball joint of the sample bulb to ensure a good seal to the socket on the manifold station. First the bulbs are evacuated by opening both the bulb (*B*) and station (*S*) stopcocks. When the system pressure has dropped to about 1 psia (or the smallest division on the Bourdon gauge),

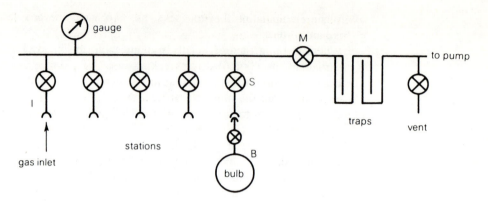

FIGURE 4.1.2   Schematic diagram of gas-handling vacuum manifold for evacuating and filling sample bulbs.

continue pumping on the bulbs for a few minutes to remove adsorbed water. It will help to warm the bulbs slightly with a heat gun while pumping. (1 psia corresponds to an absolute pressure of about 52 torr. Mechanical pumps in good condition can produce a vacuum with a background pressure of much less than 1 torr.)

3. Remove the sample bulb and thoroughly wipe off the residual stopcock grease from the ball joint. For the remainder of the experiment, care should be used in handling this bulb because small mass differences are to be determined. For this reason, a ballast bulb is used to correct the measured bulb mass for changes in the buoyancy of air caused by transient fluctuations in atmospheric pressure. This is discussed in Questions and Further Thoughts.

Rinse the outside of the bulb with acetone or methanol and carefully wipe to remove dirt or moisture. If a double-pan balance is used, carefully weigh the sample bulb, using the ballast bulb as a reference. The weighing should be repeated at least two more times by removing and replacing the sample bulb. This gives an indication of the reproducibility of this important measurement. Record these data. Leave the ballast bulb on the reference pan during the rest of the experiment. If a single-pan balance is used, first determine the mass of the ballast bulb and then that of the sample bulb. Both bulbs should be weighed several times to determine the mass reproducibility.

4. Apply a *light* coating of stopcock grease (too much grease is often worse than not enough) to the sample bulb ball joint and attach it to the vacuum rack station using the pinch clamp. Open the station *(S)* and bulb *(B)* stopcocks and wait until the rack is again pumped down. Open the gas inlet stopcock *(I)* to the N_2O_4 cylinder and evacuate the gas inlet line.

5. Carefully turn the manifold stopcock *(M)* to isolate the manifold from the trap and pump. The vacuum gauge should not show any change; if it does, there is a leak, most likely from one of the sample bulb joints. This leak must be found and eliminated before the bulbs can be filled. Open the N_2O_4 cylinder head valve to allow gas to fill the manifold and the sample bulbs. The system should be filled to a pressure of *at least* 1 atm, but not more than 5 psi (250 torr) above 1 atm. If necessary, the N_2O_4 tank can be heated with a heat gun to promote evaporation. After the bulbs are filled, each *bulb* stopcock should be turned off and the manifold stopcock *(M)* should be *slowly* opened. The gas is then pumped out of the system and is condensed in the trap(s). After the system is reevacuated, the gas inlet *(I)* and station stopcocks *(S)* close and remove the filled bulbs. Do not shut the vacuum rack down yet because at the end of the experiment the sample bulbs

will be reevacuated. Top off the Dewar flask(s) with cryogenic liquid if necessary. Thoroughly remove the stopcock grease from the bulb ball joint.

6. Place the filled bulb in the 30° temperature bath (which *must* be located in a fume hood). Record the actual temperature. It is important to prevent any water from getting into the cell opening above the stopcock. After a minute or two, open the bulb stopcock *momentarily* every 2 to 3 min until the brown vapor no longer emerges from the opening. Make sure the stopcock is securely closed and remove the bulb from the bath. Rinse the bulb with acetone or methanol and carefully wipe the outside of the bulb. Weigh the sample *and* ballast bulbs. If a double-pan balance is used, the former is again weighed relative to the latter. Repeat measurement of the sample bulb to establish reproducibility. Record these data.

7. Place the bulb in the 45°C bath and follow the procedure in step 6; repeat for the other temperatures (60° and 75°). Finally, record the barometric pressure and ambient temperature at the end of the experiment.

8. After the final weighing, attach the bulb to the vacuum rack and pump out the gas. When this is complete, shut the stopcock and remove the bulb. When all the bulbs are reevacuated, the instructor will pump out the system and shut it down.

9. If the volume of the sample bulb is not provided, fill it with deionized or distilled water by immersing the bulb tip in the water and opening the stopcock. If the bulb does not become completely filled, use a narrow-tipped (e.g., Pasteur) pipet. Remove any water from the space above the stopcock, and weigh the filled bulb. Make sure the balance you are using can withstand this weight. Record the ambient temperature.

Calculations

1. Tabulate the sample and (if measured) ballast bulb weights. Subtract the tare weight of the bulb, and list the masses of the N_2O_4/NO_2 contents at the different temperatures studied. From the ballast bulb weights (or calculations based on the displaced air weights) tabulate the buoyancy correction for each measurement and list the respective corrected N_2O_4/NO_2 weights. The corrected bulb weight is obtained by subtracting the buoyancy correction from the observed bulb weight. [Thus if the buoyancy correction is -2.0 mg (e.g., the air had become warmer), the observed bulb weight is increased by 2.0 mg.]

2. Using the value of the bulb volume, and the atmospheric pressure (average the two results if a change is recorded), calculate values of a. Determine K_P and α at the temperatures studied. Tabulate these results.

3. Plot $\ln K_P$ vs. $1/T$ and obtain the slope both graphically and by linear regression (least squares). From this, calculate $\Delta H°$ and indicate the probable error.

4. By interpolation, determine K_P at 25°C and, from this, $\Delta G°$ and $\Delta S°$. Estimate the errors in these values.

5. Compare these results with literature values. (Use the standard thermodynamic constants of formation of NO_2 and N_2O_4.)

Questions and Further Thoughts

1. Regarding the importance of the ballast bulb in these measurements, the change in the measured mass of the bulb arising from air buoyancy effects is equal to the change in air mass brought about by pressure and temperature changes. Using the ideal gas law for air and a propagation of errors treatment, show that for a mean temperature and pressure of 300 K and 1 atm, respectively, a change in the temperature of 1° and in the

pressure of 1 torr together cause a change in the measured mass of about 1.4 mg. Assume a volume of 250 mL. Comment on whether the buoyancy corrections are important in your experiment.

2. Look up the boiling point of liquid nitrogen (LN_2) and the sublimation temperature of CO_2 at 1 atm pressure. Which of the following gases would be trapped (i.e., have a vapor pressure below 1 atm) by these cryogenic media: He, Ne, Ar, CH_4, SiH_4, C_2H_4, C_2H_6, C_3H_8, C_3H_6 (propene)?

3. Why is it necessary to immerse the bulbs in baths in order of increasing temperature?

4. Describe an arrangement by which the experiment would be performed using a fixed amount of N_2O_4/NO_2 in a sample bulb by measuring the total pressure as a function of temperature. The entire pressure-measuring device would always have to be kept at a higher temperature than that of the sample bulb itself. Why?

5. In 4 above, derive the expression by which K_P is obtained from P_{tot}. What else would have to be measured?

6. One source of error in the experiment is the possibility that air could be introduced into the bulb during the venting process. If this occurred, equation (4.1.10) would be replaced by $P_{tot} = P_{NO_2} + P_{N_2O_4} + P_{air}$, or $P_{tot} > P_{NO_2} + P_{N_2O_4}$. What would be the effect on K_P and ΔG° if equation (4.1.10) in its original form were used to calculate K_P (i.e., this error went undetected)?

7. The mole fraction equilibrium constant, $K_x = x_{NO_2}^2/x_{N_2O_4}$ is related to K_P by the equation $K_x = K_P(P_{tot})^{-\Delta v}$. Since P_{tot} is known (and constant) for this experiment, K_P could be determined if K_x were evaluated. Suggest a method to determine x_{NO_2} and $x_{N_2O_4}$ from the average molecular weight obtained from the relationship, $<M> = (m_{tot})RT/PV$, where m_{tot} is the mass of gas determined in this experiment and P, V, R, and T have the usual meanings.

8. The standard heat capacities of $N_2O_4(g)$ and $NO_2(g)$ are 77.28 and 37.20 J K^{-1} mol^{-1}, respectively. Assuming constant C_p° values, by how much does the enthalpy of N_2O_4 dissociation change between the lowest and highest temperatures in the experiment? In light of this, would you be able to detect the expected change in the slope of your ln $(K_P), 1/T$ data?

Notes

1. The standard state for a gas is 1 atm and *ideal gas conditions*. Thus any nonideal behavior of the gas at 1 atm pressure must be taken into account. The expression for this "correction term" is given in I. N. Levine's "Physical Chemistry," 2nd ed., pp. 139–140.

2. The intrinsic spin of the unpaired electron contributes a magnetic moment to the molecule, and thus it becomes oriented with respect to the lines of force of an externally applied magnetic field. This phenomenon is called paramagnetism.

3. J. G. Calvert and J. N. Pitts, Jr., "Photochemistry," pp. 216–222, Wiley, New York, 1967.

4. F. S. Wettack, *J. Chem. Educ.*, 49, 556 (1972).

Further Readings

R. A. Alberty, "Physical Chemistry," 7th ed., pp. 139–149, Wiley (New York), 1987.

P. W. Atkins, "Physical Chemistry," 3rd ed., pp. 212–224, W. H. Freeman (New York), 1986.

G. N. Castellan, "Physical Chemistry," 3rd ed., pp. 233–240, Addison-Wesley (Reading, Mass.), 1983.

I. N. Levine, "Physical Chemistry," 2nd ed., pp. 163–176, McGraw-Hill (New York), 1983.

J. H. Noggle, "Physical Chemistry," pp. 266–290, Little, Brown (Boston), 1985.

F. H. Verhoek and F. Daniels, *J. Am. Chem. Soc.*, 53, 1250 (1931).

Experiment 1

Thermodynamics of a Gas Phase Reaction:
Reversible Dissociation of N_2O_4

NAME _____   DATE _____

Bulb # 1

Volume _____

Bath temperature	Density bulb weight	Ballast bulb weight
_____	_____	_____
_____	_____	_____
_____	_____	_____
_____	_____	_____
_____	_____	_____

Evacuated _____   _____   _____

Bulb # 2

Volume _____

Bath temperature	Density bulb weight	Ballast bulb weight
_____	_____	_____
_____	_____	_____
_____	_____	_____
_____	_____	_____
_____	_____	_____

Evacuated _____   _____   _____

Experiment 1

Thermodynamics of a Gas Phase Reaction: Reversible Dissociation of N_2O_4

NAME _____   DATE _____

Bulb # 1

Volume _____

Bath temperature	Density bulb weight	Ballast bulb weight
_____	_____	_____
_____	_____	_____
_____	_____	_____
_____	_____	_____
_____	_____	_____

Evacuated _____   _____   _____

Bulb # 2

Volume _____

Bath temperature	Density bulb weight	Ballast bulb weight
_____	_____	_____
_____	_____	_____
_____	_____	_____
_____	_____	_____

Evacuated _____   _____   _____

☐ Experiment 2

Heat Capacity Ratio (C_P/C_V) of Gases

Objective

To determine the ratio of constant-pressure to constant-volume heat capacities of gases (e.g., He, air, CO_2 and SF_6) by measuring the speed of sound.

Introduction

Constant-volume and constant-pressure heat capacities, C_V and C_P, respectively, are important thermodynamic properties that concern such practical matters as heat transfer, as well as the theoretical treatment of thermodynamic equations of state. These quantities are defined as

$$C_V = \left(\frac{\partial U}{\partial T}\right)_V \quad \text{and} \quad C_P = \left(\frac{\partial H}{\partial T}\right)_P. \tag{4.2.1}$$

Heat capacities are experimentally determined by measuring the change in temperature that is brought about by the absorption of a known quantity of heat by a substance under controlled conditions (such as constant volume or pressure). The numerical value of the heat capacity of a pure substance, and its temperature dependence, provides a vital link between experiment and theory (statistical mechanics).

A quantity that appears frequently in thermodynamics is the *ratio* of C_P to C_V, γ, thus:

$$\gamma \equiv \frac{C_P}{C_V}. \tag{4.2.2}$$

As an example, the expression linking the pressure and volume in a reversible *adiabatic* process for an ideal gas is PV^γ = constant. This is distinctly different from the case of an *isothermal* process for which $PV = RT$ = constant.

As will be shown below, the value of γ for a gas reflects certain structural characteristics of the gas. Knowledge of γ also allows the computation of C_V if only C_P is known (or vice versa). In this experiment, γ will be determined for several gases using a technique in which the speed of sound is measured. This method might appear to be indirect when compared, for example, with an adiabatic expansion of the gas in which the temperature change is measured consequent to a known pressure change. In the latter case, the application of the ideal gas law to PV^γ = constant provides the expression

$$P^{[(\gamma - 1)/\gamma]} \frac{1}{T} = \text{constant}. \tag{4.2.3}$$

Although γ can be determined in this manner, the speed of sound method has the advantage of illustrating some important characteristics of waves and sound propagation. The technique also involves the use of fundamental and important electronic equipment.

In order to understand the nature of this experiment, as well as the methodology of data analysis, it is important to consider the nature of sound waves. Sound is propagated through a medium by *longitudinal waves*. A longitudinal wave is a

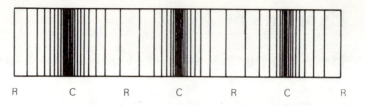

FIGURE 4.2.1   Schematic diagram of a longitudinal standing wave. The vertical lines represent constant-density fronts; C and R show positions of gas compressions and rarefactions (high and low densities), respectively.

type of periodic motion in which the displacement of the particles in the medium occurs in the *same* direction as the wave itself. This behavior is in contrast to a *transverse wave*, a common example of which is the spreading of circular ripples on the quiescent surface of water after a stone is thrown into the water. The water oscillates vertically (as the wave passes), but the wave moves outward, horizontally. There is little or no transport of the water itself along the direction of wave propagation.

A schematic diagram of a longitudinal sound wave is shown in Figure 4.2.1. For simplicity, a one-dimensional wave is depicted; one can imagine that sound, generated by an oscillating boundary at the left, is traveling to the right through a medium. The motion of the sound wave is a function of both time and space. The vertical lines indicate density contours of the medium and show that in the wake of the traveling wave there are periodic *compressions (C)* and expansions, or *rarefactions (R)*. The density of the fluid ahead of the wavefront is the undisturbed bulk density, d_0, which is intermediate between the local densities of the medium at C and R.

A general treatment of propagating sound waves through a medium results in the following expression for the *wave velocity, v_s*:[1]

$$v_s = \left(\frac{B}{d_0}\right)^{1/2}, \tag{4.2.4}$$

in which B is called the *bulk modulus* and d_0 is the bulk density of the medium. B is explicitly defined as

$$B = -V\frac{dP}{dV}. \tag{4.2.5}$$

The negative sign makes $B > 0$ because, for all media, a decrease in V causes an increase in P. It should be noted that B is a *bulk* property of the medium and is also the reciprocal of the compressibility, κ, i.e.,

$$\kappa \equiv -\frac{1}{V}\frac{dV}{dP}. \tag{4.2.6}$$

An important aspect of sound propagation is the fact that if the frequency of the sound being generated is high enough, i.e., audio frequencies which, in this experiment, are between 10^3 and 10^4 Hz (oscillations per second), the compressions and rarefactions are established very rapidly as the sound wave moves through the medium. This condition means that heat transport between the compressed and rarefied regions of the medium and the surroundings is slow relative to the creation of the compressions and rarefactions. Thus, on a local basis, the compressions and rarefactions are carried out *adiabatically*. At much lower sound frequencies, on the other hand, it is possible to imagine that heat transport between the medium and the surroundings is fast enough to allow the medium to be compressed and expanded *isothermally* (if the thermal mass of the surroundings is large enough). Accordingly, the compressibility κ can be described under constant-temperature or constant-energy conditions, and one can thus distinguish between isothermal and adiabatic compressibilities of a substance, κ$_T$, and κ$_S$, respectively.[2]

Since audio frequencies are used in this experiment, we must use the *adiabatic* (or isentropic) bulk modulus, which can be explicitly written as

$$B_S = -V\left(\frac{\partial P}{\partial V}\right)_S. \tag{4.2.7}$$

Wave Velocity, Speed of Sound, and Thermodynamics

Our goal is to express the wave velocity in terms of thermodynamic relations. To this end, we consider the medium (which up to now could be any fluid or solid) to be an *ideal gas*. This is not an unreasonable assumption because the experiment deals with gases at atmospheric pressure. Moreover, the intensity of the sound is not so great that the local pressure of the compression zone becomes excessive for the ideal gas approximation.

We start by applying the cyclic rule to the partial derivative in equation (4.2.7), thus:

$$\left(\frac{\partial P}{\partial V}\right)_S = \frac{-(\partial S/\partial V)_P}{(\partial S/\partial P)_V}. \tag{4.2.8}$$

By using the Maxwell relations, $(\partial S/\partial V)_P = (\partial P/\partial T)_S$ and $(\partial S/\partial P)_V = -(\partial V/\partial T)_S$, the derivatives on the right-hand side of (4.2.8) can be expressed as

$$\left(\frac{\partial P}{\partial V}\right)_S = \frac{(\partial P/\partial T)_S}{(\partial V/\partial T)_S}. \tag{4.2.9}$$

The fundamental relationships for dU and dH as applied to an ideal gas are

$$dU = T\,dS - P\,dV = C_V dT,$$

and $\hspace{9cm}$ (4.2.10)

$$dH = T\,dS + V\,dP = C_P dT.$$

Solving these respective expressions for dS provides

$$dS = \frac{C_V dT}{T} + \frac{P\,dV}{T},$$

and $\hspace{9cm}$ (4.2.11)

$$dS = \frac{C_P dT}{T} - \frac{V\,dP}{T}.$$

In view of the fact that a reversible adiabatic process is also isentropic ($dS = 0$), the partial derivatives needed in equation (4.2.9) can be obtained from (4.2.11):

$$\left(\frac{\partial V}{\partial T}\right)_S = \frac{-C_V}{P},$$

and $\hspace{9cm}$ (4.2.12)

$$\left(\frac{\partial P}{\partial T}\right)_S = \frac{C_P}{V}.$$

Finally, by substituting the derivatives in (4.2.12) into equation (4.2.9), and using the definition of B_S [equation (4.2.7)], we see that

$$B_S = \frac{PC_P}{C_V} = P\gamma \qquad \text{(ideal gas)}. \qquad (4.2.13)$$

The speed of sound in an ideal gas is now, from (4.2.4),

$$v_s = \left(\frac{\gamma P}{d_0}\right)^{1/2} \qquad (4.2.14)$$

Since for an ideal gas, $P = nRT/V$, and $d_0 = nM/V$, where M is the molecular weight, the wave velocity, or the speed of sound, may also be expressed as

$$v_s = \left(\frac{\gamma RT}{M}\right)^{1/2} \qquad (4.2.15)$$

This simple-looking result predicts that the speed of sound in an ideal gas: (1) is independent of the pressure and (2) varies as $T^{1/2}$ and $M^{-1/2}$. It is interesting to realize that equation (4.2.15) resembles the expression for the mean speed of molecules obtained from kinetic theory, namely, $<c> = (8RT/\pi M)^{1/2}$. Since γ is usually between 1 and 1.7, depending on the complexity of the molecule (see below), the speed of sound and mean molecular speed are comparable, roughly 10^4 cm/s.

Heat Capacity and Molecular Structure

We now discuss the interpretation of heat capacities in terms of molecular structure. First consider the meaning of C_V; it is the derivative of the internal energy with respect to temperature, see (4.2.1). From gas kinetic theory applied to hard sphere particles (masses having no "internal" structure), the only vestige of energy is kinetic, or translational, motion, and the mean translational energy, $<E_{\text{trans}}>$, (per mole) is $3/2RT$. This corresponds to the internal energy, U, and thus C_V for this monatomic gas is $3/2R$. A review of the derivation of E_{trans} will reveal that motion relative to the three Cartesian axes, x, y, and z, is isotropic, that is, movement can occur along either axis with equal probability (this ignores gravitational effects). Thus each of the three components of motion contributes $1/2RT$ to the mean translational energy and $1/2R$ to the heat capacity. Each of the available, separate, ways of manifesting energy (in this case, E_x, E_y, and E_z) is called a *degree of freedom*. More specifically, a degree of freedom is an energy component of a system (e.g., a large number of molecules) that is expressed in terms of a squared coordinate or a squared momentum. In this case, $E_x = 1/2mv_x^2 = p_x^2/2m$, etc. The *equipartition theorem*, first deduced by Maxwell, states that each degree of freedom of a mole of particles possesses the same average energy, $1/2RT$.

The relationship between C_V and C_P for an ideal gas is $C_P = C_V + R$. Therefore, C_P for a monatomic ideal gas is $5/2R$. These predictions are borne out by measurements; experimental values of C_V and C_P at ambient temperature for He are 12.45 and 20.79 J mol^{-1} K^{-1} ($1.50R$ and $2.50R$), respectively. In considering the heat capacities for molecules, however, one must take into account the fact that there are additional ways that a molecule can manifest energy other than translation, namely, rotation and vibration. The expression for the rotational (kinetic) energy of a rigid molecule is $E_{\text{rot}} = 1/2I\omega_x^2$, where ω_x is the angular velocity around the x axis and I is the moment of inertia. For a diatomic molecule, A—B, $I = \mu r^2$, where μ is the reduced mass, $m_A m_B/(m_A + m_B)$, and r is the internuclear distance. It turns out that for a linear molecule, there are two independent axes around which the rotating molecule possesses kinetic energy (e.g., the x and y axes; the z axis is taken to coincide with the internuclear coordinate). For a nonlinear molecule, there are three components of rotational energy. The consequence of this is that a linear molecule has *two* rotational degrees of freedom, while a nonlinear one possesses *three* such degrees of freedom. For a linear molecule, rotation would be expected to contribute $2(1/2)R$, or R to the (constant volume) heat capacity, while for a nonlinear molecule, a contribution of $3/2R$ to C_V is expected.

In understanding the role of molecular vibrations in this discussion of heat capacities, it is best to consider first a diatomic molecule that possesses only one vibration: the internuclear stretch. The expression for the total vibrational energy associated with this stretching mode contains two terms, $1/2\mu\ v_z^2 + 1/2kz^2$, where v_z represents the relative motion of the nuclei along the internuclear z axis, and k is called the *force constant*, a quantity that reflects the strength of the chemical bond between A and B. These two terms, therefore, represent the kinetic and potential energy components of molecular vibration. Thus there are *two* degrees of freedom associated with this (and any other) vibration, and it is accordingly expected to contribute ca. 8.31 J mol^{-1} K^{-1} to C_V.

For a polyatomic molecule containing N atoms, there are $3N$ coordinates needed to specify the locations of the atoms (an x, y, z set for each atom). This corresponds to a *total* of $3N$ degrees of freedom (translational, rotational, and vibrational). Of these, three account for the net kinetic energy of the molecule and three pertain to the rotational energy (if nonlinear; if linear, two degrees of rotational freedom are required). The remaining degrees of freedom account for the vibrational energy. This is summarized below for the degrees of freedom for an N-atom molecule.

Molecule Type	Total	Translational	Rotational	Vibrational
Linear	$3N$	3	2	$3N - 5$
Nonlinear	$3N$	3	3	$3N - 6$

As a specific example, we consider the nature of the vibrational degrees of freedom for the nonlinear triatomic molecule, H_2O. There are nine degrees of freedom, of which three are vibrational. These three vibrations involve different patterns of nuclear displacements which are illustrated in Figure 4.2.2. The arrows

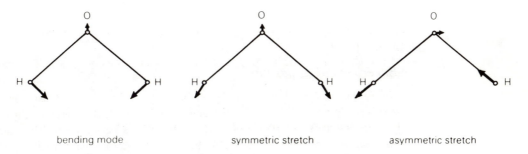

bending mode   symmetric stretch   asymmetric stretch

FIGURE 4.2.2   Representations of the three vibrational modes of water showing atomic displacements. The length and orientation of each arrow indicate the magnitude and direction of an atom's motion. In each case, the center of mass of the molecule is stationary.

indicate the directions of motion of the atoms. The first is called the bending mode in which the molecule moves in a scissors-like manner. The second is the symmetric stretch wherein the H atoms move away from (or toward) the central O atom together, i.e., in phase. The third is the asymmetric stretch in which one H atom approaches the O atom while the other moves away; this is the out-of-phase motion.

In summary, the total energy (translational and internal) of a nonlinear molecule is given by

$$U = U_{trans} + U_{rot} + U_{vib} = 3/2RT + 3/2RT + \frac{2(3N - 6)}{2}RT, \quad (4.2.16)$$

and hence the heat capacity, C_V, is expected to be

$$C_V = (3N - 3)R, \quad (4.2.17)$$

$[(3N - 2)R$ for a linear molecule]. If the system is considered to be an ideal gas for which $C_P = C_V + R$, it follows that $C_P = (3N - 2)R$ and $(3N - 1)R$ for nonlinear and linear molecules, respectively. Therefore, γ is expected to be

$$\gamma_{nonlin} = \frac{3N - 2}{3N - 3},$$

or $\qquad\qquad\qquad\qquad\qquad\qquad\qquad\qquad\qquad\qquad (4.2.18)$

$$\gamma_{lin} = \frac{3N - 1}{3N - 2},$$

for polyatomic molecules in the ideal gas approximation.

In reality, however, this prediction is not borne out by actual measurements of C_P and C_V. For example, in the case of methane, CH_4, the predicted and measured values of γ are 1.083 and 1.31, respectively. The reason for this discrepancy is that the equipartition theorem, which is based on classical mechanics, fails to account for quantum mechanical effects that are exhibited by ultramicroscopic systems such as molecules, and thus the average energy and the heat capacity expressions (4.2.16) and (4.2.17) do not always represent these quantities quantitatively. To explain the reason for this, one has to consider quantum mechanical effects and realize that vibrational, rotational, and even translational energies are characterized by discrete *quantum levels*. The spacing between these levels is sufficiently large for *vibrational* degrees of freedom that molecules are not "fully excited" at ordinary temperatures, and therefore many vibrational degrees of freedom do not provide their full contribution of RT to the average energy. In one case, namely, H_2, rotational levels are relatively far apart so that rotational motion does not contribute $1/2RT$ (per degree of freedom) at ambient temperatures. For further discussion of this very interesting topic, the reader should consult treatments of quantum mechanics and statistical mechanics (see the Notes and Further Readings).

Experimental Apparatus

A schematic diagram of the apparatus is shown in Figure 4.2.3. An earphone *(E)* is driven by an audio (sine wave) generator *(AG)* which operates at a frequency between about 2 and 10 kHz (1 kHz is 10^3 cycles per second). The microphone *(M)* converts the pressure changes that it senses into an electrical signal that is amplified by a simple integrated circuit *(AMP)*. The earphone and microphone are mounted at the ends of a movable cylindrical cavity that is filled with the gas to be studied. [Kundt (1866–1868) used a similar device to measure the speed of sound. The sound was generated by stroking a glass rod; the wave maxima and

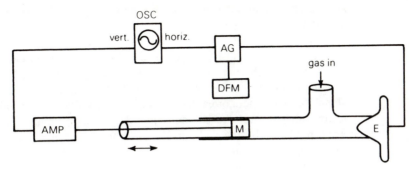

FIGURE 4.2.3   Schematic diagram of Kundt tube and associated acoustical and electronic components.

minima were located by using fine powder.] A marker attached to the moving part of the cylinder allows its displacement to be measured. The waveforms produced by the audio generator and the amplified microphone signal are displayed by an oscilloscope. The value of the audio frequency is indicated by a digital frequency meter (DFM). The gas to be studied is slowly passed through the tube; it enters through an inlet near the earphone driver and escapes through the space between the tube and the movable microphone.

Safety Precautions

☐ Safety glasses must be worn in the laboratory at all times.

☐ This experiment requires the use of gases contained at high pressures in metal cylinders. These cylinders must be securely strapped. Each cylinder is equipped with a reducing valve that delivers gas to the Kundt tube at a pressure of only a few (less than 5) psig. Do not adjust this pressure. If the pressure or flow rate is excessive, the movable microphone can be forced out of the resonance tube.

☐ This experiment must be conducted in an open, well-ventilated laboratory.

Procedure

The experiment will be performed in two parts, each demonstrating a different characteristic of periodic sound waves. In the first part, you will determine the speed of sound in the gases by measuring the distance between compressions in a standing, or resonant, wave established in the acoustical tube. See Figure 4.2.1. This is done by examining the phase relationship between the driver (earphone) and the receiver (microphone) as a function of cavity length at a fixed (but arbitrary) frequency. In the second part, the speed of sound will be obtained by determining the audio frequencies required to sustain a standing wave in a fixed cavity length.

Part I

1. Briefly purge the cavity with dry air before measurements are started. The instructor will demonstrate the use of the gas supply. After initial purging, a *very slight* trickle of air should flow through the tube. Never allow the flow rate of gas to be excessive because the movable part of the acoustical tube could be blown off. Now, and periodically throughout the experiment, read and record the ambient temperature.

2. Set the audio generator to a frequency between about 2.5 and 5.0 kHz. The digital frequency meter displays this frequency; after it remains constant, record its value.

3. The signal from the audio generator is coupled to the horizontal (or external) input of the oscilloscope, while the amplified microphone signal is fed into the vertical input. The scope is thus used in the *X-Y* mode.

4. After you adjust the horizontal *(X)* and vertical *(Y)* gain controls, a pattern will appear on the oscilloscope screen. This figure is called a Lissajous pattern, and its shape provides information about the phase relationship between the *X* signal (audio generator/earphone) and the *Y* signal (microphone/amplifier). Assuming that neither the earphone nor the microphone/amplifier distorts the sine wave produced by the audio generator, and that each input to the scope has the same amplitude, a *circle* will be produced when the two waveforms differ in phase by $\pm 90°$. (If the amplitudes are unequal, an ellipse is observed.) However, if the two input waveforms differ in phase by $0°$, a *straight line* having a slope of $+45°$ is seen; likewise, if the phase difference is $180°$, the slope is $-45°$.

5. Since the occurrence of a straight line can be determined quite accurately, the position of the microphone that produces phase lags of either $0°$ or $180°$ can be readily measured. The distance through which the microphone is displaced between two successive equal Lissajous patterns is equal to one wavelength. By carefully moving the rod to which the microphone is attached, and by reading the position of the marker attached to the rod, record the positions at which straight-line patterns are produced. Make sure the oscilloscope is properly focused so that the precision of the straight-line determination can be maximized.

6. Repeat the above covering as much microphone "travel" as possible. Determine as many "nodal positions" as possible.

7. Choose a different frequency (between 2.5 and 5 kHz) and perform another set of measurements in duplicate.

8. Now shut off the air supply and gently purge the tube with CO_2 by slowly opening the appropriate needle valve. After about 1 minute, reduce the flow rate so that only a *minimal* amount of gas flows through the tube.

9. Change the audio frequency so that the frequency meter reads between about 2.0 and 3.5 kHz. Record the value chosen and repeat the above procedure followed for air in duplicate.

10. Change to a new audio frequency between 2 and 3.5 kHz and carry out another set of readings in duplicate.

11. Change to He by following the procedure in step 8.

12. Use a frequency between 8 and 11 kHz. For this gas, you may have to increase the amplifier gain and/or the vertical display sensitivity on the scope.

13. Repeat the same procedure followed with the other gases in duplicate; if time permits, another set of readings with a different frequency should be obtained.

14. If SF_6 is to be studied, purge the cavity with this gas and perform a set of measurements as above. Frequencies between 1.5 and 2.5 kHz are appropriate.

Part II

In this part, the cavity length will be fixed and the sound frequency will be varied. A maximum in signal strength is observed when the cavity length is equal to an integral number of half wavelengths. This is because the signal strength, or intensity, is proportional to the square of the wave amplitude. See the diagram of a standing wave and its intensity pattern in Figure 4.2.4.

1. The oscilloscope is now used in the internal time base mode. Switch the time sweep to the 10-kHz (or 0.1-ms) position. Choose an arbitrary cavity length (about 50 cm) and fix the position of the microphone rod by taping it to the outer cylinder. Measure and record the length of the resonant cavity (between the earphone diaphragm and the microphone head).

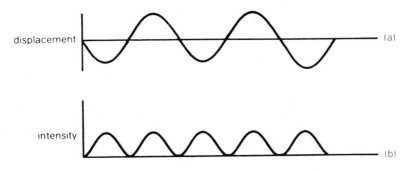

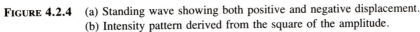

FIGURE 4.2.4 (a) Standing wave showing both positive and negative displacement.
(b) Intensity pattern derived from the square of the amplitude.

2. Change the gas back to dry air by gently purging, etc. You should observe a vertical display on the oscilloscope that corresponds to the amplified signal produced by the microphone.

3. Starting with about 2.5 kHz, adjust the frequency of the audio generator until the vertical display on the oscilloscope reaches a distinct maximum; record the frequency indicated by the digital frequency meter. Gradually increase the audio frequency until another maximum is produced on the oscilloscope and record this frequency. Continue to increase the audio frequency, recording values that produce maxima. Go to as high a frequency as possible; you may have to increase the gain of the microphone amplifier and/or the vertical display of the oscilloscope. The earphone, microphone, and amplifier combination operates less efficiently at higher frequencies.

4. Purge the tube, fill with CO_2, and repeat the above measurements.

5. Purge, fill with SF_6 (or whatever other gas is studied), and repeat.

6. Fill the tube with He and determine the frequency maxima. Start with a frequency of about 5 kHz.

Data Analysis

From the data obtained in Part I, determine the average value of the differences between the microphone positions that provided the straight-line Lissajous pattern. This average corresponds to one wavelength at the frequency used. The speed of sound is then determined from $v_s = \lambda f$, where λ is the wavelength and f is the frequency. For each gas studied, tabulate the average values of v_s thus obtained, indicating the errors. Compare with literature values.

For each of the data sets in Part II, plot the frequency at which maxima are observed vs. n, an arbitrary integer, beginning with the low-frequency data. This plot should be linear having a slope equal to $v_s/2L$, where L is the cavity length. Note that for each gas studied, the plot may deviate from linearity at lower frequencies. This appears to occur at frequencies for which 2λ approaches (and exceeds) L. Notice that the speed of sound is independent of the frequency.

From these slopes, and the measured value of L, obtain v_s values for the gases. Tabulate these values separately and compare with those obtained in Part I.

Finally, using the most reliable values of v_s for the gases, calculate γ for each gas. Comment on the values of γ with respect to the discussion in the introduction. The write-up should contain a discussion of the types of errors that enter into the two methods for measuring v_s. Indicate which method seems to provide the more accurate and/or the more precise data, and justify your conclusions.

Questions and Further Thoughts

1. According to equation (4.2.15), the speed of sound is independent of pressure. Does this (at first thought) make sense physically? What property of sound propagation changes at very low pressures (compared with moderate pressures, e.g., 1 atm)? Remember, sound does not propagate through a vacuum.

2. From equation (4.2.15) the speed of sound varies as $T^{1/2}$. Can you relate your experience in hearing sound travel across a flat, unobstructed surface (such as calm water) at both high and low temperatures with the difference in the *speed* of sound?

3. What structural information could be obtained by measuring the speed of sound through a solid? How could the speed of sound through a solid, eg., a uniform rod, be measured (imagine using an audio generator, receiver, and oscilloscope)?

4. Have you ever heard someone speak after having inhaled some helium (such as from a He-filled balloon)? If so, explain what you hear in terms of what you have learned in this experiment.

5. Predict the number of degrees of freedom and the number of translational, vibrational, and rotational modes for HCN, HCl, C_2H_2, C_2H_4.

Notes

1. D. Halliday and R. Resnick, ''Fundamentals of Physics,'' pp. 292–294, 317–321, Wiley (New York), 1981.

2. J. N. Noggle, ''Physical Chemistry,'' pp. 103, 143–144, Little, Brown (Boston), 1985.

Further Readings

A. W. Adamson, ''A Textbook of Physical Chemistry,'' 3rd ed., pp. 121–131, Academic Press (Orlando, Fla.), 1986.

P. W. Atkins, ''Physical Chemistry,'' 3rd ed., pp. 71–75, 539–541, W. H. Freeman (New York), 1986.

G. W. Castellan, ''Physical Chemistry,'' 3rd ed., pp. 77–80, 122–123, Addison-Wesley (Reading, Mass.), 1983.

I. N. Levine, ''Physical Chemistry,'' 2nd ed., pp. 58–59, 444–447, 763–764, McGraw-Hill (New York), 1983.

Experiment 2

C_P/C_V Ratio

NAME _____ DATE _____

Part I Temperature _____

Gas _____   Gas _____   Gas _____

Frequency #1 _____ kHz   Frequency #1 _____ kHz   Frequency #1 _____ kHz

Positions, cm: Positions, cm: Positions, cm:

_____ _____   _____ _____   _____ _____

_____ _____   _____ _____   _____ _____

_____ _____   _____ _____   _____ _____

_____ _____   _____ _____   _____ _____

_____ _____   _____ _____   _____ _____

_____ _____   _____ _____   _____ _____

_____ _____   _____ _____   _____ _____

Frequency #2 _____ kHz   Frequency #2 _____ kHz   Frequency #2 _____ kHz

_____ _____   _____ _____   _____ _____

_____ _____   _____ _____   _____ _____

_____ _____   _____ _____   _____ _____

_____ _____   _____ _____   _____ _____

_____ _____   _____ _____   _____ _____

_____ _____   _____ _____   _____ _____

_____ _____   _____ _____   _____ _____

(*continues*)

Part II Temperature _____

Gas _____   Gas _____   Gas _____

Cavity Length _____   Cavity Length _____   Cavity Length _____

Frequencies, kHz   Frequencies, kHz   Frequencies, kHz

____	____	____	____	____	____
____	____	____	____	____	____
____	____	____	____	____	____
____	____	____	____	____	____
____	____	____	____	____	____
____	____	____	____	____	____
____	____	____	____	____	____
____	____	____	____	____	____
____	____	____	____	____	____
____	____	____	____	____	____
____	____	____	____	____	____

Experiment 2

C$_P$/C$_V$ Ratio

NAME _____ DATE _____

Part I Temperature _____

Gas _____   Gas _____   Gas _____

Frequency #1 _____ kHz   Frequency #1 _____ kHz   Frequency #1 _____ kHz

Positions, cm: Positions, cm: Positions, cm:

_____ _____   _____ _____   _____ _____

_____ _____   _____ _____   _____ _____

_____ _____   _____ _____   _____ _____

_____ _____   _____ _____   _____ _____

_____ _____   _____ _____   _____ _____

_____ _____   _____ _____   _____ _____

_____ _____   _____ _____   _____ _____

Frequency #2 _____ kHz   Frequency #2 _____ kHz   Frequency #2 _____ kHz

_____ _____   _____ _____   _____ _____

_____ _____   _____ _____   _____ _____

_____ _____   _____ _____   _____ _____

_____ _____   _____ _____   _____ _____

_____ _____   _____ _____   _____ _____

_____ _____   _____ _____   _____ _____

_____ _____   _____ _____   _____ _____

(continues)

Part II Temperature _____

Gas _____   Gas _____   Gas _____
Cavity Length _____   Cavity Length _____   Cavity Length _____

Frequencies, kHz   Frequencies, kHz   Frequencies, kHz

_____ _____   _____ _____   _____ _____
_____ _____   _____ _____   _____ _____
_____ _____   _____ _____   _____ _____
_____ _____   _____ _____   _____ _____
_____ _____   _____ _____   _____ _____
_____ _____   _____ _____   _____ _____
_____ _____   _____ _____   _____ _____
_____ _____   _____ _____   _____ _____
_____ _____   _____ _____   _____ _____
_____ _____   _____ _____   _____ _____
_____ _____   _____ _____   _____ _____

□ Experiment 3

Real Gases

PART I: THE JOULE-THOMSON COEFFICIENT

Objective

To measure the Joule-Thomson coefficient for several gases, such as CO_2, N_2, SF_6, and He.

Introduction

The enthalpy of an ideal gas depends only on the temperature and not on the pressure because of the absence of intermolecular forces. In a real gas, however, for which such forces cannot be neglected, the enthalpy is pressure-dependent, and this property has many practical consequences.

One famous direct experimental study of intermolecular attractions was reported by Joule and Thomson (later Lord Kelvin) in 1853. Phenomenologically, it is the change in temperature accompanying the expansion of a gas that is measured in this experiment. This effect is of great practical importance and has many industrial applications. It is relevant not only in the liquefaction of gases (an ultimate consequence of intermolecular attractions) but also in the operation of the refrigerator and the heat pump. To understand the theoretical background of this experiment, refer to Figure 4.3.1.

The gas, initially at a temperature T_1, enters an insulated cylinder at a constant pressure, P_1. Under these conditions, the gas has a molar volume (V/n), V_{m1}. The gas is driven through a porous plug (which acts as a throttle) and then emerges at a lower (constant) pressure, P_2. The outflowing gas therefore has a larger molar volume, V_{m2}. The equilibrium temperature of the exiting gas is T_2. For a real gas, $T_2 \neq T_1$. The dashed lines in Figure 4.3.1 indicate the imaginary boundary of *one mole* of flowing gas. It is thus represented that $V_2 > V_1$. We will first show that this expansion is carried out at constant enthalpy (isenthalpically). Figure 4.3.1 also shows the pressure profile assumed in the system.

Because the entire system is insulated, $q = 0$ (no heat is exchanged with the

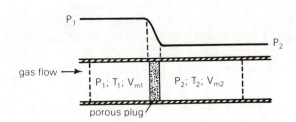

FIGURE 4.3.1   Gas flowing through an insulated tube containing a porous obstruction. The entering gas is at temperature and pressure T_1 and P_1 and has molar volume V_{m1}; the exiting gas is characterized by T_2, P_2, and V_{m2} . $P_2 > P_1$.

surroundings). The work, w_1, involved in pushing an arbitrary volume of gas (e.g., V_1) through the plug is given by

$$w_1 = -\int_{V_1}^{0} P_1 \, dV = P_1 V_1, \qquad (4.3.1)$$

where it is assumed that the driving pressure, P_1, is constant throughout the entire left-hand part of the tube. In equation (4.3.1) $w_1 > 0$ because work is done *on* the gas ($\Delta V < 0$, and work is absorbed). The work done *by* the same quantity of gas as it emerges in the right-hand side of the tube is

$$w_2 = -\int_{0}^{V_2} P_2 \, dV = P_2 V_2, \qquad (4.3.2)$$

where it is again assumed that the low pressure, P_2, is constant throughout the low-pressure region. Note that in equation (4.3.2) $w_2 < 0$ ($\Delta V > 0$) and hence work flows from the system. In the overall expansion, the net work, w, accompanying the flow of 1 mole of gas is

$$w = w_1 + w_2 = P_1 V_1 - P_2 V_2. \qquad (4.3.3)$$

Using equation (4.3.3) and the fact that $q = 0$, the change in the internal energy (per mole) of flowing gas becomes

$$\Delta U = q + w = U_2 - U_1 = P_1 V_1 - P_2 V_2. \qquad (4.3.4)$$

Equation (4.3.4) can be rearranged to give

$$U_2 + P_2 V_2 = U_1 + P_1 V_1, \qquad (4.3.5)$$

and because enthalpy, H, is defined as $U + PV$, it follows that $H_2 = H_1$ and $\Delta H = 0$; therefore, we see that the process is carried out at *constant enthalpy*. The gas can be said to undergo an irreversible, *isenthalpic, adiabatic expansion*.

As mentioned above, in this experiment one measures the temperature change $(T_2 - T_1)$ accompanying an isenthalpic expansion between known initial and final pressures. Thus, the Joule-Thomson (JT) (sometimes called Joule-Kelvin) coefficient, μ_{JT}, which is defined as

$$\mu_{JT} \equiv \lim_{\Delta P \to 0} \left(\frac{\Delta T}{\Delta P}\right)_H = \left(\frac{\partial T}{\partial P}\right)_H, \qquad (4.3.6)$$

where $\Delta T = T_2 - T_1$ and $\Delta P = P_2 - P_1$, can be evaluated. If a gas *cools* in a JT expansion, $T_2 < T_1$ and the JT coefficient is *positive* (since $P_2 < P_1$). Alternatively, if the gas warms, μ_{JT} is negative. As will be mentioned below, the JT coefficient is a function of both T and P, and therefore a gas that heats in a JT expansion under certain conditions can be made to cool if the temperature and/or pressure of the expansion are appropriately changed.

It is desirable to relate μ_{JT} to thermodynamically useful expressions. To this end, we express the enthalpy change in terms of differentials of T and P:

$$dH = \left(\frac{\partial H}{\partial T}\right)_P dT + \left(\frac{\partial H}{\partial P}\right)_T dP = C_P \, dT + \left(\frac{\partial H}{\partial P}\right)_T dP, \qquad (4.3.7)$$

where C_P, the constant pressure heat capacity, is introduced. For an isenthalpic process $dH = 0$, and $(\partial T/\partial P)_H$ is expressed from (4.3.7) as

$$\left(\frac{\partial T}{\partial P}\right)_H = \mu_{JT} = \frac{-(\partial H/\partial P)_T}{(\partial H/\partial T)_P} = \frac{-(\partial H/\partial P)_T}{C_P}. \qquad (4.3.8)$$

As mentioned above, $\mu_{JT} = 0$ for an ideal gas. Because $C_P \neq 0$, the JT coefficient is zero only if $(\partial H/\partial P)_T = 0$. We can understand this derivative to imply something about the nature of intermolecular interactions. At high pressures where the average intermolecular distance is smaller, the presence of intermolecular forces will result in a change in energy (in this case, enthalpy). For example if intermolecular *attractions* dominate, the enthalpy is lower at high pressure and heat is *absorbed* ($\Delta H > 0$) when the pressure is decreased ($\Delta P < 0$), as weakly attracted molecules are separated from each other (assuming T is kept constant). Thus in this case, $(\partial H/\partial P)_T < 0$, and from (4.3.8), $\mu_{JT} > 0$. On the other hand, if intermolecular (or atomic) *repulsions* are more significant, a decrease in pressure is accompanied by a *liberation* of energy and $\Delta H < 0$ for $\Delta P < 0$. Under these circumstances, $(\partial H/\partial P)_T > 0$ and $\mu_{JT} < 0$. At ambient temperature, the only two gases that have negative JT coefficients are He and H_2.

The relationship between attractive and repulsive molecular forces and their effect on the sign (and magnitude) of the JT coefficient can be obtained from the equation of state of the gas, e.g., the van der Waals equation. This will be illustrated below [see equations (4.3.16), (4.3.20), and (4.3.23)].

Since both μ_{JT} and C_P are measurables, the quantity $(\partial H/\partial P)_T$ can be determined quantitatively. Because the latter reflects the nature and magnitude of intermolecular interactions, it allows the correctness of an equation of state for a real gas to be evaluated. Knowledge of $(\partial H/\partial P)_T$ is also important in correcting the enthalpies of gases at high pressures to standard conditions. This is used in *bomb calorimetry* (Experiments 5 and 6).

There are practical aspects to the JT expansion. It is the basis of the Linde method of gas liquefaction. In this process, a gas is compressed to a reasonably high pressure and the heat liberated thereby is removed. The temperature is then low enough that $\mu_{JT} > 0$. The cooled, pressurized gas is then expanded adiabatically through a nozzle, and if the pressure drop is sufficiently large for the particular gas, it will cool so much that the final temperature is below its boiling point, and the gas liquefies. The liquid is drawn off and stored in heavy-walled containers. When the liquid warms up (i.e., to ambient temperature), the presure in the container can be significant. In addition, the JT effect is the basis of operation of the refrigerator and the device that operates in the reverse sense, the heat pump.

It is desirable to obtain $(\partial H/\partial P)_T$ from any particular equation of state. To achieve this, we write the fundamental Gibbs equation for dH:

$$dH = T\,dS + V\,dP. \qquad (4.3.9)$$

Constructing $(\partial H/\partial P)_T$ from (4.3.9):

$$\left(\frac{\partial H}{\partial P}\right)_T = T\left(\frac{\partial S}{\partial P}\right)_T + V, \qquad (4.3.10)$$

$(\partial S/\partial P)_T$ can be expressed in terms of measurables by using the Maxwell relation implied in $dG = -S\,dT + V\,dP$, namely,

$$\left(\frac{\partial S}{\partial P}\right)_T = -\left(\frac{\partial V}{\partial T}\right)_P. \qquad (4.3.11)$$

Substituting this result into equation (4.3.10) provides the relation

$$\left(\frac{\partial H}{\partial P}\right)_T = V - T\left(\frac{\partial V}{\partial T}\right)_P, \qquad (4.3.12)$$

which can be directly applied to an equation of state, $V = f(T,P)$.

Since many equations of state relate $P = f(T,V)$, it is useful to recast equation (4.3.12) accordingly. We can use the cyclic relation:

$$\left(\frac{\partial V}{\partial T}\right)_P = \frac{-(\partial P/\partial T)_V}{(\partial P/\partial V)_T,} \tag{4.3.13}$$

to express the JT coefficient as

$$\mu_{JT} = \frac{-1}{C_P}\left[V + \frac{T(\partial P/\partial T)_V}{(\partial P/\partial V)_T}\right]. \tag{4.3.14}$$

Equation (4.3.14) can be used to estimate μ_{JT} from an equation of state that is of the form $P = f(T,V)$. Alternatively, the adequacy of an equation of state can be tested by using its ability to predict μ_{JT} values using (4.3.14).

Calculation of μ_{JT} from Equations of State:

THE VAN DER WAALS (vdW) EQUATION: The vdW equation of state can be expressed as

$$P = \frac{RT}{V_m - b} - \frac{a}{V_m^2}, \tag{4.3.15}$$

where a and b are parameters that can be obtained from the critical constants of a gas (see below). If equation (4.3.14) is applied to the vdW equation, one obtains a result in which μ_{JT} is a function of T and V_m (or pressure). This is observed to be the case, although the pressure dependence of μ_{JT} is rather small (except at very high pressures). At low pressures (a few tens of atmospheres, i.e., large molar volumes), the expression obtained from equations (4.3.14) and (4.3.15) is

$$C_P\mu_{JT} = \frac{2a}{RT} - b \qquad \text{(vdW gas)}, \tag{4.3.16}$$

and the temperature dependence of μ_{JT} in the low-pressure regime can be obtained from equation (4.3.16). Generally, in applying this equation (4.3.16), the temperature dependence of C_P should be taken into account.

THE REDLICH-KWONG EQUATION: Another useful two-parameter equation of state is the Redlich-Kwong (RK) equation[1,2]

$$P = \frac{RT}{V_m - \underline{b}} - \frac{\underline{a}}{V_m(V_m + \underline{b})T^{1/2}}. \tag{4.3.17}$$

The $\underline{a}$ and $\underline{b}$ parameters are different from those in the vdW equation. In equation (4.3.17), for example, $\underline{a}$ has different units relative to a in equation (4.3.15). These parameters can also be obtained from the critical constants of a gas. For the RK equation,

$$\underline{a}_{RK} = \frac{R^2 T_c^{5/2}}{9(2^{1/3} - 1)P_c},$$

and

$$\underline{b}_{RK} = \frac{(2^{1/3} - 1)RT_c}{3P_c}, \tag{4.3.18}$$

while for the vdW equation, the a and b constants are

$$a_{vdW} = \frac{27R^2 T_c^2}{64P_c},$$

and   (4.3.19)

$$b_{vdW} = \frac{RT_c}{8P_c}.$$

In these equations, T_c and P_c are the critical temperature and pressure of the gas (in K and atm, respectively). By applying equation (4.3.14) to the RK equation, one obtains, in the limit of large V_m,

$$C_P\mu_{JT} = \frac{5a}{2RT^{3/2}} - b.   \quad(4.3.20)$$

THE BEATTIE-BRIDGEMAN (BB) EQUATION: The BB equation is a five-parameter equation of state that is often used for precise work; the five parameters allow the equation to track the P,V,T behavior of a real gas up to higher pressures. The BB equation is written[3]

$$P = \frac{RT(1 - E)(V_m + B)}{V_m^2} - \frac{A}{V_m^2},   \quad(4.3.21)$$

where $A = A_0(1 - a/V_m)$, $B = B_0(1 - b/V_m)$, and $E = c/(V_mT^3)$. Thus the five BB parameters are A_0, a, B_0, b, and c. The BB equation can be cast into a virial form explicit in V[4]

$$PV_m = RT + \frac{\beta}{V_m} + \frac{\gamma}{V_m^2} + \frac{\delta}{V_m^3}, \quad(4.3.22)$$

where $\beta = RT[B_0 - A_0/(RT) - c/T^3]$,

$$\gamma = RT[-B_0b + A_0a/(RT) - B_0c/T^3],$$

$$\delta = RT[B_0bc/T^3].$$

While the parameters in the vdW and RK equations can be obtained from critical constants, those of the BB equation are derived from experimental P,V,T data, and the analysis is quite complicated. The table below lists the BB constants for three of the gases used in this experiment.[5]

Beattie-Bridgeman Constants for He, N_2, and CO_2. Units are in dm^3, mol, atm, and K

	A_0	a	B_0	b	c
He	0.0216	0.05984	0.01400	0	40
N_2	1.3445	0.02617	0.05046	-0.00691	4.20×10^4
CO_2	5.0065	0.7132	0.10476	0.07235	6.600×10^5

The application of equation (4.3.14) to the BB equation (4.3.22) yields (in the limit of large V)

$$C_P\mu_{JT} = \frac{2A_0}{RT} + \frac{4c}{T^3} - B_0,   \quad(4.3.23)$$

and thus only three of the BB parameters determine μ_{JT} at low pressures. In this experiment, all three equations of state will be used to calculate μ_{JT} at the appropriate temperature. These results will be compared with the experimentally determined values.

The vdW and RK *a* and *b* constants, as mentioned above, can be determined from the critical constants of the gases. These constants are contained in the following table:[6]

Critical Constants and Heat Capacities of Gases

	V_c(dm^3 mol^{-1})	P_c(atm)	T_c(K)	C_P°(dm^3 atm mol^{-1}K^{-1})
He	0.0578	2.261	5.20	0.2052
N$_2$	0.0901	33.54	126.27	0.2874
SF$_6$	0.199	37.11	318.71	0.9626
CO$_2$	0.094	72.85	304.20	0.3663

Experimental Method

The procedure to be followed in this experiment involves the direct measurement of the temperature change that accompanies the expansion of a gas under nearly isenthalpic conditions. Thus ΔT and ΔP values are measured, and these are used to compute the JT coefficient. The inlet pressure of the gas is measured using a Bourdon gauge (see pp. 30–31), and the temperature *change* of the gas after expansion is determined directly from a pair of matched thermocouples connected in series. One thermocouple is placed in the high-pressure gas stream, and the other is located in the emerging, low-pressure, gas. See Figure 4.3.2.

μ_{JT} can be measured more accurately using a microcalorimeter in which a known and carefully controlled amount of heat is added to (or removed from) the flowing gas after it expands through a throttle in order to keep its temperature equal to that of the entering high-pressure gas. The heat is provided by a resistance wire, and the energy (heat flow) dissipated per unit time (the power P) is determined from $P = iV$, where i is the current flow through the resistor and V is the voltage across the resistor.

Apparatus

A diagram of the experimental apparatus is shown in Figure 4.3.2.[7] The JT cell is constructed from stainless-steel fittings. Stainless steel is used because of its relatively low thermal conductivity; this helps keep the process adiabatic. The system is all metal so that it can withstand the pressures used in this experiment (up to 4 to 5 atm). In order to measure the JT effect for He (for which μ_{JT} is very small), expansions from relatively high pressures are used.

The heart of the apparatus, the porous plug (throttle), is a 3/8-in.-diameter stainless-steel frit having a porosity of about 2 μm. The entire apparatus is insulated by a layer of glass wool to keep it as adiabatic as possible. Gas at high pressure is forced through the cell. It expands through the frit and is then released at ambient pressure (into the atmosphere). The high-pressure side of the cell is monitored by a pressure gauge. The thermocouple that senses the temperature of the flowing, high-pressure gas is mounted in the center of the stainless-steel cross. This thermocouple junction is rather small (ca. 0.5 mm diameter) and thus has a very small thermal mass; it can react very rapidly to any change in the entering gas temperature. The temperature of the exiting low-pressure gas is monitored by an identical thermocouple.

The thermocouples are constructed of very fine gauge wire in order to minimize heat transfer between the system and the surroundings. The thermocouple pair used is copper-constantan, which produces a potential of 39 μV per C° (at ambient

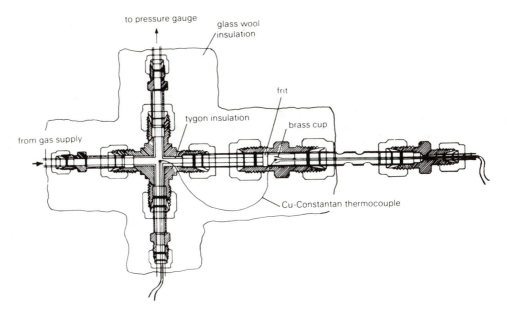

FIGURE 4.3.2   Cutaway view of the Joule-Thomson cell.

temperature). The junctions are connected in *series* (the copper wire of one thermocouple is connected to the constantan wire of the other), and thus the pair directly measures the temperature *difference* between the high- and low-pressure sides of the JT cell. The thermocouples are well matched; they produce a negligible voltage when the two junctions are equilibrated at a common temperature. The thermocouple pair is attached directly to a microvoltmeter (having a bipolar meter) that indicates the voltage difference between the two junctions. With this particular apparatus, a meter deflection to the right (positive voltage) indicates that the high-pressure-sensing thermocouple (gas inlet) is at a higher temperature than is the low-pressure-sensing thermocouple (gas outlet). A negative deflection indicates the opposite situation.

The pressure of the gas entering the JT apparatus is read from a Bourdon-type pressure gauge, a mechanical device capable of adequate accuracy. The gas first passes through a length of 1/4-in. copper tubing so that it is equilibrated to ambient temperature before it enters the JT cell.

Safety Precautions

☐ Safety glasses must be worn at all times during this experiment.
☐ This experiment requires the use of gases contained at high pressures in metal cylinders. These cylinders must be firmly strapped to a secure foundation. Each cylinder is equipped with a reducing valve that delivers gas to the JT apparatus at reduced pressures (up to 90 psig). Do not increase the gas pressure.
☐ This experiment must be performed in an open, well-ventilated laboratory.

Procedure

1. Make sure you have been instructed (and understand) how to use and read the microvoltmeter, that is, how and when to change scales and how to read the meter scales properly.

2. Before you start the experiment, read the "offset" voltage produced by the thermocouples. This corresponds to the initial temperature gradient across the thermocouples under zero-gas-flow conditions. Record this value along with the ambient temperature as indicated by the thermometer near the apparatus.

3. Set the microvoltmeter scale to the 1-V position. Starting with helium, slowly (over a period of $\sim$ 1 min) increase the pressure to 90 psig as indicated on the pressure gauge. After this pressure is reached, wait for the system to stabilize (about 30 to 60 s); read and record the pressure and voltage values. For the latter, increase the sensitivity of the meter as appropriate.

4. Record readings every $\sim$ 10 s until a constant value is approached, or until three consecutive readings are within experimental error. There may be substantial fluctuations in the voltage readings with He (possibly because of its high thermal conductivity).

5. Slowly lower the pressure to about 75 psig. This should be done over a 10-s interval. After about 30 to 60 s, record the thermocouple voltage as above.

6. Lower the pressure in increments of 10 to 15 psig, obtaining appropriate voltage readings for each pressure.

7. Continue until a final pressure of 30 to 40 psig is reached.

8. Switch the meter back to the 1-V full-scale position and change the gas to N_2. Your instructor will illustrate how to use the gas manifold. It takes only a few seconds for the gas to be purged from the JT cell.

9. Follow the same procedure for He. Slowly raise the pressure to 90 psig. Obtain voltage readings at 10 to 15 psig increments until a final pressure of 30 to 40 psig is reached.

10. Change the gas to SF_6. Because this gas is relatively expensive, an initial pressure of 50 to 60 psig will be used. Wait the minimal time needed for the voltage to stabilize after each pressure change. The final pressure should be about 15 psig.

11. Finally, change to CO_2 and bring the system to 80 to 90 psig. Repeat the procedure as above. The experiment can be stopped at 20 to 30 psig.

12. When the experiment is completed, make sure the gas cylinder head valves are closed.

Data Analysis

Convert the voltage readings to ΔT values using the conversion factor of 39 μV/C° (this is a characteristic of the *copper-constantan* thermocouple). Also convert the pressure readings to atmospheres. For each gas, tabulate these ΔT, ΔP values along with their respective estimated uncertainties.

Plot ΔT vs. ΔP separately for each gas. While one would expect these plots to be linear (since the pressure is not too great), some of these plots may show slight curvature and may not appear to extrapolate to the (0,0) origin, as would be expected. These problems arise from defects in the JT cell design that prevent the process from being carried out completely adiabatically or isenthalpically. Using all the points (or as many as you feel are justified), determine the slope of each plot. This can be taken to be the JT coefficient. Tabulate these values (in K atm^{-1}) and their uncertainties along with those calculated from the three equations of state discussed above. Finally compare these values with those found in the literature, giving appropriate citation(s).

Questions and Further Thoughts

1. Suppose the apparatus produced a voltage offset of 3 μV before gas started to flow through the JT cell. What temperature difference does this correspond to?

2. The JT coefficient is an intensive property. Can one get more cooling from a given gas expanding between a certain pressure difference with a JT apparatus by letting the gas flow for a longer period of time?

3. Suppose a JT apparatus is to function as a refrigerator. What characteristics (i.e., JT cell design, quantity of gas flowing, etc.) determine how much *power,* i.e., the amount of heat removed from the surroundings of the expanding gas per second, can be realized from such a device?

4. Suggest an explanation of why the ΔT vs. ΔP plots sometimes do not go through the origin. Should this effect depend on the nature of the gas being studied?

5. The JT inversion temperature is the temperature at which μ_{JT} changes sign (i.e., $\mu_{JT} = 0$). Calculate this point for each gas studied using the vdW and RK equations of state, i.e., equations (4.3.16) and (4.3.20).

6. Derive equations (4.3.16) and (4.3.20) using the relation in (4.3.14) and equations (4.3.15) and (4.3.17), respectively.

Notes

1. O. Redlich and J. N. S. Kwong, *Chem. Rev., 44,* 233 (1949).
2. I. N. Levine, ''Physical Chemistry,'' 2nd ed., pp. 200–206, McGraw-Hill (New York), 1983.
3. J. A. Beattie and O. C. Bridgeman, *J. Am. Chem. Soc., 49,* 1665 (1927).
4. G. W. Castellan, ''Physical Chemistry,'' 3rd ed., pp. 47–48, Addison-Wesley (Reading, Mass.), 1983.
5. J. A. Beattie and O. C. Bridgeman, *J. Am. Chem. Soc., 50,* 3133 (1928); J. A. Beattie, *Phys. Rev., 35,* 543 (1930).
6. Data taken from J. A. Dean, ed., ''Lange's Handbook of Chemistry,'' 13th ed., pp. 9–180, 9–4, McGraw-Hill (New York), 1985, except for SF_6, for which the heat capacity is from K. Bier, G. Maurer, and H. Sand, *Ber. Bunsenges. Phys. Chem., 84,* 430 (1980).
7. A. M. Halpern and S. Gozashti, *J. Chem. Educ., 63,* 1001 (1986).

Further Readings

R. A. Alberty, ''Physical Chemistry,'' 7th ed., pp. 43–45, Wiley (New York), 1987.

P. W. Atkins, ''Physical Chemistry,'' 3rd ed., pp. 66–68, W. H. Freeman (New York), 1986.

J. de Heer, ''Phenomenological Thermodynamics,'' pp. 91–95, 136–140, Prentice-Hall (Englewood Cliffs, N.J.), 1986.

J. H. Noggle, ''Physical Chemistry,'' pp. 106–110, Little, Brown (Boston), 1985.

See also references 2 and 4, above.

Experiment 3

Joule-Thomson Coefficient

NAME _____ DATE _____

Gas _____   Gas _____   Gas _____

Pressure (psig) _____   Pressure (psig) _____   Pressure (psig) _____

Voltage, μV   ___ ___   Voltage, μV   ___ ___   Voltage, μV   ___ ___

___ ___ ___ ___   ___ ___ ___ ___   ___ ___ ___ ___

___ ___ ___ ___   ___ ___ ___ ___   ___ ___ ___ ___

Pressure (psig) _____   Pressure (psig) _____   Pressure (psig) _____

Voltage, μV   ___ ___   Voltage, μV   ___ ___   Voltage, μV   ___ ___

___ ___ ___ ___   ___ ___ ___ ___   ___ ___ ___ ___

___ ___ ___ ___   ___ ___ ___ ___   ___ ___ ___ ___

Pressure (psig) _____   Pressure (psig) _____   Pressure (psig) _____

Voltage, μV   ___ ___   Voltage, μV   ___ ___   Voltage, μV   ___ ___

___ ___ ___ ___   ___ ___ ___ ___   ___ ___ ___ ___

___ ___ ___ ___   ___ ___ ___ ___   ___ ___ ___ ___

Joule-Thomson Coefficient

NAME _____   DATE _____

Gas _____

Pressure (psig) _____
Voltage, μV   ____ ____
____ ____ ____ ____ ____
____ ____ ____ ____ ____

Pressure (psig) _____
Voltage, μV   ____ ____
____ ____ ____ ____ ____
____ ____ ____ ____ ____

Pressure (psig) _____
Voltage, μV   ____ ____
____ ____ ____ ____ ____
____ ____ ____ ____ ____

Gas _____

Pressure (psig) _____
Voltage, μV   ____ ____
____ ____ ____ ____ ____
____ ____ ____ ____ ____

Pressure (psig) _____
Voltage, μV   ____ ____
____ ____ ____ ____ ____
____ ____ ____ ____ ____

Pressure (psig) _____
Voltage, μV   ____ ____
____ ____ ____ ____ ____
____ ____ ____ ____ ____

Gas _____

Pressure (psig) _____
Voltage, μV   ____ ____
____ ____ ____ ____ ____
____ ____ ____ ____ ____

Pressure (psig) _____
Voltage, μV   ____ ____
____ ____ ____ ____ ____
____ ____ ____ ____ ____

Pressure (psig) _____
Voltage, μV   ____ ____
____ ____ ____ ____ ____
____ ____ ____ ____ ____

☐ Experiment 4

Real Gases

PART II: THE SECOND VIRIAL COEFFICIENT

Objective

To measure the second virial coefficient of CO_2 or SF_6 at different temperatures.

Introduction

That the ideal gas law,

$$PV = nRT, \qquad (4.4.1)$$

is an abstraction is evident when one considers the underlying assumptions made: that molecules are point masses, and that they undergo only elastic collisions. Since we know these to be inherently false, it may be surprising how well the ideal gas law seems to work. At higher pressures and/or lower temperatures, however, finite molecular volumes and intermolecular forces are considerable, and the expected deviations from (4.4.1) become too large to be ignored.

One logical and systematic way in which deviations from ideal gas behavior can be expressed mathematically is to measure the state properties, $P, V,$ and $T,$ of n moles of a gas at equilibrium, and to determine the extent to which the PV/nRT quotient deviates from unity. The dimensionless expression PV/nRT is called the *compressibility factor,* and is denoted as Z. Since the extent to which Z differs from unity depends on the pressure [or alternatively, at a given temperature, the reciprocal molar volume, $1/(V/n)$], Z can be expressed as a power series in either of these state variables. Thus,

$$Z = \frac{PV}{nRT} = 1 + B_2\left(\frac{n}{V}\right) + B_3\left(\frac{n}{V}\right)^2 + B_4\left(\frac{n}{V}\right)^3 + \cdots, \qquad (4.4.2)$$

where $B_2, B_3, \ldots,$ are called the second, third, $\ldots$ *virial coefficients.* (The term virial here indicates a power series.) They are functions of temperature and actually relate to the *simultaneous interactions* of two, three, four $\ldots$ molecules, respectively. It can be understood, therefore, that the higher-order virial coefficients become significant only at smaller molar volumes (i.e., higher pressure).

The virial coefficients can be calculated from theoretical concepts involving statistical mechanics and knowledge of the intermolecular potential energy function appropriate to the particular molecular system. This is a case in which thermodynamics, which deals empirically with macroscopic systems, can be linked with microscopic entities, molecules.

Z can also be expressed as a power series in the pressure (it often being a more convenient state variable):

$$Z = 1 + A_2P + A_3P^2 + A_4P^3 + \cdots, \qquad (4.4.3)$$

where the temperature-dependent $A_2, A_3, A_4, \ldots$ are also virial coefficients. If equation (4.4.2) is solved for P and this expression is substituted in equation

(4.4.3), the coefficients of (n/V) of equal powers in the two expressions can be equated. Thus,

$$B_2 = A_2 RT \qquad B_3 = (A_2^2 + A_3)R^2 T^2,$$

and (4.4.4)

$$B_4 = (A_2^3 + 3A_2 A_3 + A_4)R^3 T^3.$$

For most gases at moderate pressures (below ca. 50 to 100 atm), the squared and higher terms can be neglected, and equation (4.4.3) reads

$$Z = \frac{PV}{nRT} = 1 + A_2 P. \qquad (4.4.5)$$

This is a one-parameter equation of state (considering R as a constant). If the van der Waals equation of state (1873)

$$P = \frac{RT}{V/n - b} - \frac{a}{(V/n)^2}, \qquad (4.4.6)$$

is cast into a virial form in either (V/n) or P (see references 1 and 2), and the results are compared with equation (4.4.2) or (4.4.3), respectively, it becomes evident that

$$B_2 = b - \frac{a}{RT},$$

and (4.4.7)

$$A_2 = \frac{1}{RT}\left(b - \frac{a}{RT}\right).$$

The temperature dependence of B_2 and A_2 here is explicit. Expressions for the higher virial coefficients can also be obtained in terms of a and b. Thus the virial coefficients can be estimated from the van der Waals a and b constants. Alternatively, a and b can be determined from the temperature dependence of B_2 (or A_2).

The Beattie-Bridgeman (BB) equation (1927), is a five-parameter equation of state:

$$P = \frac{RT(1 - E)(V/n + B)}{(V/n)^2} - \frac{A}{(V/n)^2}, \qquad (4.4.8)$$

where $A = A_0[1 - a/(V/n)]$, $B = B_0[1 - b/(V/n)]$, and $E = c/[(V/n)T^3]$. Thus the five parameters are A_0, B_0, a, b, and c. This equation, which, because of the five parameters, works well over a wider pressure range, can be cast into a virial form (i.e. power series) in which the second coefficient is

$$B_2 = B_0 - \frac{A_0}{RT} - \frac{c}{T^3}. \qquad (4.4.9)$$

Here, only three of the BB parameters appear.

Once A_2 (or B_2) is determined, the equation of state

$$PV_m = RT + B_2 P, \qquad (4.4.10)$$

where V_m is the molar volume [see equation (4.4.5)] can be used to obtain certain real gas properties such as the fugacity coefficient (γ), internal energy (U), $C_P - C_V$, and the Joule-Thomson coefficient. As an example of how the simple equation of state (4.4.10) can be used, we will consider the calculation of the fugacity coefficient. The molar Gibbs free energy (chemical potential), μ, for a gas under ideal conditions ($P \rightarrow 0$ or $V \rightarrow \infty$) is expressed as

$$\mu^{id} = \mu°(T) + RT \ln\left(\frac{P}{P°}\right), \qquad (4.4.11)$$

where P is the pressure in atmospheres ($P° = 1$ atm) and $\mu°$ is the standard state chemical potential of the gas at 1 atm pressure and under "ideal gas conditions."

At higher pressures where gas imperfection cannot be ignored, equation (4.4.11) is modified to express the chemical potential as

$$\mu = \mu°(T) + RT \ln\left(\frac{f}{f°}\right), \qquad (4.4.12)$$

where f is called the *fugacity* of the gas. In other words, the fugacity of a gas is a quantity whose logarithm in equation (4.4.12) represents the actual chemical potential of the gas. Thus as $P \rightarrow 0$, $f \rightarrow P$, and if we take the standard state pressure to be 1 atm, the dimensions of fugacity must also be in atmospheres in order to make equation (4.4.12) quantitatively correct.

We wish to find out how the fugacity depends on pressure (so we can use a particular equation of state to determine f). At constant temperature, the pressure dependence of the chemical potential is simply $d\mu = V_m dP$. Using this expression along with (4.4.12), we get

$$d\mu = RT \, d(\ln f) = V_m \, dP. \qquad (4.4.13)$$

In principle, this result could be integrated to get $\ln f$ as a function of P. The problem is with the lower boundary condition: for $P = 0$ (ideal gas), $f = P$, and thus $\ln f$ is not finite. We can get around this problem by expressing the fugacity as a factor, γ, times the pressure:

$$f = \gamma P, \qquad (4.4.14)$$

where γ is called the *fugacity coefficient*. There is an analogy here between the fugacity coefficient (for gases) and the activity coefficient (usually used for solutions). We recognize that, in the limit of zero pressure, $\gamma \rightarrow 1$. Substituting equation (4.4.14) into (4.4.13), we get [after using $d(\ln P) = dP/P$]

$$RT \, d(\ln \gamma) = \left(V_m - \frac{RT}{P}\right) dP. \qquad (4.4.15)$$

This equation is used for calculating fugacity coefficients. After integration (using the dummy variable P') between $P' = 0$ (where $\ln \gamma = 0$) and $P' = P$, (4.4.15) yields

$$\ln \gamma = \frac{1}{RT} \int_0^P \left(V_m - \frac{RT}{P'}\right) dP'. \qquad (4.4.16)$$

The integrand in equation (4.4.16) is obtained from an equation of state. Notice that the RT/P' term in (4.4.16) would appear to be troublesome as $P' \rightarrow 0$; however, in this limit, $V_m = RT/P'$ and the integrand vanishes.

From the simple equation of state presented in equation (4.4.10), $V_m - RT/P = B_2$, and using this result in equation (4.4.16) and integrating between 0 and P gives finally

$$\ln \gamma = \frac{B_2 P}{RT}. \qquad (4.4.17)$$

EXPERIMENTAL DESIGN

Virial coefficients are not particularly easy to measure. One must work with systems at high pressures where gas nonideality is significant. However, if the pressure is too high (> 100 atm), not only is specialized equipment needed, but the third virial coefficient becomes nonnegligible. Moreover, at low temperatures where nonideality is greater (relative to high temperatures), gas adsorption on the inner surface of the container becomes more of a problem. Values of Z for common, unreactive gases under the most practical experimental conditions (ambient temperature) less than 20 to 30 atm, are usually greater than 0.8. Thus, one is measuring a relatively small effect.

In determining the second virial coefficient (A_2) experimentally, the most straightforward procedure would be to measure Z as a function of pressure and, assuming a linear relation to hold (up to moderate pressures), to plot Z vs. P [as implied in equation (4.4.5)], thereby obtaining A_2 as the slope. While P, V, and T can be easily measured, the determination of n—the number of moles of gas— is not straightforward. A gravimetric technique is inappropriate because the mass of vapor in the system is a very small fraction of the total mass of the container (which is constructed of heavy-gauge metal so that it is capable of withstanding high pressures). An indirect method is therefore needed to determine n, and an approach developed by Burnett in 1936[3] and recently modified by Baskett and Matthews[4] is particularly straightforward. It involves filling a bomb (the *sample cylinder*) with the gas to be studied at a moderately high pressure (ca. 20 atm), reading the pressure, and then withdrawing a small amount of gas so that it fills another, somewhat larger, container (the *expansion cylinder*) at a known, low pressure (less than 1 atm). The ideal gas law can then be applied to the gas in the expansion vessel, and thus the number of moles of gas withdrawn can be determined. A schematic diagram of the experimental arrangement is shown in Figure 4.4.1.

The mathematical treatment of the above experimental method that leads to an expression of $Z = f(P)$ follows. Let V^e and V^s be the volumes of the expansion and sample cylinders, respectively. P^e and P^s are the corresponding gas pressures in these containers. The experiment is carried out at a controlled temperature, T. Initially, the sample cylinder is filled (at high pressure) with n_0 moles of the gas to be studied. A small amount of the gas is transferred to the expansion cylinder

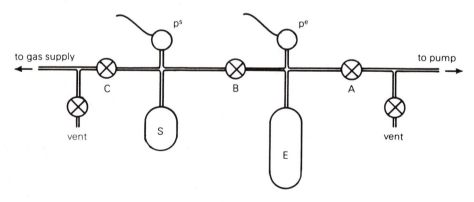

FIGURE 4.4.1   Schematic diagram of gas filling and handling apparatus. S and E are the sample and evacuation cylinders, and P^s and P^e are the respective pressure-sensing devices. The function of valves A, B, and C is described in the text.

which has been previously evacuted; the pressure then rises to P_1^e, which is not allowed to exceed $\simeq 1$ atm. In this way, the ideal gas law can be justifiably applied to the gas in the expansion cylinder. The pressure in the sample cylinder falls from P_0^s to P_1^s. From the ideal gas law, the number of moles of gas *transferred* is

$$n_1 = \frac{P_1^e V^e}{RT}.$$

The expansion cylinder is then pumped out and refilled with another $\simeq 1$ atm amount of gas from the sample cylinder. The resulting pressures are now P_2^e and P_2^s, and the number of moles transferred this time is $n_2 = P_2^e V^e / RT$. The expansion cylinder is reevacuated and the cycle repeated until finally, in the mth transfer, the pressure in the sample cylinder drops to about 1 atm (or lower), i.e., $P_m^e = P_m^s$.

In order to proceed further, we must determine the number of moles of gas *remaining in the sample cylinder* after a given transfer. This is achieved as follows. The initial number of moles of gas in the sample cylinder, n_0, is equal to the number of moles of gas remaining after the mth expansion *plus* the total number of moles delivered via the individual transfers to the expansion cylinder; thus,

$$n_0 = \frac{1}{RT}(P_m^e V^s + P_1^e V^e + P_2^e V^e + \cdots + P_m^e V^e),$$

or   (4.4.18)

$$n_0 = \frac{1}{RT}\left(P_m^e V^s + V^e \sum_{i=1}^{m} P_i^e\right),$$

where P_i^e is the pressure in the expansion cylinder after the ith expansion. Thus the number of moles of gas in the sample cylinder that remains after r expansions have been carried out is

$$n_r = n_0 - \sum_{i=1}^{r} n_i,$$ (4.4.19)

where n_i is the number of moles of gas withdrawn in the ith expansion: $n_i = P_i^e V^e / RT$. Substituting n_i and n_0 [equation (4.4.18)] into (4.4.19) results in

$$n_r = \frac{1}{RT}\left(P_m^e V^s + V^e \sum_{i=1}^{r} P_i^m - V^e \sum_{i=1}^{r} P_i^e\right),$$

or   (4.4.20)

$$n_r = \frac{1}{RT}\left(P_m^e V^s + V^e \sum_{i=r+1}^{m} P_i^e\right).$$

Using this result for n_r, we can now write the compressibility factor

$$Z_r = \frac{P_r^s V^s}{n_r RT} = \frac{P_r^s V^s}{P_m^e V^s + V^e \displaystyle\sum_{i=r+1}^{m} P_i^e},$$ (4.4.21)

and this simplifies, after dividing by V_s, to

$$Z_r = \frac{P_r^s}{P_m^e + (V^e/V^s)\displaystyle\sum_{i=r+1}^{m} P_i^e}.$$ (4.4.22)

Equation (4.4.22) is the desired result and is the computational basis of this experiment. The measurables are $\{P_i^s\}$, the set of pressure readings in the sample cylinder, and $\{P_i^e\}$, the corresponding set of pressures in the expansion cylinder. The summation in equation (4.4.22) goes between the $(r+1)$th and the final, mth, expansion, where r is a running index. The volume ratio, V^e/V^s, is separately measured by performing a gas expansion at low pressure.

Sources of Error

As mentioned above, virial coefficients are difficult to measure. While more sensitivity can be achieved by measuring Z at higher gas pressures (just as the Joule-Thomson coefficient can be determined more readily at higher pressure differences), the maximum practical pressure in this experiment is ca. 20 atm. Given this constraint, it would appear that there are only two intrinsic error sources: those of the primary measurables—the high and low pressures of the sample and expansion cylinders (assuming that the entire system is always at thermal equilibrium).

A detailed examination of equation (4.4.22), the source of the $Z_r(P_r^s)$ determinations, reveals that two components to that equation are particularly important: the volume ratio, (V^e/V^s), and the final, residual pressure, P_m^e. The volume ratio is determined from pressure measurements (see below). *It is important that these two quantities be determined as accurately as possible.*

An error analysis of equation (4.4.22) indicates that a 5 percent (positive) error in the volume ratio causes the *intercept* of the Z_r vs. P_r^e to be too low by ca. 5 percent even though the slope of the plot is not significantly affected. Thus intercepts that are different from unity can be observed without having serious error in the value of A_2 (as long as the plot is linear). An error in P_m^e has a more profound effect on the data, especially at the lowest values of P_r. Downward curvature at low pressures can be caused by a value of P_m^e that is too large. These facts should be borne in mind when the data are analyzed. This is a case where linear regression should not be blindly performed using *all* of the data points.

Safety Precautions

☐ Safety goggles must be worn in the laboratory at all times.

☐ This experiment requires the use of gases contained at high pressures in metal cylinders. These cylinders must be securely strapped to a firm foundation. Each cylinder is equipped with a reducing valve that delivers gas to the apparatus at a pressure of ca. 300 psig. Do not increase this pressure.

☐ Before starting the experiment, be sure you know the general procedure, i.e., the sequence of manipulating the valves.

☐ Be extremely cautious in admitting gas from the (high-pressure) sample cylinder to the expansion cylinder. If the expansion cylinder is overpressurized (> 1 atm), the pressure transducer can be forced from the apparatus, causing damage.

☐ Only one person should operate the gas-handling valves at one time.

☐ The vacuum pump outlet must be connected to an active exhaust system to avoid letting pump oil mist into the laboratory.

☐ This experiment must be performed in an open, well-ventilated room.

Procedure

CAUTION: This experiment involves high pressures; goggles must be worn. Although the system is periodically tested, leaks may develop. If you hear gas escaping from the apparatus, notify your instructor at once. When metering in gas from the sample cylinder to the evacuation cylinder, be sure not to allow the pressure to exceed 1 atm (otherwise the low-pressure transducer could be brought out of calibration).

A diagram of the apparatus is shown in Figure 4.4.1. The cylinders should be immersed in the constant temperature bath (stable to $\pm 0.2°C$) as fully as possible. Record the bath temperature. Be sure the power supplies to the pressure transducers as well as the digital voltmeters are turned on for *at least* 15 min.

1. First evacuate the entire system by opening valves A and B. If the gas inlet line is to be evacuated also, valve C should be opened as well. Consult your instructor on this point. The pressure transducers (or Bourdon gauges) should indicate nominal "0" pressure. Often, however, these devices show nonzero readings, and these "offset" values should be subtracted from the actual pressure readings acquired in the experiment. After the system has been pumped down for a few minutes, record the offset values of the two transducers.

2. Isolate the system by closing valves A and B (and C as well if the inlet line was evacuated). Fill the sample cylinder by admitting the gas (CO_2 or SF_6) to the inlet line and then by gradually opening valve C until a pressure of about 20 atm is indicated by the high-pressure transducer (P^s) [*MAKE SURE VALVE B IS CLOSED*]. Close valve C securely but *do not overtighten*. Record this pressure after it stabilizes (after 1 to 2 min).

3. Fill the evacuated expansion cylinder with a quantity of the gas. Gradually open valve B (no more than 1/4 revolution) until the low-pressure transducer (P^e) indicates about 1 atm (e.g., 760 ± 50 torr). Then make sure valve B is firmly closed, but do not overtighten. Because the expansion is not, initially, isothermal, there will be temperature (and pressure) fluctuations in the two cylinders. After both P^s and P^e have stabilized, record these pressures.

4. Evacuate the expansion cylinder by opening valve A. Wait for P^e to read "0," and then shut valve A. Fill the expansion cylinder as in step 3.

5. Continue to meter in quantities of gas from the sample cylinder by repeating steps 3 and 4 until, after the *m*th transfer, the pressure in the expansion cylinder fails to reach a full 1 atm and thus $P^e_m = P^s_m$. Record the equilibrated pressure. This value must be obtained as accurately as possible.

6. The cylinder volume ratio, V^e/V^s, must be determined. It is particularly important that this quantity be obtained as accurately as possible (see Sources of Error, above). Pump out the system by opening valves A and B. After the transducers read 0, isolate the sample cylinder by shutting valve B. Shut off the vacuum pump and let in enough air through the vent so that the expansion cylinder is filled to about 1 atm pressure. Close valve A and record this pressure (P_1) after it has stabilized. Open valve B and let the air expand into the evacuated sample cylinder. Record the final (equilibrated) pressure (P_2) indicated by the *low*-pressure transducer (it is more accurate). Treating the air as an ideal gas under these circumstances, we have

$$P_1 V^e = P_2(V^e + V^s).$$

Thus the needed volume ratio is

$$\frac{V^e}{V^s} = \frac{P_2}{P_1 - P_2}.$$

Calculations and Data Analysis

1. Tabulate the corresponding sets of high- and low-pressure readings, $\{P_i^s\}$ and $\{P_i^e\}$. List the value of the final, equilibrated pressure, P_m^e. Using the respective offset values for the two pressure transducers, correct the observed readings by subtracting these values from *each* reading. Tabulate the corrected data.

Calculate and list the value of the volume ratio, (V^e/V^s). Assign a running index (r) to each of the P^s, P^e data starting with 0 (there will only be a P^s value for this case—the initial one).

2. Calculate values of Z_r using equation (4.4.22); tabulate, and plot Z_r vs. P_r^s. The plot should be linear and have an intercept of unity. Preferably using linear regression, obtain the value of A_2, and from it, calculate B_2.

3. Calculate values of B_2 from the van der Waals and Beattie-Bridgeman equations and compare them with your data. For CO_2, the van der Waals constants (all in units of dm^3, moles, and atm) are

$$a = 3.59 \qquad b = 0.0427,$$

and the Beattie-Bridgeman constants (dm^3 and atm units) are

$$A_0 = 5.0065 \qquad B_0 = 0.10476$$

$$a = 0.07132 \qquad b = 0.07235 \qquad c = 6.6 \times 10^5.$$

4. Using your data, calculate the fugacity and the fugacity coefficient of the gas under the *initial* conditions in the sample cylinder.

Questions and Further Thoughts

1. Is it possible for a gas to have a positive value of A_2 at room temperature? If so, which gas comes to mind?
2. Recalling that $(\partial H/\partial P)_T = -C_P\mu_{JT} = V - T(\partial V/\partial T)_P$, show that

$$\mu_{JT} = \frac{1}{C_P[T(dB_2/dT) - B_2]}$$

for a gas whose behavior is represented by equation (4.4.10). Note that if $B_2(T)$ were obtained, the inversion temperature could be determined.
3. What are the advantages of having $V^e > V^s$? Are there any advantages to having $V^e < V^s$?
4. Why must the entire apparatus (so far as possible) be immersed in a constant temperature bath?
5. What are the accuracy (and precision) limiting components of the experimental setup?
6. In this experiment, it is assumed that the total number of moles of gas is constant throughout the series of expansions. What would the data look like if there were a small leak in the sample cylinder?
7. Look up the value of A_3 (or B_3) for CO_2 at ambient temperature. At what initial pressure would the effect of the second virial term have a measurable impact on the value of Z determined in this experiment? Assume that Z_0 can be determined with a precision of 1 percent.

Notes

1. I. N. Levine, "Physical Chemistry," 2nd ed., pp. 199–202, McGraw-Hill (New York), 1983.
2. G. W. Castellan, "Physical Chemistry," 3rd ed., pp. 34–49, Addison-Wesley (Reading, Mass.), 1983.
3. E. S. Burnett, *J. Appl. Mech.*, *58*, A136 (1936).
4. W. P. Baskett and G. P. Matthews, *J. Chem. Educ.*, *62*, 353 (1985).

Further Readings

R. A. Alberty, ''Physical Chemistry,'' 7th ed., pp. 8–20, Wiley (New York), 1987.

P. W. Atkins, ''Physical Chemistry,'' 3rd ed., pp. 26–32, 590–592, W. H. Freeman (New York), 1986.

J. H. Noggle, ''Physical Chemistry,'' pp. 9–14, Little, Brown (Boston), 1985.

Note: Values of virial coefficients can be found in J. H. Dymond and E. B. Smith, ''The Virial Coefficients of Pure Gases and Mixtures,'' Clarendon Press (Oxford), 1980.

Experiment 4

The Second Virial Coefficient

NAME _____   DATE _____

Gas _____   Bath temperature _____

Offset values: High pressure _____   Low pressure _____

Sample cylinder pressure reading   Expansion cylinder pressure reading

_____	_____	_____	_____
_____	_____	_____	_____
_____	_____	_____	_____
_____	_____	_____	_____
_____	_____	_____	_____
_____	_____	_____	_____
_____	_____	_____	_____
_____	_____	_____	_____
_____	_____	_____	_____
_____	_____	_____	_____
_____	_____	_____	_____
_____	_____	_____	_____
_____	_____	_____	_____

V^e/V^s Data: Initial Pressure _____   Final pressure _____

Experiment 4

The Second Virial Coefficient

NAME _____   DATE _____

Gas _____   Bath temperature _____

Offset values: High pressure _____   Low pressure _____

Sample cylinder pressure reading   Expansion cylinder pressure reading

_____   _____   _____   _____

_____   _____   _____   _____

_____   _____   _____   _____

_____   _____   _____   _____

_____   _____   _____   _____

_____   _____   _____   _____

_____   _____   _____   _____

_____   _____   _____   _____

_____   _____   _____   _____

_____   _____   _____   _____

_____   _____   _____   _____

_____   _____   _____   _____

_____   _____   _____   _____

V^e/V^s Data: Initial Pressure _____   Final pressure _____

☐ PART FIVE

Thermochemistry

☐ Experiment 5

Bomb Calorimetry: Heat of Formation of Naphthalene or Sucrose

Objective

To determine the standard heat of formation of a pure substance (e.g., naphthalene or sucrose) from its heat of combustion.

Introduction

One of the most useful collections of thermodynamic information is the large number of standard heats of formation of chemical compounds. From these data, and through the application of Hess's law, the standard heats of reaction of numerous chemical processes can be calculated. The experimental determination of the heat of formation of a material is an important contribution to thermodynamics, and the measurement must therefore be performed with great attention to detail in order to achieve the desired precision and accuracy.

In this experiment, the standard heat of formation of a common hydrocarbon, naphthalene, $C_{10}H_8$ (or sucrose, $C_{12}H_{22}O_{11}$), will be determined from its measured heat of combustion. The enthalpy of the desired reaction, i.e.,

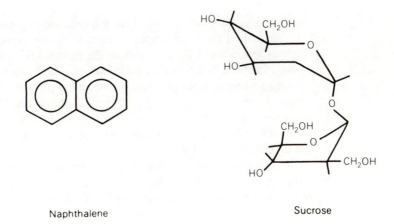

Naphthalene   Sucrose

95

$$10C(\text{graphite}) + 4H_2(g) \rightarrow C_{10}H_8(s) \qquad (\text{naphthalene}), \qquad (5.5.1)$$

or

$$12C(\text{graphite}) + 11H_2(g) + 11/2\ O_2 \rightarrow C_{12}H_{22}O_{11} \qquad (\text{sucrose}), \qquad (5.5.1a)$$

could not be determined simply by carrying out the reaction directly between the elements. The chances of such a reaction actually producing the desired product would be remote, even if the reactants were combined in the correct stoichiometric ratio.

A solution to this problem is to measure the heat of another reaction that *exclusively* involves naphthalene or sucrose as one of the reactants. An example of such a reaction is the combustion of naphthalene or sucrose [equation (5.5.2)] which, if carried out properly, produces *known* products under specific conditions. Thus the heat flow accompanying a controlled, stoichiometrically known reaction is measured. The pertinent reactions in this experiment are

$$C_{10}H_8(s) + 12O_2(g) \rightarrow 10CO_2(g) + 4H_2O(l), \qquad (5.5.2)$$

or

$$C_{12}H_{22}O_{11}(s) + 12O_2 \rightarrow 12CO_2(g) + 11H_2O(l). \qquad (5.5.2a)$$

Once the heat (enthalpy) of the combustion reaction is determined, the heat of formation of naphthalene (or sucrose) may be obtained by applying Hess's law (1840) to reactions (5.5.2) or (5.5.2a). For the former case,

$$\Delta H_f[C_{10}H_8(s)] = 10\Delta H_f[CO_2(g)] + 4\Delta H_f[H_2O(l)] - \Delta H_{\text{comb}}, \qquad (5.5.3)$$

where ΔH_{comb} is the enthalpy of combustion, i.e., the heat of reaction (5.5.2), considered to take place at constant temperature. It should be noted that the enthalpies on the right-hand side of equation (5.5.2) are negative. The desired heat of formation, obtained from equation (5.5.3), is the *difference* between two terms, one set obtained from thermodynamic tables and the other from the combustion measurement. This fact has a significant effect on the magnitude of the error with which ΔH_f can be determined.

Because it is of utmost importance that the reaction be carried out as cleanly as possible (that is, that the reactants and products be exactly as specified in the written reaction), the combustion reaction is carried out with a large excess of oxygen. This ensures that after the reaction there are no unburned naphthalene (or sucrose) or other by-products of incomplete combustion such as elemental carbon (graphite). The large excess of O_2 also causes the combustion to take place very rapidly. The reaction is initiated by suddenly raising the temperature of the fuel (the material to be combusted). This is accomplished by causing a very hot wire, energized via an electric current, to fall onto the fuel sample.

Thus one of the experimental requirements in carrying out this reaction is that high pressures must be sustained in the reactor. For this reason, the reaction is carried out in a heavy-walled metal vessel called, appropriately, a bomb. In addition to being capable of safely withstanding pressures of several hundred atmospheres, this reactor must be constructed of (or lined with) a chemically inert material because highly corrosive products such as nitric acid are commonly formed (see below). The experimental apparatus, called a *bomb calorimeter,* will be described in detail below.

The experimental measurable in this reaction is the heat flow under *constant volume* conditions. From the first law, $\Delta U = q + w$, we have for the case of *only* gas expansion/compression, or $P-V$ work, $\Delta U = q + \int P_{\text{ext}}\, dV$. Hence,

$$\Delta U = q_V, \tag{5.5.4}$$

and the measured heat flow in the bomb is equal to the change in internal energy for the reaction. Since the objective is to determine the *enthalpy* of the reaction, we apply the definition of enthalpy, namely, $H \equiv U + PV$ and write

$$\Delta H = \Delta U + \Delta(PV), \tag{5.5.5}$$

where $\Delta(PV)$ is the difference of the PV value of the products of (5.5.3) from that of the reactants. Because the volume of the gaseous constituents of the reaction is so much larger than the volume of solids and liquids, $\Delta(PV)$ can be determined from the change in mole numbers of gas in the reaction. [The volume of the bomb is approximately 0.3 dm^3, and the volumes of naphthalene (or sucrose) used and (liquid) water produced are $\sim$ 0.8 and 0.6 cm^3, respectively.] Therefore, using the ideal gas law, we can write

$$\Delta(PV) \approx \Delta n_{\text{gas}} RT. \tag{5.5.6}$$

The enthalpy of reaction can thus be obtained from the measured heat flow at constant volume, ΔU, through (5.5.6) and the balanced equation for the combustion reaction. However, it is desirable to obtain the *standard* heat of reaction, $\Delta H°$, in which the reactants and products are all under standard conditions: 1 atm pressure, 298.15 K, and all gases are in the hypothetical "ideal gas state." Three adjustments must be made to convert ΔH to $\Delta H°$. The first of these, ΔH_1, is the correction term to standard temperature and is obtained from

$$\Delta H_1 = \int_{T°}^{T} \Delta C_P \, dT', \tag{5.5.7}$$

where ΔC_P is the difference between the heat capacities of the products and the reactants and T is approximated as the mean temperature of the system before and after reaction (the reaction is controlled to prevent a temperature change of more than 2 to 3°C). ΔH_1 is usually very small because the reaction is generally carried out at ambient temperature and because ΔC_P is usually small.

The second enthalpy correction, ΔH_2, brings the *solid* reactant (naphthalene or sucrose) and *liquid* product (water) from their high-pressure condition in the bomb (ca. 30 atm) to the standard state, 1 atm. The needed relation, $(\partial H/\partial P)_T$, is obtained from the fundamental expression $dH = T \, dS + V \, dP$, thus:

$$\left(\frac{\partial H}{\partial P}\right)_T = T\left(\frac{\partial S}{\partial P}\right)_T + V. \tag{5.5.8}$$

Using the Maxwell relation, $(\partial S/\partial P)_T = -(\partial V/\partial T)_P$, equation (5.5.8) becomes

$$\left(\frac{\partial H}{\partial P}\right)_T = -TV\alpha + V, \tag{5.5.9}$$

in which the coefficient of thermal expansion, $\alpha = V^{-1}(\partial V/\partial T)_P$ is used. [If V is the molar volume of the solid (or liquid), then the expression yields the pressure dependence of its enthalpy per mole.] Integration of (5.5.9) between P, the mean pressure of the system (e.g., 30 atm), and 1 atm can be carried out by assuming that these materials are incompressible so that

$$\Delta H_2 = (TV\alpha - V)(1 - P), \tag{5.5.10}$$

ΔH_2, like ΔH_1, is usually very small.

The third adjustment is the correction to gas "ideality," ΔH_3, which is applied to O_2 and CO_2. It is obtained from the total differential of H:

$$dH = \left(\frac{\partial H}{\partial T}\right)_P dT + \left(\frac{\partial H}{\partial P}\right)_T dP. \tag{5.5.11}$$

Using the definitions for the constant-pressure heat capacity, $(\partial H/\partial T)_P = C_P$, and the Joule-Thomson coefficient, $\mu_{JT} = (\partial T/\partial P)_H$, the desired result is

$$\left(\frac{\partial H}{\partial P}\right)_T = -C_P\mu_{JT}. \tag{5.5.12}$$

Integration of (5.5.12) between the (mean) bomb pressure, P atm, and 0 atm, the latter representing the ideal gas condition, gives (for a pressure-independent μ_{JT} and C_P):

$$\Delta H_3 = PC_P\mu_{JT}. \tag{5.5.13}$$

Although ΔH_3 is also small in comparison with the (molar) heat of combustion of most substances, it cannot be neglected in precise work.

The Calorimeter

The apparatus used in this experiment is called a bomb calorimeter. It consists of the *bomb* in which the combustion is carried out, and the container called the bucket in which the bomb is placed. The bomb is constructed of highly resistant stainless steel. The calorimeter bucket, which is chromium-plated to retard heat exchange through radiation, contains a known quantity of water to absorb the heat liberated by the bomb during the combustion.

In this particular apparatus, called an *adiabatic* calorimeter, the bucket is jacketed. Water flows through this outer vessel, and the temperature of this water is monitored and controlled after the sample is ignited to match the temperature of the water in the calorimeter bucket. This arrangement ensures that minimal heat flow takes place between the calorimeter and the surroundings because the temperature differential is kept as small as possible; hence the term adiabatic. (Recall Newton's law of cooling, which says that the rate of heat exchange is proportional to the temperature difference between the system and the surroundings.) In a static jacket (or isothermal) calorimeter, the bucket is well insulated (e.g., by a polyurethane jacket) but allows a small amount of heat to flow from the bomb after ignition. The adiabatic calorimeter is more complex (and expensive) and is capable of higher calorimetric precision.

The terms adiabatic and isothermal, used in this context, might be misleading. Of course, any calorimeter should be adiabatic. By keeping the surroundings of the calorimeter at (or very near) the changing temperature of the calorimeter itself, the system is "truly" adiabatic because the heat flow approaches zero when the temperature gradient is infinitesimal. In an insulated-wall calorimeter (e.g., a Dewar flask), the temperature of the surroundings is constant (hence the term isothermal) relative to the changing temperature inside the calorimeter. Heat flow in this case is retarded because of the reduced conductivity with the surroundings. This type of calorimeter might be called quasiadiabatic.

The temperatures of the calorimeter bucket and the jacket are determined using highly precise mercury thermometers. When used with a magnifying glass, these thermometers can be read to $\pm 0.01°C$. Alternatively, a pair of platinum resistors (RTDs), thermocouples, or thermistors can be used.

A cross-sectional diagram of the Parr adiabatic calorimeter is shown in Figure 5.5.1. Notice that there are stirrers that gently agitate the water in the bucket and jacket. This mixing is deliberately kept to a level low enough to avoid mechanical (Joule-type) heating.

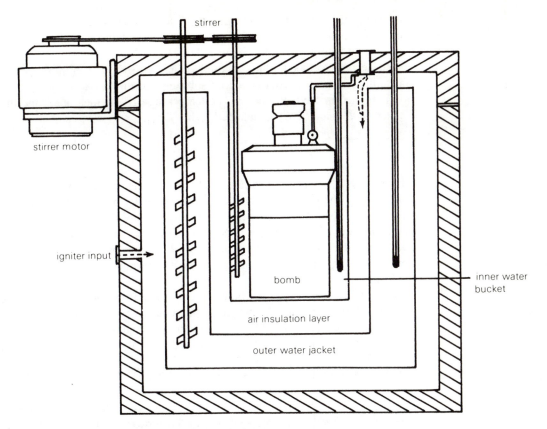

FIGURE 5.5.1   Cutaway view of adiabatic calorimeter.

Combustion of the Sample

The sample, if a solid, is usually introduced to the bomb in the form of a compressed pellet. Combustion is externally triggered by sending an electric current through a thin-gauge nichrome wire that is situated just above (or touches) the pellet. This fuse wire is connected to a voltage source (ca. 20 V ac) by means of electrodes that pass through the bomb via high-pressure seals. When current passes through the nichrome fuse, it becomes very hot. The hot wire expands, touching (or dropping onto) the sample, and this is usually sufficient to induce the combustion of the sample in the presence of the high O_2 pressure. Volatile liquid samples are often placed in a gelatin capsule to facilitate weighing. The capsule is then positioned in the fuel container so that it is in contact with the nichrome fuse. Nonvolatile liquid samples can be placed (and weighed) directly in the fuel container.

Calibration of the Calorimeter

Because absolute values of heat flow are required in this experiment, the calibration of the apparatus is a crucial part of the procedure. Calibration can be performed in two ways, electrically or chemically. In the first method, a known amount of heat is introduced to the fully loaded calorimeter (filled bomb, bucket, and water) by passing a measured amount of electric current through a resistor for a definite time. The power dissipated by the resistor is $P = iV$ where V is the applied voltage (if V is in volts and i in amperes, P is given in Joules/s, or watts).

The energy flow, which is presumed to be totally thermal, is iVt, where t is the time. The temperature rise ΔT is measured, and the calorimeter heat capacity, C, is calculated from $C = q/\Delta T = iVt/\Delta T$.

The second method (which is used in this experiment) employs a chemical standard. A known amount of a pure substance having a known heat of combustion is ignited in the calorimeter. The temperature rise is measured, and the calorimeter *heat capacity, C,* is obtained from

$$C = \frac{Q_{\text{tot}}}{\Delta T},$$
(5.5.14)

where Q_{tot} is the total amount of heat liberated in the calorimeter as a result of the combustion. The calorimeter heat capacity essentially provides the conversion between the measured temperature change and the absolute heat flow that caused that temperature rise *for the particular calorimeter*. The two most significant components of the calorimeter that determine its heat capacity are the mass of the bomb (ca. 8 kg) and the mass of the water that surrounds the bomb in the bucket (ca. 2 kg).

The total heat absorbed is obtained principally from the heat liberated by the combustion standard sample, which in this experiment is pure benzoic acid supplied in nominal 1-g pellet form. This quantity is equal to $Q_m m$, where Q_m is the heat released per gram of benzoic acid and m is the mass of the benzoic acid sample. $Q_m = 26{,}435$ Jg^{-1}.

Two small correction terms must be added to the numerator of equation (5.5.14). These terms must be used with *any* combustion reaction in the calorimeter to account (1) for the heat released by the combustion of part of the nichrome wire, and (2) for the heat associated with the formation of nitric acid. The latter is produced when the N_2 in the air initially contained in the bomb is oxidized (in the presence of water):

$$N_2(g) + 5/2O_2(g) + H_2O(l) \rightarrow 2HNO_3(aq).$$

Calculation of Heat Flow

In carrying out a combustion reaction, whether for a calibration or for an actual thermochemical measurement, the corrected heat flow is expressed as

$$Q_{\text{tot}} = Q_m m + q_1 + q_2,$$
(5.5.15)

where $Q_m m$ is the heat liberated by the actual combustion of the sample and q_1 and q_2 are the two correction terms mentioned above. q_1 is obtained by measuring the *change* in length of the nichrome fuse after ignition and multiplying this value by the conversion factor. For the No. 34 gauge nichrome wire used in this experiment, this value (supplied by the manufacturer) is 9.6 J cm^{-1}; hence

$$q_1 = 9.6(L_0 - L) \quad \text{(J)},$$
(5.5.16)

where L_0 and L are the initial and final lengths of the nichrome fuse, respectively, in centimeters.

The determination of the correction term for the formation of nitric acid is made by titrating the aqueous residue in the bomb with a standardized Na_2CO_3 solution to a methyl red (or methyl orange) end point. If this solution is made $0.0725N$, 1 mL corresponds to an amount of HNO_3 produced in combustion equivalent to 4.18 J. Hence the expression for q_2 is

$$q_2 = 4.18V_b \quad \text{(J)},$$
(5.5.17)

where V_b is the number of milliliters of $0.0725N$ base needed to reach the end point.

Safety Precautions

☐ Safety goggles must be worn at all times during this experiment.

☐ A gas cylinder containing O_2 at high pressure is used in this experiment. The cylinder must be strapped to a secure foundation.

☐ The bomb is filled to high pressure (e.g., 20 atm). Make sure the gas line is correctly and tightly connected to the bomb, and that the bomb is firmly closed before filling.

☐ When carrying the bomb, support it by the bottom; use two hands.

☐ Do not place *anything* in the bomb other than the sample you are instructed to use. Do not use more fuel than indicated in the Procedure.

☐ If the charged bomb shows considerable bubbling after it is placed in the water-filled calorimeter bucket, consult the instructor. The bomb may have to be vented, resealed, and refilled.

☐ After combustion, vent the bomb slowly, preferably in a fume hood.

Experimental Procedure

CAUTION: Bomb calorimetry involves the use of high pressures. Safety goggles MUST be worn. Make sure you are aware of and understand the pressurizing, ignition, and depressurizing steps described below. Use two hands when transporting the pressurized bomb from the O_2 cylinder to the calorimeter; support the bomb from the bottom.

Calibration

1. Weigh (to the nearest 0.1 mg) a 1-g benzoic acid pellet (supplied by the stockroom). Place the pellet in the metal pan (called the fuel capsule) which is supported by the ring-shaped electrode; the pellet could be situated upright, resting against the side of the fuel capsule, or placed directly on the bottom of the fuel capsule.

2. Cut a 10-cm length of nichrome ignition wire and connect it to the bomb electrodes. Make sure there are no kinks in the wire. Insert the ends of the fuse wire through the holes in the electrodes so that only a few millimeters protrude. Slide down the contact sleeves to ensure good electrical contact. The shape and position of the fuse wire are important. The wire should be formed into a loop that is positioned *just above* the pellet. See Figure 5.5.2. The loop formed by the ignition wire should touch the side (or top) of the pellet. It is sometimes helpful to use a pair of forceps to shape and position the fuse loop. Finally, make sure the pellet is secure enough so that it will not move when the bomb is assembled and moved. Take a few minutes to carry out this procedure correctly; a complete combustion depends on the correct fuse/sample attachment.

3. Add 1 mL of deionized water to the bomb using a pipet. This provides a medium for the HNO_3 that is produced. It also promotes the condensation of the water formed in the combustion.

4. Inspect the O-ring in the cap to make sure that it has no nicks or dirt on its surface. Close the bomb by gently turning the screw cap down *firmly* by hand. Make sure you know how to seal the bomb; consult the instructor.

5. Attach the O_2 inlet line to the fitting on the bomb; charge to a pressure of 30 atm (ca. 450 psig—see pp. 28–29 for the definition of psig) by cautiously

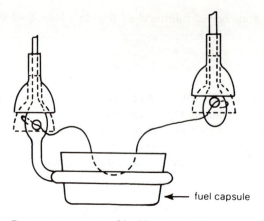

← fuel capsule

FIGURE 5.5.2   Proper arrangement of ignition wire with respect to the fuel capsule.

turning the filling knob. It may take 10 to 15 s until the bomb is filled; you can hear when the gas flow has stopped. When the bomb is filled, close the filling valve and purge the filling line using the small lever. The gas fitting on the bomb head automatically seals closed when the filling line is purged. The instructor will help you with the first run.

6. Add exactly 2 L of deionized water (a 2-L volumetric flask is provided) to the calorimeter bucket. The water should be about 1 to 2° below room temperature. The same amount of water must be added in all the runs in this experiment. (For higher accuracy, the *mass* of water delivered to the bucket should be determined. An error of 1 mL corresponds to 4.18 J error in the calculated heat flow.) Holding the bomb by the head of the venting valve nut, submerge the bomb in the calorimeter bucket. Avoid jarring the bomb. Make sure that there are no gas bubbles escaping from the bomb seal. *If you see gas escaping,* **DO NOT CONTINUE**. Inform the instructor; the bomb might have to be purged and refilled.

7. Carefully place the calorimeter bucket in the outer jacket. Be sure that the flexible metal couplings to the electrodes are clean and dry so that good contact is made. Attach the ignition voltage supply wire to the lug on the top of the bomb. Close the calorimeter by swinging the cover to the right and lowering the cam lever. Make sure that it seats properly and that the water pump and stirrer couplings are in place. Next, lower the two thermometers into the apparatus.

8. Switch on the motor that drives the circulating and mixing pulleys. Adjust the hot/cold water flow through the jacket so that the temperatures are within ±0.01°F (±0.005°C) of one another. Make sure the reading lens is attached to the calorimeter thermometer. It is needed to achieve the optimal precision.

9. Record the temperature of the calorimeter (to 0.01°F). Ignite the sample by pressing the button on the ignition module. Keep the button depressed *until the indicator light goes out* (a few seconds). The calorimeter temperature should start to rise within 1 min.

10. Carefully manipulate the hot/cold water flowing through the jacket to keep the temperature as close as possible to the rising temperature of the calorimeter. It is important to do this especially as the temperature levels off. Record the temperature of the calorimeter periodically, at about 15- to 30-s intervals, until it has become constant; this will take 10 to 15 min.

11. Turn off the motor, *raise the thermometers,* and wait for 15 to 30 s to allow the water to drain from the calorimeter lid. Open the top of the calorimeter jacket and gently lift out the inner bucket. Remove the bomb and slowly relieve

the pressure by opening the needle valve. This should be done over a 1-min interval.

12. Remove the screw cap and inspect the inside of the bomb; make sure that there has been a "clean" combustion. There should be no unburned sample or soot. The run must be considered a failure if the combustion was incomplete. Wash the interior surface of the bomb using deionized water. Quantitatively transfer the washings to an Erlenmeyer flask, add a few drops of indicator, and titrate with the standard Na_2CO_3 solution. Record the number of milliliters of alkali used as well as its normality.

13. Remove the remaining pieces of fuse wire from the bomb electrode posts and measure the combined length.

REPEAT: Clean the inside of the bomb and repeat the standardization procedure using another benzoic acid pellet.

Check the results of the two standardization runs for internal consistency. The ratios of the temperature increase to the mass of benzoic acid used for the two runs should be *within 1 percent* of each other.

Sample Combustion—A: Naphthalene

1. On the triple-beam balance, weigh out about 0.9 g of naphthalene. Prepare a pellet of this sample using the pellet press. The instructor will help. *Do not overcompress the sample.* The idea is to apply enough pressure to bind the naphthalene particles together so that the pellet will not break apart while being handled. If the pellet is too hard, it will fragment upon ignition, and combustion may not be complete.

2. After the pellet is prepared, weigh it to within 0.1 mg.

3. Attach a 10-cm length of fuse wire to the electrodes (see above). Place the naphthalene pellet in the fuel capsule; shape and position the fuse wire so that the loop is just above the pellet.

4. Assemble the bomb, charge with 30 atm O_2, and carry out a combustion following the same procedure used in the standardization.

Repeat at least once; check the results for consistency.

Sample Combustion—B: Sucrose

1. Accurately weigh about 1.5 g of sucrose directly in the metal fuel capsule. Secure the fuel capsule on the electrode ring.

2. Attach a 10-cm length of fuse wire to the electrodes (see above). Position the fuse wire so the bottom of the loop is a few millimeters above the sucrose sample.

3. Assemble the bomb, charge with 30 atm of O_2, and carry out a combustion following the same procedure as in the standardization.

Repeat at least once; check the results for consistency.

Data Analysis

1. Determine the heat capacity (also called the energy equivalent) of the bomb calorimeter in J °C^{-1} using $C = Q_{tot}/\Delta T$. The specific heat of combustion of benzoic acid is 26.43 kJ/g. In determining q_2, use the conversion factor appropriate to the actual normality of the Na_2CO_3 used in the titration.

2. Using the same procedure, determine the heat of combustion of naphthalene (or sucrose) in kJ mol^{-1}.

3. Convert this value to standard conditions by using the correction term for gas ideality, ΔH_3 [see equation (5.5.13)]. Heat capacities and Joule-Thomson coefficients for O_2 and CO_2 are shown below.

	C_P (J mol^{-1} K^{-1})	μ_{JT} (K° atm^{-1})
O_2	29.72	0.31
CO_2	37.14	1.10

The C_P values are for 298 K and 1 atm pressure. Assume both C_P and μ_{JT} to be pressure-independent. Notice that ΔH_3 is a molar quantity; thus be sure to take into account the stoichiometric coefficients of the gases in the balanced combustion reaction. As an example, consider the correction for O_2 in the combustion reaction (5.5.2). The correction term needed to bring O_2 from high pressure (P_h) to ideal gas conditions (P_{id}) is obtained by *adding* the following quantity to equation (5.5.2); see also equation (5.5.13):

$$12 O_2 \ (P_h) \rightarrow 12 O_2 \ (P_{id}) \qquad \Delta H_3 = 12 C_P \, \mu_{JT}(0 - 30).$$

The correction term involving CO_2 must also be considered.

4. Determine the standard heat of formation of naphthalene (or sucrose) from the result in step 3 using the standard heats of formation of $CO_2(g)$ and $H_2O(l)$. Compare this result with the literature value.

5. Perform an error analysis of the standard heat of formation of naphthalene (or sucrose).

6. Estimate the compressibility factor for O_2 at 30 atm and 300 K from the van der Waals equation (consult your textbook) and, using the volume of the bomb (ca. 0.36 dm^3), determine the mole ratio of O_2 to sample (benzoic acid or naphthalene) and compare with the stoichiometric ratio for the combustion. If you have not studied compressibility factors yet, use the ideal gas law as an approximation.

Questions and Further Thoughts

1. The conversion equivalent required to take into account the formation of HNO_3 was given as 4.18 J mL^{-1} if Na_2CO_3(aq) is $0.0725N$; see equation (5.5.17). From this information, calculate ΔH for the combustion reaction associated with the formation of HNO_3(aq).

2. A student performs a calorimeter calibration using an electric heater. When a voltage of 30 V is applied to the heater for 3.0 min, a constant current of 2.5 A is observed. The calorimeter temperature increases from 20.00 to 25.59°C. Calculate C_P for the calorimeter.

3. What is the major source of error in measuring heats of combustion?

4. To obtain a larger effect that one is measuring (and hence lower the relative error), a larger sample could be used. What is the disadvantage of this approach in this particular experiment?

5. Suppose the sample had a 1 mole percent impurity that had a heat of combustion 10 percent larger than that of the compound being studied, e.g., naphthalene or sucrose. Assuming that the measured heat of combustion of the mixture is a mole-fraction-weighted average of the components, what is the effect of this impurity on the determined *heat of formation* of the particular compound?

6. In this experiment only two trials (each) for the calorimeter calibration and sample combustion are possible in the time available. How many runs (calibration and sample combustion) would be required in order to reduce the random error in the measured heat of combustion by a factor of three? See pp. 4–5.

Further Readings

R. A. Alberty, "Physical Chemistry," 7th ed., pp. 53–64, Wiley (New York), 1987.

P. W. Atkins, "Physical Chemistry," 3rd ed., pp. 79–90, W. H. Freeman (New York), 1986.

G. N. Castellan, "Physical Chemistry," 3rd ed., pp. 138–145, Addison-Wesley (Reading, Mass.), 1983.

I. N. Levine, "Physical Chemistry," 2nd ed., 1983, pp. 142–146, McGraw-Hill (New York), 1983.

J. N. Noggle, "Physical Chemistry," pp. 251–256, Little, Brown (Boston), 1985.

H. A. Skinner, ed., "Experimental Thermochemistry; Measurement of Heats of Reaction," IUPAC Commission of Chemical Thermodynamics, vol. 2, Interscience (New York), 1956–1962.

R. C. Wilhoit, *J. Chem. Educ.*, *44*, A-629 (1967).

Experiment 5

Heat of Combustion

NAME _____   DATE _____

Trial number	Weight of benzoic acid	T_1	T_2	Volume Na_2CO_3	Length wire
1	_____	_____	_____	_____	_____
2	_____	_____	_____	_____	_____
3	_____	_____	_____	_____	_____
4	_____	_____	_____	_____	_____

- -

Trial number	Weight of sample	T_1	T_2	Volume Na_2CO_3	Length wire
1	_____	_____	_____	_____	_____
2	_____	_____	_____	_____	_____
3	_____	_____	_____	_____	_____
4	_____	_____	_____	_____	_____

Experiment 5

Heat of Combustion

NAME _____ DATE _____

Trial number	Weight of benzoic acid	T_1	T_2	Volume Na_2CO_3	Length wire
1	_____	_____	_____	_____	_____
2	_____	_____	_____	_____	_____
3	_____	_____	_____	_____	_____
4	_____	_____	_____	_____	_____

- -

Trial number	Weight of sample	T_1	T_2	Volume Na_2CO_3	Length wire
1	_____	_____	_____	_____	_____
2	_____	_____	_____	_____	_____
3	_____	_____	_____	_____	_____
4	_____	_____	_____	_____	_____

□ Experiment 6

Bomb Calorimetry: The Resonance Energy of Benzene

First read Experiment 5.

In this experiment, the heat of combustion of the macrocyclic compound, 1,5,9-trans,-trans,-cis-cyclododecatriene (CDDT) will be measured in the bomb calorimeter. The structure of this compound, which is a trimer of butadiene (C_4H_6), a common petroleum feedstock, is shown below.

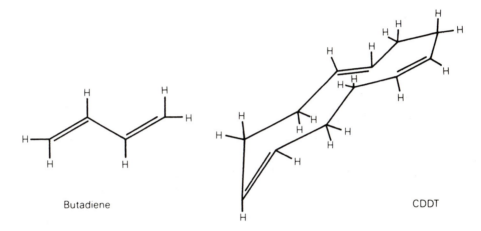

Butadiene

CDDT

The combustion reaction is

$$C_{12}H_{18}(l) + (33/2)O_2(g) \longrightarrow 12CO_2(g) + 9H_2O(l).$$

CDDT contains three C—C double bonds, nine C—C single bonds, six C—H bonds, and six CH_2 groups. An interesting comparison can be made between CDDT and two simple molecules, benzene (C_6H_6) and cyclohexane (C_6H_{12}).[1] If benzene is simply considered to possess three C—C double bonds, three C—C single bonds, and six C—H bonds, and cyclohexane to consist of six C—C bonds and six CH_2 groups, then the heat of formation of CDDT should be equal to the sum of the heats of formation of benzene and cyclohexane. Verify this by counting the number and types of bonds in benzene and cyclohexane and compare with CDDT.

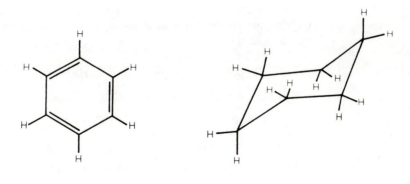

Benzene

Cyclohexane

This bond energy analysis pertains to the gas phase properties of the three molecules (to avoid discrepancies due to differences in the heats of vaporization). Because CDDT contains the same number of C and H atoms as the benzene and cyclohexane "constituents," the heat of combustion of CDDT should be equal to the sum of the heats of combustion of benzene and cyclohexane (again, all hydrocarbons being in the gas phase).

This bond energy approach to thermochemical reactions assumes that all the bond "groups" are isolated, or noninteracting. For example, it would be assumed that benzene is 1,3,5-cyclohexatriene, i.e., as it is literally depicted above. While the bond group analysis may be valid for CDDT and cyclohexane, we know it cannot be true for benzene, which can be represented by a resonance hydrid between the two Kekulé structures:

Benzene is, in fact, thermodynamically more stable than three isolated C—C double bonds and three C—C single bonds in a cyclic system; it is *not* cyclohexatriene. The increased stability of benzene relative to (hypothetical) cyclohexatriene is called the *resonance energy* and is associated almost entirely with the delocalization of the six π-electrons occupying the (six) carbon $2p_z$ atomic orbitals. There are several experimental manifestations of this resonance energy, including thermochemical (e.g., the heat of hydrogenation of benzene vis-à-vis that of cyclohexene), spectroscopic (ultraviolet absorption spectrum of benzene vis-à-vis ethylene or cyclohexene), and chemical reactivity (e.g., the rate of bromination of benzene as compared with ethylene).

The resonance energy can be determined from the following hypothetical reaction:

$$C_6H_6(g) \quad \text{(hexatriene)} \longrightarrow C_6H_6(g) \quad \text{(benzene)} + E_{res}, \quad (5.6.1)$$

where E_{res} is the resonance energy. Considering the previous discussion about the comparison of bond energies and combustion enthalpies, we can also define the resonance energy of benzene as the enthalpy of the following reaction:

$$C_6H_6(g) \quad \text{(benzene)} + C_6H_{12}(g) \quad \text{(cyclohexane)} \longrightarrow$$
$$C_{12}H_{18}(g) \quad \text{(CDDT)}. \quad (5.6.2)$$

This reaction can be expressed in terms of the following thermodynamic cycle involving the combustion enthalpies of the *liquid* hydrocarbons and their heats of vaporization:

a. $\quad C_6H_6(g) \quad\quad \longrightarrow \quad\quad C_6H_6(l) \quad\quad -\Delta H_{vap}$ (benzene)

b. $C_6H_6(l) + 15/2O_2(g) \longrightarrow 6CO_2 + 3H_2O(l) \quad \Delta H_{comb}$ (benzene)

c. $\quad C_6H_{12}(g) \quad\quad \longrightarrow \quad\quad C_6H_{12}(l) \quad\quad -\Delta H_{vap}$ (cyclohexane)

d. $C_6H_{12}(l) + 9O_2(g) \longrightarrow 6CO_2(g) + 6H_2O(l) \quad \Delta H_{comb}$ (cyclohexane)

e. $12CO_2(g) + 9H_2O(l) \longrightarrow C_{12}H_{18}(l) + 33/2O_2(g) \quad -\Delta H_{comb}$ (CDDT)

f. $\quad C_{12}H_{18}(l) \quad\quad \longrightarrow \quad\quad C_{12}H_{18}(g) \quad\quad \Delta H_{vap}$ (CDDT)

The enthalpy of step e is measured in this experiment; the others can be obtained from the literature. Hence the enthalpy of reaction (5.6.2), the resonance energy of benzene, can be determined.

Safety Precautions

☐ Safety goggles must be worn during this experiment.

☐ The use of a high-pressure oxygen supply is required. Make sure the gas cylinder is securely strapped to a firm support. Be sure that the gas connections are properly attached to the bomb before you charge it with oxygen.

☐ Do not place anything in the bomb except the fuel (combustible) used in this experiment.

☐ In transporting the filled, charged bomb, use two hands; support the bomb from the bottom.

☐ CDDT has a slight odor. Be careful not to drop it or spill it in the laboratory. The experiment should be performed in an open, well-ventilated laboratory. If the fumes are objectionable, consult your instructor.

Procedure

1. *First read the Procedure in Experiment 5.*

2. Standardize the calorimeter by performing duplicate measurements using benzoic acid pellets.

3. Because CDDT is a low-vapor-pressure liquid, it can be placed in the bomb directly in the fuel capsule. First remove the fuel capsule from the round electrode and clean thoroughly with acetone and water. Dry with a clean paper towel. Weigh the empty capsule to within 0.1 mg; then add about 0.7 g of CDDT (ca. 40 drops). Reweigh ($\pm$0.1 mg), and replace the capsule on the electrode ring.

4. Cut a 10-cm length of fuse wire and attach the ends to the electrodes. The fuse wire should form a loop that extends to within about 1/8 in. of the surface of the CDDT in the capsule. A pair of forceps can be used to manipulate the fuse wire. Take time with this step; the proper attachment and adjustment of the ignition wire is vital to achieving a "clean" combustion.

5. Follow the same procedure outlined on pp. 101–102 for closing, charging, and loading the bomb. Perform combustion measurements in duplicate.

Data Analysis

1. Determine the heat equivalent of the bomb calorimeter (and the approximate uncertainty).

2. Obtain the standard (molar) heat of combustion of CDDT (and the approximate uncertainty). Use the correction term, ΔH_3, to bring the data to ideal gas conditions. See the discussion in Experiment 5. Be sure to use the appropriate stoichiometric coefficients for O_2 and CO_2 in the balanced combustion equation for CDDT. See Experiment 5, Data Analysis, step 3.

3. Look up (or calculate) the standard heats of combustion of benzene and cyclohexane as well as their heats of vaporization. The vapor pressure of CDDT has been measured over a range of temperatures by Rauh et al.,[2] and they report

$$\log_{10}(P_{\text{vap}}) = \frac{-3552}{T} + 10.861$$

The heat of vaporization can be obtained from this expression.

4. From the above data, calculate the resonance energy of benzene. Report the uncertainty. Compare with "literature values"; see your physical chemistry or organic chemistry texts.

5. Estimate the compressibility factor for O_2 at 30 atm and 300 K from the van der Waals equation (consult your textbook) and, using the volume of the bomb (ca. 0.36 dm^3), determine the mole ratio of O_2 to sample (benzoic acid or CDDT) and compare with the stoichiometric ratio for the combustion. If you have not studied compressibility factors yet, use the ideal gas law as an approximation.

Questions and Further Thoughts

1. The conversion equivalent required to take into account the formation of HNO_3 was given as 4.18 J mL^{-1} if Na_2CO_3(aq) is $0.0725N$; see equation (5.5.17). From this information, calculate ΔH for the combustion reaction associated with the formation of HNO_3(aq).

2. A student performs a calorimeter calibration using an electric heater. When a voltage of 30 V is applied to the heater for 3.0 min, a constant current of 2.5 A is observed. The calorimeter temperature increases from 20.00 to 25.59°C. Calculate C_P for the calorimeter.

3. What is the major source of error in measuring heats of combustion?

4. To obtain a larger effect that one is measuring (and hence lower the relative error), a larger sample could be used. What is the disadvantage of this approach in this particular experiment?

5. Suppose the sample had a 1 mole percent impurity with a heat of combustion 10 percent larger than that of CDDT. Assuming that the measured heat of combustion of the mixture is a mole-fraction-weighted average of the components, what is the effect of this impurity on the determined *heat of combustion* of CDDT and the resonance energy of benzene?

6. In this experiment only two trials (each) for the calorimeter calibration and sample combustion are possible because of time limitations. How many runs (calibration and sample combustion) would be required in order to reduce the random error in the measured heat of combustion by a factor of three? See pp. 4–5.

Notes

1. M. Pickering, *J. Chem. Educ.*, *59*, 318, (1982).
2. H.-J. Rauh, W. Geyer, H. Schmidt, and G. Geisler, *Z. phys. Chem.* (Leipzig), *253*, 43, 1973.

Further Readings

R. A. Alberty, "Physical Chemistry," 7th ed., pp. 53–64, Wiley (New York), 1987.

P. W. Atkins, "Physical Chemistry," 3rd ed., pp. 79–90, W. H. Freeman (New York), 1986.

G. N. Castellan, "Physical Chemistry," 3rd ed., pp. 138–145, Addison-Wesley (Reading, Mass.), 1983.

I. N. Levine, "Physical Chemistry," 2nd ed., pp. 142–146, McGraw-Hill (New York), 1983.

J. N. Noggle, "Physical Chemistry," pp. 251–256, Little, Brown (Boston), 1985.

H. A. Skinner, ed., "Experimental Thermochemistry; Measurement of Heats of Reaction," IUPAC Commission of Chemical Thermodynamics, vol. 2, Interscience (New York), 1956–1962.

R. C. Wilhoit, *J. Chem. Educ.*, *44*, A-571, A-629 (1967).

Experiment 6

Heat of Combustion

NAME _____ DATE _____

Trial number	Weight of benzoic acid	T_1	T_2	Volume Na_2CO_3	Length wire
1	_____	_____	_____	_____	_____
2	_____	_____	_____	_____	_____
3	_____	_____	_____	_____	_____
4	_____	_____	_____	_____	_____

Trial number	Weight of sample	T_1	T_2	Volume Na_2CO_3	Length wire
1	_____	_____	_____	_____	_____
2	_____	_____	_____	_____	_____
3	_____	_____	_____	_____	_____
4	_____	_____	_____	_____	_____

Experiment 6

Heat of Combustion

NAME _____ DATE _____

Trial number	Weight of benzoic acid	T_1	T_2	Volume Na_2CO_3	Length wire
1	_____	_____	_____	_____	_____
2	_____	_____	_____	_____	_____
3	_____	_____	_____	_____	_____
4	_____	_____	_____	_____	_____

Trial number	Weight of sample	T_1	T_2	Volume Na_2CO_3	Length wire
1	_____	_____	_____	_____	_____
2	_____	_____	_____	_____	_____
3	_____	_____	_____	_____	_____
4	_____	_____	_____	_____	_____

□ PART SIX

Thermodynamics of Solutions: Electrochemistry

□ Experiment 7

The Entropy of Mixing

Objective

To determine the ideal entropy of mixing from electrochemical measurements.

Introduction

The chemical potential of component i in a liquid phase solution may be determined from the relationship

$$\mu_i = \mu_i^{\bullet}(T,P) + RT \ln a_i = \mu_i^{\bullet}(T,P) + RT \ln \gamma_i x_i, \qquad (6.7.1)$$

where $\mu_i^{\bullet}$ is the molar free energy (chemical potential) of the ith component in its pure form at a temperature T and pressure P and $a_i, \gamma_i,$ and x_i are, respectively, the activity, activity coefficient, and mole fraction of this component in the solution. Equation (6.7.1) is based on a similar expression derived for ideal gas mixtures in which P_i, the partial pressure of the ith component (in atmospheres), replaces x_i.

An ideal solution may be defined as one for which $\gamma_i = 1$ for all components in all compositions. Ideal behavior is usually rationalized on the basis of equivalent intermolecular interactions between *different* components vis-à-vis a *single* component. For example, in a binary mixture of A and B, the A-B interaction energy is assumed to be equal to that of A-A and B-B. When an ideal solution is formed, the enthalpy changes associated with overcoming A-A and B-B attractions in the pure components are assumed to be of equal magnitudes (and opposite sign) relative to the new attractions between A and B. This energy balance results in a net enthalpy change of zero; thus it is rationalized that ΔH_{mix} (ideal) = 0.

The free energy change that takes place when two components, A and B, are mixed can be determined from the expression

$$\Delta G_{\text{mix}} = n_A \mu_A + n_B \mu_B - n_A \mu_A^{\bullet} - n_B \mu_B^{\bullet}, \qquad (6.7.2)$$

where n_A and n_B are the respective mole numbers of A and B that are combined. Substitution of equation (6.7.1) for the chemical potential in a solution into this expression gives

$$\Delta G_{\text{mix}} = n_A RT \ln \gamma_A x_A + n_B RT \ln \gamma_B x_B, \qquad (6.7.3a)$$

which, for the ideal case, becomes

$$\Delta G_{\text{mix}} = n_A RT \ln x_A + n_B RT \ln x_B, \qquad (6.7.3b)$$

(because $\gamma_A = \gamma_B = 1$). From (6.7.3b) it follows that the general expression for the free energy of mixing of N pure components to form an ideal solution is

$$\Delta G_{\text{mix}} = RT \sum_{i=1}^{N} (n_i \ln x_i). \qquad (6.7.4)$$

The determination of ΔG_{mix} for a solution is usually carried out by measuring the partial pressures of the mixed components in the vapor phase in equilibrium above the solution. For an ideal solution, this would correspond to an experimental confirmation of Raoult's law, $P_{\text{tot}} = \sum_{i}(x_i P_i^{\bullet})$, where $P_i^{\bullet}$ is the vapor pressure of *pure* component i.

An expression for the ideal entropy of mixing may be obtained from equation (6.7.3b) by using the fundamental relation, $\Delta S = -(\partial G / \partial T)_P$, or

$$\Delta S_{\text{mix}} = -R \sum_{i=1}^{N} n_i \ln x_i. \qquad (6.7.5)$$

Entropies of mixing are likewise usually obtained from vapor pressure measurements.

In this experiment, the free energy of mixing will be determined directly from *electrochemical* measurements.[1] The materials being mixed are not, themselves, pure substances, but are aqueous solutions of electrolytes: potassium hexacyanoferrate(II) and (III), i.e., $K_4Fe(CN)_6$ and $K_3Fe(CN)_6$. These will be abbreviated HCFeII and HCFeIII, respectively. Equation (6.7.1) cannot, strictly speaking, be used to represent the chemical potential of either of these components in the mixture because the $x_i = 1$ reference state is not a pure compound. Instead, the experiment is designed so that the reference solutions of each of the hexacyanoferrate species are at the same composition, m_0 molal. In this case, the chemical potentials of these reference solutions are given by the equations

$$\mu_{\text{II}} = \mu_{\text{II}}^{\circ} + RT \ln (m_0 \gamma_{\text{II}}), \qquad (6.7.6a)$$

and

$$\mu_{\text{III}} = \mu_{\text{III}}^{\circ} + RT \ln (m_0 \gamma_{\text{III}}), \qquad (6.7.6b)$$

where γ_{II} and γ_{III} are the activity coefficients of HCFeII and HCFeIII in the reference (m_0) states. When these solutions are *mixed*, the chemical potentials become

$$\mu_{\text{II}}' = \mu_{\text{II}}^{\circ} + RT \ln (m_{\text{II}} \gamma_{\text{II}}') = \mu_{\text{II}}^{\circ} + RT \ln (m_0 x_{\text{II}} \gamma_{\text{II}}'), \qquad (6.7.7a)$$

$$\mu_{\text{III}}' = \mu_{\text{III}}^{\circ} + RT \ln (m_{\text{III}} \gamma_{\text{III}}') = \mu_{\text{II}}^{\circ} + RT \ln (m_0 x_{\text{III}} \gamma_{\text{III}}'), \qquad (6.7.7b)$$

where γ_{II}' and γ_{III}' now refer to the activity coefficients of the HCFe species in the mixed state. Substituting these equations into equation (6.7.1) gives

$$\Delta G_{\text{mix}} = RT\left\{ n_{\text{II}}\left[\ln x_{\text{II}} + \ln\left(\frac{\gamma'_{\text{II}}}{\gamma_{\text{II}}}\right) \right] + n_{\text{III}}\left[\ln x_{\text{III}} + \ln\left(\frac{\gamma'_{\text{III}}}{\gamma_{\text{III}}}\right) \right] \right\} \quad (6.7.8)$$

Because the ionic strengths[2] of the mixed and reference states are nearly identical, the activity coefficient ratios, $\gamma'_{\text{II}}/\gamma_{\text{II}}$ and $\gamma'_{\text{III}}/\gamma_{\text{III}}$, may be approximated as unity.[3] We are thus assuming that the hexacyanoferrate species experience the same activity effects (e.g., solvation) in the mixed state as they do in their respective reference states. This aqueous system is therefore treated as a "pseudo ideal," two-component mixture with its free energy of mixing given by the relationship

$$\Delta G_{\text{mix}} = n_{\text{II}}RT \ln x_{\text{II}} + n_{\text{III}}RT \ln x_{\text{III}}, \quad (6.7.9)$$

an expression identical in form to equation (6.7.3b).

Electrochemical Measurements

One hypothetical method by which ΔG_{mix} could be determined for this system would be to construct an electrochemical cell consisting of 0.1 *m* HCFeII reference solution in one half cell, an equal volume of 0.1 *m* HCFeIII reference solution in the other half cell and allow electrons to flow *reversibly* in the external circuit. (We take the term reversible to denote a process that is carried out at an infinitesimal rate so that each point along the path represents the system arbitrarily close to equilibrium.) Refer to Figure 6.7.1. The cell potential would run down as HCFeII in the left half cell was replaced by HCFeIII, *and,* at the same time, HCFeIII in the right half cell was replaced by HCFeII. (Notice that when an electron is transferred from HCFeII to HCFeIII, the two species "switch identity.") As electron flow continues and the system eventually reaches equilibrium, the cell potential, E, and hence the free energy difference, ΔG, goes to zero. At this final state, each half cell would contain a 50/50 mixture of HCFeII and HCFeIII. Because this process involves the net mixing to equal volumes of the two HCFe reference solutions, the work dissipated by the load (i.e., expended by the motor in Figure 6.7.1) would be a direct measure of ΔG_{mix}. Actually, the free energy change in this case is equal to the (negative of the) maximum useful, or reversible, work: $\Delta G = -w_{\text{rev}}$.

Unfortunately, an *infinite* amount of time is required to complete such a reversible change because at each stage along the current flow (mixing) process, the system would be in a state infinitesimally close to equilibrium. A more practical

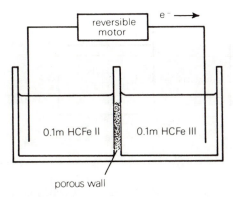

FIGURE 6.7.1　0.1 *m* solutions of HCFeII and HCFeIII separated by a porous plug. Electrodes immersed in each solution are coupled to a load (motor).

approach is to examine a series of electrochemical cells in which the ratio of the HCFeII/HCFeIII molalities in the two half cells is varied but in which the *total* molality in each is maintained at m_0. In effect, this approach examines the mixing process *statically* at a series of reversible stages along the mixing path, and the data are mathematically combined to determine the final result. In order to understand how this can be done, it is necessary to take a closer look at the thermodynamics of the system.

The differential of the Gibbs free energy is

$$dG = -S\,dT + V\,dP + dw_{rev}, \tag{6.7.10}$$

where dw_{rev} is the maximum amount of useful, non-PV work that accompanies the process. For the case in which this work is electrical,

$$dw_{rev} = -FE_{rev}\,dn,$$

where dn is the number of moles of charge that flows under the electrical potential, E_{rev}, and F is the Faraday constant (96,487 coulombs mol^{-1}—an SI unit). At constant temperature and pressure, equation (6.7.10) becomes

$$dG = -FE_{rev}\,dn. \tag{6.7.11}$$

The net free energy change accompanying a finite process is obtained by integrating equation (6.7.11). In the Fe^{2+}/Fe^{3+} example considered here, this corresponds to the free energy of mixing of the two species, thus

$$\Delta G_{mix} = \int_{initial}^{final} dG = \int_{initial}^{final} -FE_{rev}\,dn. \tag{6.7.12}$$

Figure 6.7.2 shows the initial and final states pertinent to the hypothetical example portrayed in Figure 6.7.1. These schematic half cells help in establishing the integration limits needed in equation (6.7.12). If the molality of Fe^{3+} in the initial left-hand half cell, $m_{Fe^{3+}}$, is used as the variable that represents the extent of reaction (mixing), we see that, initially, $m_{Fe^{3+}} = 0$, and finally, $m_{Fe^{3+}} = 0.05$. We could equivalently have used $m_{Fe^{2+}}$ in the right-hand half cell as the extent of reaction variable.

As mentioned above, the "mixing" of the Fe^{2+} and Fe^{3+} species is brought about by external charge (electron) flow between the half cells, and, in fact, as the charge flow (mixing) nears completion, E_{rev} falls to zero. Thus E_{rev} in equation (6.7.12) is a function of n. To obtain a numerical value for ΔG_{mix}, we need to express n as a function of the cell composition; this is done as follows.

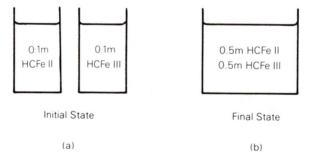

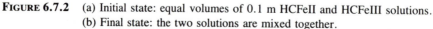

FIGURE 6.7.2   (a) Initial state: equal volumes of 0.1 m HCFeII and HCFeIII solutions.
(b) Final state: the two solutions are mixed together.

Suppose we consider the generalized situation represented in Figure 6.7.2; the initial and final molalities of each Fe species before and after mixing are m_0 and $m_0/2$, respectively. The number of moles of electrons required to bring the molality of Fe^{3+} contained in W kg of solvent in the left-hand half cell from 0 to $m_{Fe^{3+}}$ is $n = Wm_{Fe^{3+}}$. Since the sum of Fe^{2+} and Fe^{3+} molalities in each cell is constant, i.e., m_0, the mole fraction of Fe^{3+} that we are using as the extent-of-reaction variable is $x_{Fe^{3+}} = m_{Fe^{3+}}/m_0$. Combining these two relations, we have

$$n = Wm_0 x_{Fe^{3+}},$$

and thus

$$dn = Wm_0\, dx_{Fe^{3+}}. \tag{6.7.13}$$

Again, note that we could have used the molality of Fe^{2+} in the right-hand half cell as the extent-of-reaction variable. In order to avoid specifying which species we are monitoring, we can generalize equation (6.7.13) and write

$$dn = Wm_0\, dx_1, \tag{6.7.14}$$

where x_1 is the mole fraction of any *one* of the HCFe species in the system. Replacing dn in equation (6.7.12) by that shown in (6.7.14) gives us the desired result:

$$\Delta G_{mix} = -FWm_0 \int_0^{0.5} E_{rev}\, dx_1, \tag{6.7.15}$$

where the initial and final values of x_1 are 0 and 0.5, respectively, for complete mixing of Wm_0 moles of HCFeII and HCFeIII.

The Electrochemical Cell Potential

How does one measure E_{rev}? In practice, it is not possible to use the system represented in Figures 6.7.1 and 6.7.2. This is because there must be a finite amount of both Fe species in each half cell in order for there to be a measurable cell potential. This limitation, however, does not invalidate, or even compromise, the rigor of the methodology described above. In fact, it is an important part of the methodological design of the experiment that the cell molalities, denoted as m_1 and m_2, are *complementary*. That is, $m_{Fe^{2+}}(\text{left}) = m_{Fe^{3+}}(\text{right})$ *and* $m_{Fe^{3+}}(\text{left}) = m_{Fe^{2+}}(\text{right})$. Experimentally, E_{rev} can be measured using a device that has an extremely high resistance (e.g., $> 10^9\ \Omega$) so that a negligible amount of current flows from the cell. (See the discussion on pp. 34–35.)

The redox couple between the hexacyanoferrate complexes is indicated by the following half reactions:

$$Fe(CN)_6^{4-} \longrightarrow Fe(CN)_6^{3-} + e^-, \tag{6.7.16}$$

and

$$Fe(CN)_6^{3-} + e^- \longrightarrow Fe(CN)_6^{4-}, \tag{6.7.17}$$

and the cell for the overall process is schematically represented as

$$Pt_L|Fe(CN)_6^{4-}(m_1),\ Fe(CN)_6^{3-}(m_2)\|Fe(CN)_6^{4-}(m_2),\ Fe(CN)_6^{3-}(m_1)|Pt_R. \tag{6.7.18}$$

In this expression, Pt indicates that inert platinum electrodes couple the electroactive iron species to the external circuit and $\|$ denotes a salt bridge (or a porous barrier) that permits internal charge migration [e.g., $K^+(\text{aq})$] to compensate for external electron flow. Because the potential of cell (6.7.18) arises from the tendency of the $Fe(CN)_6^{4-}$ [or $Fe(CN)_6^{3-}$] molalities in each half cell to equalize by mixing, E_{rev} is a *direct* measure of the driving force of mixing for the particular complementary cell composition. The important advantages of using complementary solutions will be made evident below.

The entropy of mixing is obtained from the Gibbs-Helmholtz equation, $\Delta G = \Delta H - T\,\Delta S$, applied to equation (6.7.15):

$$\Delta S_{mix} = \frac{-\Delta G_{mix}}{T} = \frac{FWm_0}{T}\int_0^{0.5} E_{rev}(x_1)\,dx_1, \qquad (6.7.19)$$

by assuming that $\Delta H_{mix} = 0$. This equation represents the experimental basis for measuring the entropy of mixing.

It is desirable to test the theory and experimental design presented above and evaluate ΔS from equation (6.7.18). This can be done by expressing the cell potential in terms of the analytical concentrations of the hexacyanoferrate species. Following the usual convention for expressing cell potentials, $E = E_R - E_L$, and using the Nernst equations for each half cell,

$$E_R = E_R^\circ - \frac{RT}{F}\ln\left[\frac{a_{II}(m_2)}{a_{III}(m_1)}\right],$$

and

$$E_L = E_L^\circ - \frac{RT}{F}\ln\left[\frac{a_{II}(m_1)}{a_{III}(m_2)}\right],$$

the expression for the cell potential is

$$E = \frac{RT}{F}\ln\left[\frac{a_{II}(m_1)a_{III}(m_1)}{a_{III}(m_2)a_{II}(m_2)}\right],$$

where $a_{II}(m_1)$ denotes the activity of the HCFeII at a molality m_1, etc. Expressing each of the activities in terms of the molal activity coefficient and molality [cf. equation (6.7.1)], e.g., $a_{II}(m_1) = \gamma_{II}(m_1)[m_1/m_0]$, the cell potential becomes

$$E = \frac{RT}{F}\ln\left(\frac{m_1^2}{m_2^2}\right) + \frac{RT}{F}\ln\left[\frac{\gamma(m_{1,II})\gamma(m_{1,III})}{\gamma(m_{2,II})\gamma(m_{2,III})}\right]. \qquad (6.7.20)$$

Because complementary hexacyanoferrate solutions are used, the two half cells have nearly equal ionic strength; this permits us to approximate the activity coefficient ratio in equation (6.7.20) as unity. Equation (6.7.20), therefore, now simplifies to[5]

$$E = \frac{2RT}{F}\ln\left(\frac{m_1}{m_2}\right). \qquad (6.7.21)$$

Recalling that $m_1 = m_0 x_1$, $m_2 = m_0 x_2$, and $x_1 + x_2 = 1$, equation (6.7.21) can be rewritten as

$$E = \frac{2RT}{F}\ln\left(\frac{1 - x_1}{x_1}\right). \qquad (6.7.22)$$

Using this expression for E_{rev} in equation (6.7.19) produces the final result:

$$\Delta S_{mix} = 2RWm_0\int_0^{0.5}\ln\left(\frac{1 - x_1}{x_1}\right)dx_1. \qquad (6.7.23)$$

Although E is infinite at $x_1 = 0$ or 1, the integral in equation (6.7.23) is finite and has the value of $\Delta S_{\text{mix}} = 2Rm_0 \ln 2$ (this should be verified). This evaluation of equation (6.7.23) provides a theoretical check of the experimental data [cf. equation (6.7.19)]. If equation (6.7.23) is examined for the case in which two components, each m_0 molal and $W = 1$, are mixed, it agrees with the result obtained from pseudoideal solution theory [see equation (6.7.5)];

$$\Delta S_{\text{mix}} = -R[m_0 \ln (0.5) + m_0 \ln (0.5)].$$

It should be noted, finally, that although the heat of mixing for an ideal solution is zero, we only require in the treatment above that the enthalpy of mixing of the complementary Fe^{2+}/Fe^{3+} couple be independent of composition (x_1).

Safety Precautions

☐ Safety glasses must be worn while performing this experiment.
☐ Make sure you have been shown proper pipeting procedures. *Never* pipet by mouth.
☐ If any of the solutions come into contact with the skin, immediately wash with copious amounts of water.

Procedure

Prepare 0.100 m solutions of K_3FeCN_6 and K_4FeCN_6. (Remember, molality is moles of solute per *kilogram* solvent.) About 1 L of each is sufficient. A set of volumetric flasks and pipets is provided. [The hexacyano-iron(III) complex undergoes slow hydrolysis to form aquo complexes; this process is photochemically accelerated, and thus the solutions should not be stored in the presence of light for extended periods of time.] Prepare a series of about 8 to 10 complementary Fe^{2+}/Fe^{3+} solutions in which the ratio of Fe^{2+}/Fe^{3+} ranges from about 99/1 to about 52/48 (or some other appropriate ratio near 1/1). You will thus have a total of 16 to 20 solutions. These should be labeled immediately after preparation. Avoid pipet contamination. An example of a pair of complementary solutions is 10 mL of 0.1 m Fe^{2+} diluted with 0.1 m Fe^{3+} to a volume of 50 mL, and 10 mL of 0.1 m Fe^{3+} diluted with 0.1 m Fe^{2+} to a total volume of 50 mL.

Suggested addition volumes for each half of the complementary solutions are (into 50-mL volumetric flasks): 0.5, 1, 2, 4, 5, 10, 20, and 24 mL. For example, a particular pair of solutions would contain (1) 4 mL of 0.1 m Fe^{2+} and 46 mL of 0.1 m Fe^{3+}; and (2) 4 mL of 0.1 m Fe^{3+} and 46 mL of Fe^{2+}.

After rinsing the cell and electrodes thoroughly with distilled water, pour the components of a complementary solution into the sides of the cell and insert the electrodes. (This will be unnecessary if the electrodes are sealed through the bottom of the cell.) Place the cell in the constant temperature bath (near ambient temperature) and carefully install the salt bridge so that it couples the electrolytes in the two sides of the cell. (This is unnecessary, of course, if the half cells are coupled via a porous barrier, e.g., a glass frit.) After the cell equilibrates, read the cell potential using a potentiometer or high-impedance ($> 10^9$ Ω) millivolt-meter of sufficient accuracy (± 0.1 mV). Repeat the measurement after a few seconds to verify the stability of the cell voltage.

Remove the salt bridge (if necessary) and the cell from the bath and discard the electrolytes. Rinse the cell with a few milliliters of the components of the next solution to be measured and then fill. Repeat the above procedure until the cell potentials of all the Fe^{2+}/Fe^{3+} solutions have been measured.

Data Analysis

Tabulate the cell potentials for each of the solutions along with the molality ratio, e.g., $m_{Fe^{2+}}/m_{Fe^{3+}}$.

Plot the cell potential vs. mole fraction of Fe^{2+} (or Fe^{3+}) as implied in equation (6.7.19).

Determine the area under the $E(x_1)$ vs. x_1 plot. Since E is undefined at $x_1 = 0$, an indirect technique for obtaining the complete area must be used. One possibility is to divide the x_1 axis into arbitrary (but convenient) parts: e.g., 0.1, 0.2, 0.3, 0.4, and 0.5. Using a planimeter or other appropriate device or technique, measure the cumulative areas of these segments under the $E(x_1)$ curve starting from $x_1' = 0.5$. Plot these cumulative areas vs. x_1 and extrapolate smoothly to $x_1 = 0$ to obtain the total area required.

The area under the $E(x_1)$ curve (in square graph paper units) must be converted to absolute units (volts). This can be easily done by measuring the area of a rectangle of convenient size on the graph paper and by equating this area (e.g., cm^2) with that corresponding to the absolute area of the rectangle [i.e., (ΔE) times (Δx_1)]. After converting the area under the $E(x_1)$ curve into volts, use (6.7.19) to obtain J mol^{-1} K^{-1} from the appropriate values of m_0 and T.

Tabulate your measured $E(x_1)$ values along with those calculated from (6.7.21). Compare your measured entropy of mixing value with that obtained from (6.7.5) in which two components, each m_0 molal, are mixed.

Considering the uncertainties in T, E, and m, estimate the error in the ΔS_{mix} that you have measured.

Data Analysis Using a Microcomputer

Create an E vs. x_1 data file. Using a numerical integration routine, successively determine the areas between $x_1 = 0.5$ to 0.4, 0.5 to 0.3, 0.5 to 0.2, and 0.5 to 0.1.

Now create an area vs. x_1 data file. Fit these data to the appropriate polynomial using a curve fitting algorithm. The equation thus generated can be extrapolated to determine the value of the area at $x_1 = 0$; this corresponds to the integral in equation (6.7.19). Since the data are already entered in volts, calibration is unnecessary.

Considering the uncertainties in T, E, and m, estimate the error in ΔS_{mix}.

Questions and Further Thoughts

1. Why is it impractical to measure cell potentials of solutions in which the Fe^{2+}/Fe^{3+} molality ratio is too small (or large), e.g., 1:1000?
2. Why is it meaningless to perform a "blank" check of the cell and voltage-reading system by measuring the potential of "pure" deionized (or distilled) water? Try it and see!
3. How would one readily test the cell, its leads, and the voltage-reading device under "null" conditions, i.e., when there should be nominally zero volts?
4. Is it necessary to assume that the enthalpy of mixing of m_0 molal Fe^{3+} and m_0 Fe^{2+} cyanates is zero? Is this assumption reasonable?
5. In obtaining equation (6.7.20), it is assumed that $E_R^\circ = E_L^\circ$. Why is this justified?
6. Perform a dimensional analysis of the right-hand side of equation (6.7.19) to obtain J mol^{-1} K^{-1}.
7. Show that the right-hand side of equation (6.7.23) is equal to $2Rm_0W_s \ln 2$.
8. Derive expressions for the ionic strength of each of the parts of a complementary HCFeII and HCFeIII solution of total molality m_0 and component molality m_1. Your

result should show identical ionic strengths for $m_1 = 0.5\, m_0$. Why? What are the most disparate values of ionic strength in the series of complementary solutions that you used in the experiment? (See reference 2.)

Notes

1. N. J. Selley, *J. Chem. Educ.*, **49**, 212 (1972).
2. The molality scale *ionic strength*, I_m, is defined as

$$I_m = 1/2 \sum_i m_i z_i^2,$$

where the sum is over all ions in solution having molality, m_i, and charge, z_i. In the present example, the charges of the hexacyanoferrate ions are -3 and -4.
3. The geometric mean of the cation and anion activity coefficients, e.g., for $K_3Fe(CN)_6$, $[(\gamma_+)(\gamma_-)^3]^{1/4}$, called the *mean ionic activity coefficient*, $\gamma_\pm$, can be approximated for dilute electrolytes from the Debye-Hückel theory as

$$\log \gamma_\pm \simeq -0.510 z_+ |z_-| I_m^{1/2},$$

where z_+ and z_- are the cation and anion charges and I_m is the molal ionic strength (see reference 2).
4. For a derivation of the Nernst equation, consult your physical chemistry textbook, or see I. N. Levine, "Physical Chemistry," 2nd ed., pp. 389–395, McGraw-Hill (New York), 1983.
5. The cell potential represented in equation (6.7.21) is based on the Nernst equation and is strictly valid only in the absence of other electrochemical complications. One possible complication arises because the presence of electrolytes of differing species or concentration across a diffusive boundary causes a potential difference called a liquid junction potential, E_J. In this experiment involving complementary HCFeII and HCFeIII solutions, E_J is very small because the effect of the imbalance in K^+ concentrations on E_J is small. In addition, the contribution to E_J caused by the difference in iron(II) and (III) cyanate concentrations is probably very small. See the discussion of E_J on pp. 138–139. Also, consult J. H. Noggle, "Physical Chemistry," pp. 414–416, Little, Brown (Boston), 1984.

Further Readings

R. A. Alberty, "Physical Chemistry," 7th ed., pp. 245–264, Wiley (New York), 1987.

P. W. Atkins, "Physical Chemistry," 3rd ed., pp. 259–267, W. H. Freeman (New York), 1986.

G. W. Castellan, "Physical Chemistry," 3rd ed., pp. 371–383, Addison-Wesley (Reading, Mass.), 1983.

I. N. Levine, "Physical Chemistry," 2nd ed., pp. 369–396, McGraw-Hill (New York), 1983.

J. H. Noggle, "Physical Chemistry," pp. 395–401, Little, Brown (Boston), 1984.

Experiment 7

Entropy of Mixing

NAME _____ DATE _____

Bath temperature _____

Solution number	Volume Fe^{2+}	Volume Fe^{3+}	Cell voltage	
_____	_____	_____	_____	_____
_____	_____	_____	_____	_____
_____	_____	_____	_____	_____
_____	_____	_____	_____	_____
_____	_____	_____	_____	_____
_____	_____	_____	_____	_____
_____	_____	_____	_____	_____
_____	_____	_____	_____	_____
_____	_____	_____	_____	_____
_____	_____	_____	_____	_____
_____	_____	_____	_____	_____
_____	_____	_____	_____	_____
_____	_____	_____	_____	_____
_____	_____	_____	_____	_____

Entropy of Mixing

NAME _____ DATE _____

Bath temperature _____

Solution number	Volume Fe^{2+}	Volume Fe^{3+}	Cell voltage	
_____	_____	_____	_____	_____
_____	_____	_____	_____	_____
_____	_____	_____	_____	_____
_____	_____	_____	_____	_____
_____	_____	_____	_____	_____
_____	_____	_____	_____	_____
_____	_____	_____	_____	_____
_____	_____	_____	_____	_____
_____	_____	_____	_____	_____
_____	_____	_____	_____	_____
_____	_____	_____	_____	_____
_____	_____	_____	_____	_____
_____	_____	_____	_____	_____
_____	_____	_____	_____	_____

□ **Experiment 8**

Thermodynamics of an Electrochemical Cell

PART I: FREE ENERGY, ENTHALPY, AND ENTROPY OF REACTION

Objective

To determine the enthalpy and entropy changes for a chemical reaction from the temperature dependence of a reversible cell potential.

Introduction

Direct measurements of the free energy change associated with a chemical reaction (or phase change) can be made from studies of the equilibrium composition of the system. The relationship between the *standard* free energy change for a process (i.e., under unit activity conditions), $\Delta G°$, and the equilibrium constant, K, is

$$\Delta G° = -RT \ln K. \tag{6.8.1}$$

There is, however, another fundamental relation that links the free energy change for a process with a measurable quantity. This relationship, as we shall see below, can be applied to oxidation-reduction (redox) reactions. In general, a redox reaction can be expressed as an electron transfer between an electron donor, D, and an acceptor, A, the half reactions for which are

$$D \longrightarrow D^{n+} + ne^- \qquad \text{(oxidation)},$$

and

$$A + ne^- \longrightarrow A^{n-} \qquad \text{(reduction)}, \tag{6.8.2}$$

where n is the number of electrons transferred. The overall electron transfer reaction is the sum of these half reactions:

$$D + A \longrightarrow D^{n+} + A^{n-} \tag{6.8.3}$$

The reaction must take place in a medium that supports (dissolves) the ionic species. Usually this solvent is water, and therefore it is implied that all species are aquated, i.e., solvated by water. If the redox reaction is carried out in such a way that the electron transfer does not take place *directly* between D and A but rather through an external conductor, the system is called an *electrochemical* (or *galvanic*) cell. The electrical potential developed by the cell is a direct measurement of the useful work that can be produced by the reaction. If the conditions are controlled so that the reaction is carried out *reversibly,* which, in this case, means that a *negligible* amount of current is drawn from the system (i.e., electron flow is minimal), then the maximum useful work (and also the maximum cell potential) is produced. The reversibility condition also means that the reaction is on the verge of being driven in the *reverse* direction; thus, if a potential of opposite polarity were applied to the electrochemical cell, the reaction could be driven in the nonspontaneous direction.

The familiar battery provides a useful example. If the voltage of a battery is measured under zero-load conditions (basically no current flow), that voltage represents the maximum useful work that the battery can perform. If, however, the battery is to perform this work in finite time, that is, produce finite power (energy/time), a measurable current must flow through an external circuit, and this will cause the battery voltage to drop.

The relationship between the free energy change associated with an electron transfer reaction and the cell potential is obtained as follows. The infinitesimal change in the Gibbs free energy for a process in which *electrical work* is done is

$$dG = -S\,dT + V\,dP + dw_{\text{rev}},$$

where dw_{rev} represents an infinitesimal amount of useful, non-*P-V* work that can be obtained from the process. In the case of a charge transfer reaction in which an amount of charge dq is transferred under an electrical potential E, we have

$$dw_{\text{rev}} = E_{\text{rev}}\,dq.$$

If the electron transfer reaction takes place at constant temperature and pressure, and if the process takes place so slowly that the cell potential is constant throughout, the resulting change in free energy is $\Delta G = E_{\text{rev}}Q$, where Q is the total amount of charge transferred. Since ΔG corresponds to the free energy change when *n moles of electrons* flow, we have

$$\Delta G = -nFE_{\text{rev}}, \tag{6.8.4}$$

where F (called the Faraday constant) is equal to the charge in coulombs (the SI unit) associated with one mole of electrons; $F = 96{,}487$ coulombs mol^{-1}. It should be noted that in equation (6.8.4), the cell potential, E_{rev}, is an *intensive* quantity because its value is independent of the amount of material that makes up the cell. On the other hand, the useful work is extensive: larger batteries can do more work than smaller ones even though they develop the same potential. The factor that converts the intensive property to the extensive one is n, the number of moles of charge (electrons) that actually flows in the particular chemical cell.

Equation (6.8.4) holds only when the cell voltage represents the *reversible* process, i.e., when an infinitesimal amount of current flows. This condition is represented mathematically as

$$\lim_{i \to 0} E = E_{\text{rev}}. \tag{6.8.5}$$

In the discussion that follows, we will denote the reversible cell potential simply as E. The experimental justification for this will be presented below.

We can combine (6.8.4) with the Gibbs equation, $\Delta G = \Delta H - T\,\Delta S$ to obtain

$$nFE = -\Delta G = -\Delta H + T\,\Delta S. \tag{6.8.6}$$

If this expression is differentiated with respect to T at constant P and n, one obtains

$$nF\left(\frac{\partial E}{\partial T}\right)_P = -\left(\frac{\partial \Delta H}{\partial T}\right)_P + T\left(\frac{\partial \Delta S}{\partial T}\right)_P + \Delta S. \tag{6.8.7}$$

Since $(\partial \Delta H/\partial T)_P = \Delta C_P$ and $(\partial \Delta S/\partial T)_P = \Delta C_P/T$, equation (6.8.7) simplifies to

$$nF\left(\frac{\partial E}{\partial T}\right)_P = \Delta S. \tag{6.8.8}$$

Thus the isobaric temperature dependence of the (reversible) cell potential provides a measure of ΔS for the overall reaction.

The enthalpy change for the reaction, ΔH, can also be expressed in terms of cell potential measurements by combining equations (6.8.8) and (6.8.6):

$$\Delta H = nF\left[T\left(\frac{\partial E}{\partial T}\right)_P - E\right]. \tag{6.8.9}$$

Equations (6.8.6), (6.8.8), and (6.8.9) are of historical importance because they were used by T. W. Richards (1902) to determine the temperature dependence of ΔG and ΔH from reversible cell potentials. These data were then used by Nernst in the development of his heat theorem (1906), which, in turn, led to formulation of the third law of thermodynamics.

The Chemical Cell

The redox reaction studied in this experiment is the electron transfer between zinc and lead. The particular system is one that has been studied in detail by LaMer and coworkers.[1] The reduced forms of the metals are present not as the free solids but as liquid solutions in mercury (i.e., amalgams). Moreover, the oxidized form of lead, Pb^{2+} (aq), is in equilibrium with *solid $PbSO_4$*, a sparingly soluble salt. The rationale for this situation will be made clear below. The oxidized form of zinc exists as $ZnSO_4$, which is very soluble in water, and this property allows the Zn^{2+} (aq) composition to be an experimental variable.

The overall cell reaction is

$$Zn(Hg)(s) + PbSO_4(s) \longrightarrow Zn^{2+}(m) + SO_4^{2-}(m) + Pb(Hg)(s). \tag{6.8.10}$$

In view of the above discussion, the schematic cell diagram is

$$Pt|Zn(Hg)|Zn^{2+}(m),SO_4^{2-}(m)\|PbSO_4(s)|Pb(Hg)|Pt, \tag{6.8.11a}$$

where | indicates a phase boundary (i.e., metal electrode–liquid amalgam interface) and ‖ denotes a connection or junction between the half cells that allows charge migration to take place but prevents bulk mass transport. This coupling can be made either by a salt bridge consisting of a concentrated electrolyte (e.g., $1M$ KCl) contained in a viscous medium (e.g., agar or gelatin) or via a finely sintered glass frit.

The cell as described thus far has one important drawback, the imbalance in electrolyte composition on either side of the salt bridge (or glass frit). One side has m molal $ZnSO_4$ and the other has the very small concentration of aqueous lead sulfate arising from the finite solubility of $PbSO_4$. Whenever two media having different ionic compositions are in contact (directly, or indirectly through a salt bridge or porous plug), an electrical potential develops across that boundary. This is called a *liquid junction potential*. This voltage, although small, will contribute to the observed cell potential, and thus the electrochemical measurements will not reflect the *intrinsic* thermodynamic properties of the cell reaction (6.8.10). The liquid junction potential is discussed in more detail at the end of this experiment. To minimize this complication, the aqueous medium of the Pb side of the cell is *also* made m molal in $ZnSO_4$. Since the Zn^{2+} plays no role in establishing the Pb(Hg) half cell potential, its presence will substantially "neutralize" the concentration imbalance, which would cause the liquid junction potential but will not substantially affect the thermodynamics of the cell. (The very small contribution of ions arising from the dissociation of solid $PbSO_4$, however, has to be taken into account at very low $ZnSO_4$ molality.) Thus the actual schematic cell diagram is

$$Pt|Zn(Hg)|Zn^{2+}(m),SO_4^{2-}(m)\|Zn^{+2}(m),SO_4^{-2}(m)|PbSO_4(s)|Pb(Hg)|Pt.$$
$$\tag{6.8.11b}$$

The relationship between the measured cell potential, E, and the cell composition is given by the *Nernst equation* (1889):

$$E = E° - \frac{RT}{nF} \ln Q, \tag{6.8.12}$$

where $E°$ is the *standard* cell potential (the cell potential under standard conditions: see below and Experiment 9) and Q is the *activity quotient*. The derivation of this equation can be found in the texts cited in Further Readings. It should be noted that in reaction (6.8.10), all species are at unit activity except for $ZnSO_4(m)$. Although it might appear that the metals Zn and Pb are not present in their elemental forms but as solutions in liquid mercury, and hence their activities would not be unity, we are nevertheless justified in making this simplifying assumption. This point is discussed at the end of this experiment. Therefore,

$$Q = (a_{Zn^{2+}})(a_{SO_4^{2-}}), \tag{6.8.13}$$

where $a_{Zn^{2+}}$ is the activity of Zn^{2+}(aq), etc. If we express activities as the product of the molality and (molal) activity coefficient, Q becomes

$$Q = (m_{Zn^{2+}})(m_{SO_4^{2-}})(\gamma_+)(\gamma_-). \tag{6.8.14}$$

Because the cation and anion activity coefficients cannot be individually determined, they are lumped together as a product, $(\gamma_\pm)^2$. The activity coefficient is an indication of the nonideality of the electrolyte solution. In the limit of infinitely dilute solutions, the activity coefficient approaches unity. It is useful to define the (algebraic) mean ionic activity coefficient, $\gamma_\pm$, as $[(\gamma_+)(\gamma_-)]^{1/2}$. If we equate the Zn^{2+} and SO_4^{2-} molalities (since they are produced by $ZnSO_4$, and the SO_4^{2-} produced by the $PbSO_4$ is negligible), the activity quotient is now expressed as $Q = (m^2)(\gamma_\pm)^2$, and substitution into equation (6.8.12) results in

$$E = E^0 - \frac{RT}{nF} \ln [(m)^2(\gamma_\pm)^2]. \tag{6.8.15}$$

Equation (6.8.15) allows the composition dependence of the cell potential to be determined if values of $E°$ and $\gamma_\pm$ are known. The cell potential is equal to $E°$, the standard cell potential, when $Q = 1$, i.e., *unit activities* (and $T = 298.15$ K and $P = 1$ atm). The problem is that when the $ZnSO_4$ composition is 1 molal, the activity coefficient is not unity; hence, unit molality does not correspond to unit activity. Thus $E°$ must be determined by an extrapolation of measured E values. The particular strategy for doing this is described in Experiment 9.

Safety Precautions

☐ Safety glasses with side pieces must always be worn in the laboratory.
☐ If you are preparing or handling the Zn and Pb amalgams, work in a fume hood. Use gloves.
☐ If you prepare the $ZnSO_4$ solutions, make sure you have been shown proper pipeting techniques. *Never* pipet by mouth.

Procedure

A diagram of the cell is shown in Figure 6.8.1. It is an H cell in which a fine porosity glass plug separates the two half cell compartments. This arrangement allows the internal migration of charge to compensate for whatever (miniscule) current flows in the external circuit necessary for the potential measurement.

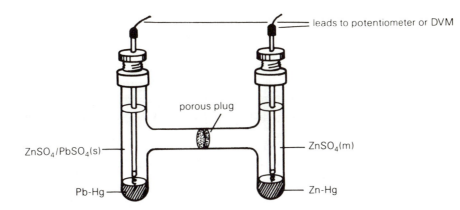

leads to potentiometer or DVM

porous plug

ZnSO$_4$/PbSO$_4$(s)

ZnSO$_4$(m)

Pb-Hg

Zn-Hg

FIGURE 6.8.1   H cell with porous plug and removable electrodes.

1. The Zn and Pb amalgams have been previously prepared (the procedure is described at the end of this experiment). Although the amalgams contain mercury (which itself is toxic), they are not particularly harmful (consider the amalgams used to fill dental caries). The electrodes are constructed of platinum wire (which is chemically unreactive) and are immersed in the amalgams during the experiment. The Pt wire is soldered to conducting leads which are connected to the voltage-measuring device. The solder joints and the nature and length of the conducting leads are designed to be as similar as possible to avoid possible junction potentials (caused by mismatches in conducting media).

2. In order that the measured cell potential will actually reflect the maximum useful work (free energy change) obtainable in the reaction, the cell must be operated reversibly. This means that a minimal amount of current should flow through the cell. To this end, either of two types of voltage-reading devices can be used (see Part II C): (1) a potentiometer, or (2) an accurate and sensitive high-resistance digital voltmeter (DVM). The former is capable of measuring cell potentials reversibly *and* with very high accuracy and precision (often to six significant figures). It is a null-reading device that applies an opposing potential to the cell, thereby limiting the current drawn to nearly zero. The latter, a DVM with (at least) a four-digit readout, can measure potentials accurately with a sensitivity of ca. $\pm$ 0.1 mV. Moreover if the DVM has an input impedance of at least 10^7 Ω, a satisfactorily small current is drawn and a true, reversible voltage reading can be made. A highly desirable feature of the DVM is that the voltage measurement is nearly instantaneous. Balancing the potentiometer takes more time—and patience. If a potentiometer is to be used in the experiment, your instructor will demonstrate its use.

3. Each side of the H cell should be filled with the same ZnSO$_4$ solution (the actual value is not important, but 0.02 m is satisfactory). Record the actual ZnSO$_4$ molality used. The Pb half cell *must* have a finite amount of solid PbSO$_4$ present if the ZnSO$_4$ solution used to fill the Pb half cell does not contain solid PbSO$_4$. Be careful to prevent PbSO$_4$-saturated ZnSO$_4$ solution from entering the zinc half cell. Pure ZnSO$_4$ solution must be added to the Zn half cell. The levels of each of the solutions should be roughly equal and extend about 1 cm above the cross-piece of the H cell. The glass frit *must* be in full contact with solution on both sides.

4. Insert the electrodes into the cell. Each electrode is labeled according to whether it is used with the Pb or Zn half cell. Make sure that the Pt tips are fully immersed in the amalgams, but *do not force them down* (the Pt wires can break off).

5. Clamp the cell securely in the 0° bath and wait for several minutes for the cell to equilibrate. If a potentiometer is used, the cell leads can be attached to the input terminals (the Pb half cell is *positive* relative to Zn). If a DVM is used, connect the cell leads to the meter *only* during a measurement. Measure and record the cell potential several times over a few minutes to be sure that it has stabilized. If there is a problem in stability, *gently* manipulate the electrodes in the amalgams; tapping the cell lightly also helps. Avoid doing this unless necessary.

6. After a stable potential is recorded, move the cell to the next bath and repeat the above procedure. The cell potential should be measured at about 10° intervals up to about 50°C.

Data Analysis

1. Tabulate the mean values of the cell potentials at the different temperatures studied. Indicate the uncertainties in these values. Plot E vs. T and determine ΔS using linear regression; estimate the error in ΔS.

2. Using the value of ΔS and the E data, determine ΔH and report its error. Comment on whether or not ΔH appears to be temperature-dependent.

3. By interpolation, obtain the value of E, and therefore ΔG, at 25.0°C. Tabulate these data and report their respective errors.

4. At 25°C, the value of the mean ionic activity coefficient, $\gamma_\pm$, for $ZnSO_4$ at 0.02 m is 0.298 (other values for $\gamma_\pm$ for different molalities are presented below). Using this information, calculate a value of $E°$. Also determine $\Delta S°$ and $\Delta H°$ [assume that $(\partial E/\partial T)_P \simeq (\partial E°/\partial T)_P$]. Compare $\Delta H°$, $\Delta S°$, and $E°$ with the respective values obtained in step 2. Finally, tabulate the standard thermodynamic data and compare with the literature values (consult, for example, the "Handbook of Chemistry and Physics").

Further Comments and Information

The Liquid Junction Potential

The system chosen for study in Experiments 8 and 9 consists of two half cells containing identical concentrations of aqueous $ZnSO_4$ but differing in the fact that the cathode is saturated with $PbSO_4$. Thus rewriting the cell diagram (6.8.11a) we have

$$Zn|ZnSO_4 \ (m)\|ZnSO_4 \ (m)|PbSO_4 \ (s)|Pb.$$

The ‖ symbol indicates a liquid junction across which ions must pass in order to compensate for the charge that flows in the external circuit. A potential will develop at this liquid junction as a result of the transport of ions if a gradient in chemical potential (i.e., concentration) exists. The cell notation above is written so that anions will move to the left (toward the anode) and cations will move to the right (toward the cathode). It follows that the pertinent liquid junction transport reactions are

$$t^- SO_4^{2-} \ (m) \quad \longleftarrow \quad t^- SO_4^{2-} \ (m),$$
$$\text{Left-hand side} \qquad\qquad \text{Right-hand side}$$

and

$$t^+ Zn^{2+}\ (m) \longrightarrow t^+ Zn^{2+}\ (m),$$

Left-hand side   Right-hand side

$t^{\pm}$ are the transport (or transference) numbers, the fraction of current carried by an ion. Note that because $PbSO_4$ is not present on the left-hand side, Pb^{2+} ions are not involved in the generation of a junction potential. Hence little or no junction potential exists for this system because the concentrations and, therefore, the chemical potentials of the transported species on either side of the cell partition are virtually equal.

Amalgam Electrodes

As indicated in the Introduction, an amalgam electrode is formed when a metal is dissolved in mercury. If the resulting mixture is a solution consisting only of a single liquid phase, it follows that the potential of the electrode relative to the pure metal should be a function of the concentration of the metal in the amalgam. For example, the potential of a univalent metal, M, dissolved in mercury would be given by

$$E_{M(Hg)} = E_{M^+/M} - \frac{RT}{F} \ln a_M,$$

where a_M, the activity of the metal in the amalgam, would depend on its concentration. In fact, the potentials of the zinc and lead amalgams used in this experiment do not display this anticipated concentration dependence. It has been demonstrated that Zn amalgam exhibits the same potential as a *pure* Zn electrode.[2] Furthermore, it has been observed that the 6 percent Zn amalgam used in this experiment consists of two phases between 0 and 50°C and is thus *not* a pure, single-phase solution.

The Pb amalgam demonstrates a potential that is approximately 0.006 V higher than that of pure Pb.[3] The potential of this "two-phase" system (the solid phase being Pb_2Hg) shows no concentration dependence between 2 and 66 percent Pb.[1] The convenience of using Zn and Pb amalgams as the reduced metal electrodes relative to the pure metals far outweighs any small errors introduced by these materials.

Preparation of Amalgams

The following procedure can be used to prepare the zinc and lead amalgams. Using a mortar and pestle, grind a quantity of the metal (granulated zinc or small pieces of lead) with mercury. The metal composition should be about 6 percent (by weight). This must be done in a fume hood. A ball mill can also be used. After the grinding is started, add a small amount of $1M$ sulfuric acid to cover the compounded metals. The acid prevents the formation of an oxide film on the surface that would retard the amalgamation. After the lead or zinc has dissolved in the mercury, rinse the amalgam with deionized or distilled water several times. The amalgams can be stored under N_2. Subsequent purification can be achieved by adding warm (50°C) water to the solid and shaking vigorously. The water is decanted and the washing is repeated until the rinsings are clear. The zinc amalgam is very viscous at ambient temperature but becomes more fluid and easier to purify and handle above about 50°C (see the discussion above).

Mean Ionic Activity Coefficients for Aqueous $ZnSO_4$ at 298 K*

Molality	$\gamma_\pm$
0.0005	0.780
0.001	0.700
0.002	0.608
0.005	0.477
0.01	0.387
0.02	0.298
0.05	0.204

*Calculated from data in reference 1.

Questions and Further Thoughts

1. It was pointed out above that the standard reduction potential (srp) for free lead, i.e.,

$$Pb^{2+} + 2e^- \longrightarrow Pb, \qquad E_1^\circ$$

is nearly equal to that of amalgamated lead,

$$Pb^{2+} + Hg + 2e^- \longrightarrow Pb(Hg). \qquad E_2^\circ$$

That is, $E_1^\circ \simeq E_2^\circ$. However, the srp of lead as considered in this experiment is actually represented as

$$PbSO_4(s) + Hg + 2e^- \longrightarrow Pb(Hg) + SO_4^{2-}, \qquad E_3^\circ$$

and E_3° is definitely not equal to E_1°. Its value is "offset" relative to E_1° by the free energy associated with the equilibrium

$$PbSO_4(s) = Pb^{2+}(aq) + SO_4^{2-}(aq).$$

E_1° is -0.1263V; adjust this value to obtain E_3° by using the equations, $\Delta G^\circ = -RT \ln K_{sp} = -nFE^\circ$, where K_{sp} is the solubility product of $PbSO_4$. Compare this value of E_3° with the literature value (consult a table of srp's). At 298 K, $K_{sp}(PbSO_4) = 1.58 \times 10^{-8}$.

2. Propose an interpretation of the values of the enthalpy and entropy of the overall reaction, equation (6.8.1), studied in this experiment, i.e., why are ΔS and ΔH positive (negative)?

3. The junctions between the Pt electrodes and the amalgams, as well as between the Pt wires and the conducting leads to the voltage-reading device, have not been considered in terms of their contributions to the measured potential. Why can these junction potentials be justifiably ignored? Are there any circumstances under which this complication might have to be taken into account?

Notes

1. I. A. Cowperthwaite and V. K. LaMer, *J. Am. Chem. Soc., 53,* 4333 (1931).
2. J. W. Clayton and W. C. Vosburgh, *J. Am. Chem. Soc. 58,* 2093 (1936).
3. W. R. Carmody, *J. Am. Chem. Soc., 51,* 2905 (1929).

Further Readings

A. W. Adamson, "A Textbook of Physical Chemistry," 3rd ed., pp. 499–520, Academic Press (Orlando, Fla.), 1986.

R. A. Alberty, "Physical Chemistry," 7th ed., Wiley (New York), 1987.

P. W. Atkins, "Physical Chemistry," 3rd ed., pp. 252–253, 270–273, W. H. Freeman (New York), 1986.

G. N. Castellan, "Physical Chemistry," 3rd ed., pp. 371–392, Addison-Wesley (Reading, Mass.), 1983.

J. de Heer, "Phenomenological Thermodynamics," pp. 168–170, Prentice-Hall (Englewood Cliffs, N.J.), 1986.

I. N. Levine, "Physical Chemistry," 2nd ed., pp. 265–270, 187–400, McGraw-Hill (New York), 1983.

J. H. Noggle, "Physical Chemistry," pp. 372–374, 395–417, Little, Brown (Boston), 1985.

Electrochemical Cell

NAME _____   DATE _____

Temperature   Cell voltage _____

_____   _____ _____ _____ _____
                                                              _____ _____ _____ _____

_____   _____ _____ _____ _____
                                                              _____ _____ _____ _____

_____   _____ _____ _____ _____
                                                              _____ _____ _____ _____

_____   _____ _____ _____ _____
                                                              _____ _____ _____ _____

_____   _____ _____ _____ _____
                                                              _____ _____ _____ _____

Experiment 8

Electrochemical Cell

NAME _____   DATE _____

Temperature   Cell voltage _____

_____   _____ _____ _____ _____

                                   _____ _____ _____ _____

_____   _____ _____ _____ _____

                                   _____ _____ _____ _____

_____   _____ _____ _____ _____

                                   _____ _____ _____ _____

_____   _____ _____ _____ _____

                                   _____ _____ _____ _____

_____   _____ _____ _____ _____

                                   _____ _____ _____ _____

□ Experiment 9

Thermodynamics of an Electrochemical Cell

PART II: STANDARD CELL POTENTIAL AND ACTIVITY COEFFICIENTS

Objective

To determine the standard cell potential and activity coefficients of an electrochemical system.

Introduction

This experiment will examine the same electrochemical cell as studied in Experiment 8. Read the Introduction of that experiment. The problem in obtaining the value of $E°$ for the electrochemical cell is the requirement that the $ZnSO_4$ be at *unit activity*.[1] Equation (6.9.1) illustrates this point [see also equations (6.8.12)–(6.8.15)]:

$$E_{cell} = E° - \frac{RT}{2F} \ln m_{Zn^{2+}} - \frac{RT}{2F} \ln m_{SO_4^{2-}} - \frac{RT}{F} \ln \gamma_{\pm}, \qquad (6.9.1)$$

where $m_{Zn^{2+}}$ and $m_{SO_4^{2-}}$ are the molalities of the indicated ions, and $\gamma_{\pm}$ is the mean ionic activity coefficient. We will begin with the assumption that the zinc and sulfate molalities are equal; i.e., $m_{Zn^{2+}} = m_{SO_4^{2-}} = m$. This will be valid if the concentration of the sulfate ion that arises from the solubility of $PbSO_4$ is small relative to that which comes from the $ZnSO_4$. This is an acceptable approximation for all but the most dilute $ZnSO_4$ solutions. If the $ZnSO_4$ molality were to be experimentally set at 1 molal, the activity coefficient would *not* be unity and thus $E°$ could not be directly evaluated unless $\gamma_{\pm}$ were known. On the other hand, $\gamma_{\pm}$ approaches unity in the limit of zero molality. However, at very low molalities E is difficult to measure. Moreover, the first two logarithm terms in equation (6.9.1) become indeterminate as $m \rightarrow 0$. The strategy that is employed in this experiment is to group the measurables in equation (6.9.1) to one side of the equation:

$$E_{cell} + \frac{RT}{2F} (\ln m_{Zn^{2+}} + \ln m_{SO_4^{2-}}) = E° - \frac{RT}{F} \ln \gamma_{\pm}, \qquad (6.9.2)$$

or

$$E_{cell} + \frac{RT}{F} \ln m = E° - \frac{RT}{F} \ln \gamma_{\pm}. \qquad (6.9.2a)$$

The left-hand side of equation (6.9.2a) is defined as $E°'$. If $E°'$ were plotted versus m and the graph extrapolated to $m = 0$ (y intercept), equation (6.9.2a) reveals that $E°'$ would be equal to $E°$ because $\ln \gamma_{\pm} \rightarrow 0$ in this limit. While an extrapolation method is sound in principle, the molality dependence of $\gamma_{\pm}$ must be considered more explicitly. To this end, we discuss a very important concept in electrolyte solutions called the Debye-Hückel (D-H) theory (1923). This theory provides a relationship between $\gamma_\pm$ (actually $\ln \gamma_\pm$), a quantity called the *ionic strength, I,* and the mean ionic radii of the cation and anion in question. The D-H theory holds that

$$\ln \gamma_\pm = \frac{-Az_+|z_-|I^{1/2}}{1 + 2BrI^{1/2}}, \tag{6.9.3}$$

where A and B contain physical constants that depend on the solvent medium as well as the temperature, z_+ and z_- are the respective charges of the cation and anion, r is the mean of the ionic radii of the cation and anion in aqueous solution, and I, the ionic strength, is defined as

$$I = 1/2\sum_i z_i^2 m_i, \tag{6.9.4}$$

in which z_i and m_i are the charge and molality of the *i*th ion in the solution. The sum is over *all* ionic species, whether or not they are involved in the chemical reaction being studied. Thus the value of I depends on the electrolyte concentration in the system. In this experiment where there are no extraneous ionic species in the system, the sum in equation (6.9.4) involves only m molal $ZnSO_4$. This will not be the case for more dilute solutions in which the finite solubility of $PbSO_4$ must be considered. This problem is discussed in Further Comments. Values of A and B for aqueous solution near room temperature are shown below.[2]

T, °C	A (kg/mol)$^{1/2}$	$10^{-9}B$ (kg/mol)$^{1/2}$m^{-1}
20.0	1.161	3.273
25.0	1.171	3.281
30.0	1.181	3.290

For aqueous Zn^{2+} and SO_4^{2-}, pertinent ionic radii are empirically determined to be 0.6 and 0.4 nm (10^{-9}m), respectively.[3] Hence, in equation (6.9.3) $r = 5 \times 10^{-10}$ m.

For very dilute solutions [i.e., where $2BrI^{1/2} \ll 1$; cf. equation (6.9.3)], the D-H equation becomes

$$\ln \gamma_\pm \simeq -Az_+|z_-|I^{1/2}. \tag{6.9.5}$$

This is known as the *Debye-Hückel limiting law* and is found to be reasonably valid for $I < 0.01\ m$.

In this experiment, $z_+ = |z_-| = 2$, and I is $4m$. Combining equations (6.9.5) and (6.9.2) under the condition of low electrolyte concentration provides

$$E^{\circ\prime} = E^\circ + \frac{8RTA}{F}m^{1/2}, \tag{6.9.6}$$

The simplification expressed by equation (6.9.6) is very useful because it suggests that a plot of $E^{\circ\prime}$ vs. $m^{1/2}$ is linear (for $m \lesssim 0.01$). This mathematical treatment thus provides an approach for obtaining E° by extrapolation.

It would be inappropriate to use the D-H theory, alone, to obtain an *experimental* value of E°. However, in this discussion we are only inferring that $\ln \gamma$ is proportional to $m^{1/2}$ at low molality. Thus equation (6.9.6) represents a graphical method for obtaining E°, i.e., by extrapolation of the $E^{\circ\prime}$ data. From equation (6.9.6) it can be seen that the limiting slope of $E^{\circ\prime}$ vs. $m^{1/2}$ is *constant* ($8RTA/F$). By contrast, the limiting slope of $E^{\circ\prime}$ vs. m would be infinite. That is, $dE^{\circ\prime}/dm = (4RTA/F)m^{-1/2}$, and this quantity increases without limit as $m \longrightarrow 0$.

The plot of $E^{\circ\prime}$ vs. $m^{1/2}$ is actually curved because the linear relationship between $\ln \gamma_{\pm}$ and $m^{1/2}$ is valid only at very low molality. A more rigorous approach could be taken by using the full D-H expression for $\ln \gamma_{\pm}$, [(equation (6.9.3)], assuming the constants A, B, and r to be valid, and plotting $E^{\circ\prime}$ vs. $\ln \gamma_{\pm}$. Computer-assisted data analysis makes this approach feasible. This method should also provide a more reliable value of E°.

Once a value of E° has been obtained from extrapolation of the experimental data ($E^{\circ\prime}$), the full D-H equation can be tested. That is, $\ln \gamma_{\pm}$ can be experimentally determined from equation (6.9.2), i.e.:

$$\ln \gamma_{\pm} = \frac{F}{RT} (E^{\circ} - E^{\circ\prime}), \qquad (6.9.7)$$

and then compared with values obtained from the D-H equation (6.9.3).[4] This can be done either graphically or by using selected measured (or interpolated) $E^{\circ\prime}$ data.

An iterative method for calculating m values at low $ZnSO_4$ concentrations where the finite solubility of $PbSO_4$ must be taken into account is presented below in Further Comments.

Most handbooks contain the standard reduction potential for the cell studied in this experiment, and it is interesting to compare this E° value with those of the free metals. It has been demonstrated[5] that the reduction potentials of free and amalgamated zinc:

$$Zn^{2+} + 2e^- = Zn, \qquad E_1^{\circ}$$

and

$$Hg + Zn^{2+} + 2e^- = Zn(Hg), \qquad E_2^{\circ}$$

are nearly identical ($E_1^{\circ} \simeq E_2^{\circ}$). There is, however, a significant difference in the standard reduction potentials of free lead and the lead half reaction in the system studied in this experiment:

$$Pb^{2+} + 2e^- = Pb, \qquad E_3^{\circ}$$

and

$$Hg + PbSO_4(s) + 2e^- = Pb(Hg) + SO_4^{2-}. \qquad E_4^{\circ}$$

Thus $E_3^{\circ} \neq E_4^{\circ}$. These points are discussed in Experiment 8 (Further Comments). Most of the difference is due to the equilibrium involving the saturated $PbSO_4$ system. The contribution of this free energy to the standard potential can be evaluated from $E^{\circ} = (RT/nF)\ln K$, where K is the activity product of $PbSO_4$ at 25°C, namely, 1.58×10^{-8} (see Experiment 8, Questions and Further Thoughts).

Safety Precautions

☐ Safety glasses must always be worn in the laboratory.

☐ If you prepare the Zn and Pb amalgams, work in a fume hood and wear gloves.

☐ When working with the amalgams, be careful to avoid spills. Waste amalgams should be stored in a labeled container.

☐ If you prepare the $ZnSO_4$ solutions, make sure you have been shown proper pipeting techniques. Never pipet by mouth.

☐ If any of the solutions come into contact with the skin, promptly wash the affected area with tap water.

□ N₂ is dispensed from a high-pressure cylinder. It must be strapped to a secure foundation.

□ A reducing valve controls the N₂ pressure to a few psi above ambient pressure (less than 5 psig). Do not increase this pressure.

Procedure

Cell potentials are to be measured as a function of $ZnSO_4$ molality at constant temperature (25°C). The same H cell used in Experiment 8 is used in this experiment. As before, Pt electrodes are used to couple the amalgams to the external conductors. The two half cells are joined by a fine-porosity frit. Because aerated water has an oxygen concentration of ca. 3×10^{-4} *m* at 25°C and 1 atm pressure, dissolved oxygen can compete with the reduction of $PbSO_4$ (at the lowest $ZnSO_4$ concentrations) due to the reaction:

$$1/2 \ O_2 + H_2O + 2e^- \longrightarrow 2OH^-.$$

For this reason, dissolved oxygen must be removed from the solutions. This is common practice in careful electrochemical work and is conveniently accomplished by bubbling a nonreactive gas through the solutions before measurement. N₂ or Ar can be used for this purpose. Figure 6.9.1 shows a pair of cell-deaerating needles.

1. If the cell is not assembled, refer to the procedure in Experiment 8. The amalgams should be purified just before the experiment is carried out by washing them with hot (ca. 50°C) deionized water until the rinsings are clear. This can be done directly in the cell or, preferably, in an Erlenmeyer flask. Be sure to use the Zn- and Pb-specified electrodes in each respective half cell consistently throughout the experiment. Remove any liquid that might be present in the cell. Using the Zn-labeled pipet, add a few milliliters of the 5.00×10^{-4} *m* $ZnSO_4$ to the Zn side of the cell. Discard the liquid and rinse again. Fill the cell with solution so that it is just above the upper part of the crosspiece containing the frit. Likewise, use the Pb-labeled pipet to rinse and fill the Pb side of the cell with the $PbSO_4$-saturated 5.00×10^{-4} *m* $ZnSO_4$ solution provided. Throughout the experiment, be sure not to contaminate the Zn side of the cell with $PbSO_4$; *keep the pipets separate*.

2. There should be a small amount of solid $PbSO_4$ in the Pb side of the cell. The solutions must next be deaerated. Swivel the degassing needles over the tops

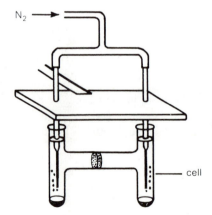

FIGURE 6.9.1   Filled H cell with mounted deaerating needles.

of the cell and carefully lower them into the solutions. The tips of the needles should be about 1/2 cm above the amalgams. *Cautiously* allow N_2 (or Ar) to flow until *gentle* bubbling takes place. Do not allow the bubbling to splatter solution out of the cell. After about 3 min carefully lift up and swivel the degassing needles away.

3. Quickly immerse the electrodes into the solutions so that the Pt tips are totally immersed in the amalgams. Do not force them down. Each electrode can be gently tapped to hasten electrical equilibrium. The electrodes should be labeled Zn and Pb. Make sure each electrode is placed in its respective half cell.

4. After about 1 min read the cell potential with a potentiometer or DVM. Repeat and record readings periodically until they are consistent. Do not disturb the electrodes; this will cause the potential to fluctuate. When the potential has converged, remove the electrodes and place them in the holding rack. Leave the cell in the bath.

5. Using the same Zn and Pb pipets, remove the solutions and replace with the solutions of next higher concentration. Deaerate as above, and place the electrodes in the cell. Read and record the cell potential until it converges.

6. Repeat the measurements with solutions in order of increasing concentration. A total of seven $ZnSO_4$ compositions will be used; these correspond to molalities of 5×10^{-4}, 1×10^{-3}; 2×10^{-3}, 5×10^{-3}, 0.01, 0.02, and 0.05. The actual values of the concentrations (and uncertainties) will be provided by the instructor.

Data Analysis

1. Tabulate the converged cell potentials for the different $ZnSO_4$ compositions studied.

2. Calculate the ionic strength of each of the cells studied. For the three lowest $ZnSO_4$ compositions, the contribution of SO_4^{2-} from the slightly soluble $PbSO_4$ will have to be added to the $ZnSO_4$. An iterative procedure can be used to calculate $m_{Zn^{2+}}$ and $m_{SO_4^{2-}}$ for these concentrations [compare equations (6.9.2) and (6.9.2a)]. This is outlined in Further Comments.

3. Tabulate values of $m_{Zn^{2+}}$, $m_{SO_4^{2-}}$, for each of the compositions studied.

4. Tabulate and plot $E^{\circ\prime}$ vs. $m^{1/2}$. By smoothly extrapolating the plot to $m^{1/2} = 0$, determine the value of E°.

5. Using the values of E°, $m_{Zn^{2+}}$, and $m_{SO_4^{2-}}$, determine values of $\ln \gamma_{\pm}$ for several concentrations; use equation (6.9.7).

6. Calculate $\ln \gamma_{\pm}$ values from the D-H theory, equations (6.9.3) and (6.9.5), and compare with the experimental data.

Further Comments

In the expression for the cell potential, $m_{Zn^{2+}}$ and $m_{SO_4^{2-}}$ refer to the molalities of zinc and sulfate ions present in the Zn(Hg) and Pb(Hg) sides of the cell, respectively. The SO_4^{2-} arises from the $ZnSO_4$ that is present in the solution in the Pb(Hg) cell, as well as from the ionization of the sparingly soluble $PbSO_4$. The latter contribution must be accounted for in the most dilute $ZnSO_4$ systems ($m < 0.005$ m).

From the solubility product of $PbSO_4$ at 25°C,

$$K_{sp} = (m_{Pb^{2+}})(m_{SO_4^{2-}})(\gamma_{\pm})^2 = 1.58 \times 10^{-8}. \tag{6.9.8}$$

If x is defined as the molality of the Pb^{2+} *and* SO_4^{2-} which arise from the $PbSO_4$ solubility, and m is the molality of $ZnSO_4$ present in the solution, then from equation (6.9.8):

$$K_{sp} = (a_{Pb^{2+}})(a_{SO_4^{2-}}) = x(m + x)\,\gamma_{\pm}^2, \qquad (6.9.9)$$

$\gamma_{\pm}$ can be determined from the D-H relation, equation (6.9.5). Equation (6.9.9) can then be solved for x, the molal solubility of $PbSO_4$ in the presence of m molal $ZnSO_4$.

$$x = -m \frac{\pm (m^2 + 4K_{sp}/\gamma_{\pm}^2)^{1/2}}{2}. \qquad (6.9.10)$$

The total SO_4^{2-} molality pertinent to equations (6.9.1) and (6.9.2) is $(m + x)$. Thus $E^{\circ\prime}$ is obtained using the formal $ZnSO_4$ molality for Zn^{2+} (m) and the corrected SO_4^{2-} molality $(m + x)$.

1. As a first approximation, calculate $\gamma_{\pm}$ from the relationship in $\gamma_{\pm} = -4.86(4m_{ZnSO_4})^{-1/2}$ (i.e., begin by assuming the contribution of $PbSO_4$ to be negligible).

2. Calculate x from equation (6.9.10) using this value for $\gamma_{\pm}$.

3. Use this value of x to calculate a new ionic strength from

$$I = 4(m_{ZnSO_4} + x),$$

and use this to calculate an improved value for $\gamma_{\pm}$.

4. Calculate an improved value of x using equation (6.9.10).

5. Repeat steps 4 and 5 until successive values of x are in agreement.

6. As will be seen, this is necessary only for $m < 0.005$ molal.

Questions and Further Thoughts

See Experiment 8.

1. Suppose the Zn side of the cell was contaminated by $PbSO_4$ (s). How would this affect the value of the cell potential?

2. How would you remove solid $PbSO_4$ that adheres to the surface of a cell or is in a hard-to-reach spot?

3. According to Noggle (see above, p. 404), "Measuring the standard emf of each electrode presents unique problems . . . every number is a story . . . ." What do you think is meant by this comment?

Notes

1. I. A. Cowperthwaite and V. K. LaMer, *J. Am. Chem. Soc.*, **53**, 4333 (1931).
2. Values calculated from I. N. Levine, "Physical Chemistry," 2nd ed., p. 267, McGraw-Hill (New York), 1983.
3. I. M. Klotz, "Chemical Thermodynamics," p. 331, Prentice-Hall (Englewood Cliffs, N.J.), 1950.
4. It should be noted that the D-H theory, equation (6.9.3), does not appear to work well for aqueous $ZnSO_4$. This was pointed out in reference 1. If the mean ionic radius, r, is determined from equation (6.9.3) using experimental values of $\gamma_{\pm}$, *negative* values are obtained.
5. W. J. Clayton and W. C. Vosburgh, *J. Am. Chem. Soc.*, **58**, 2093 (1936).

Further Readings

P. W. Atkins, "Physical Chemistry," 3rd ed., pp. 269–270, W. H. Freeman (New York), 1986.

I. N. Levine, "Physical Chemistry," 2nd ed., pp. 265–270, 387–400, McGraw-Hill (New York), 1983.

J. H. Noggle, "Physical Chemistry," pp. 372–374, 395–417, Little, Brown (Boston), 1985.

See also citations in Experiment 8.

Electrochemical Cell

NAME _____   DATE _____

ZnSO$_4$ molality: _____

E_{cell} (V):   _____

_____

_____

_____

_____

_____

ZnSO$_4$ molality: _____

E_{cell} (V):   _____

_____

_____

_____

_____

_____

ZnSO$_4$ molality: _____

E_{cell} (V):   _____

_____

_____

_____

_____

_____

ZnSO$_4$ molality: _____

E_{cell} (V):   _____

_____

_____

_____

_____

_____

ZnSO$_4$ molality: _____

E_{cell} (V):   _____

_____

_____

_____

_____

_____

ZnSO$_4$ molality: _____

E_{cell} (V):   _____

_____

_____

_____

_____

_____

Temperature: _____

Experiment 9

Electrochemical Cell

NAME _____   DATE _____

$ZnSO_4$ molality: _____ $ZnSO_4$ molality: _____

E_{cell} (V):   _____ E_{cell} (V):   _____

               _____   _____

               _____   _____

               _____   _____

               _____   _____

               _____   _____

$ZnSO_4$ molality: _____ $ZnSO_4$ molality: _____

E_{cell} (V):   _____ E_{cell} (V):   _____

               _____   _____

               _____   _____

               _____   _____

               _____   _____

               _____   _____

$ZnSO_4$ molality: _____ $ZnSO_4$ molality: _____

E_{cell} (V):   _____ E_{cell} (V):   _____

               _____   _____

               _____   _____

               _____   _____

               _____   _____

               _____   _____

Temperature: _____

□ PART SEVEN

Thermodynamics of Phase Equilibrium

□ Experiment 10

Mutual Solubilities of Liquids in a Binary Two-Phase System

Objective

To construct the mutual solubility curve of a binary two-phase liquid system (e.g., 1-butanol/water or methanol/cyclohexane).

Introduction

The determination of the mutual solubility of liquids in a two-phase system is of great practical importance. Although a number of analytical techniques can be used to obtain this information, one procedure that is both conceptually and operationally simple and does not require the removal of liquid samples for analysis (which might change the equilibrium compositions) was described by A. E. Hill in 1923.[1] This elegant approach (which Hill called a "thermostatic method") is based on a volumetric technique and requires only knowledge of the *bulk* composition of the system, that is, the total mass of each component in the mixture. It is assumed that the two liquids are in equilibrium in a two-phase system and that the phase rule therefore applies.

When two pure liquids, *A* and *B,* are combined, a two-phase system may be formed at a given temperature and pressure. Consider two samples of this binary system that have different bulk quantities of *A* and *B*. Assume that each of the two samples, at the same temperature and pressure, is at equilibrium. See Figure 7.10.1.

If m_A and m_B are, respectively, the *bulk* masses of components *A* and *B* in one sample, and m_A' and m_B' are the bulk quantities in the other sample, then it follows that

$$m_A = d_{A1}V_1 + d_{A2}V_2 \quad \text{and} \quad m_A' = d_{A1}V_1' + d_{A2}V_2', \quad (7.10.1)$$

$$m_B = d_{B1}V_1 + d_{B2}V_2 \quad \text{and} \quad m_B' = d_B V_1' + d_{B2}V_2', \quad (7.10.2)$$

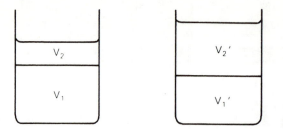

FIGURE 7.10.1   Two samples of a two-phase, two-component system containing arbitrary amounts of each component. Each solution is at the same temperature and pressure.

where V_1 and V_2 are the volumes of the lower and upper phases of one sample, and V_1' and V_2' refer to the volumes of the respective phases in the other sample. (The assignment of upper and lower phases is arbitrary.) d_{A1} and d_{A2} are the equilibrium concentrations (e.g., in g mL^{-1}) of component A in phases 1 and 2, respectively. Likewise, d_{B1} and d_{B2} are the concentrations of B in phases 1 and 2. Since the samples are at the same temperature and pressure, the phase rule requires that the equilibrium compositions be the same for both samples (that is, $d_{A1} = d_{A1}'$, $d_{A2} = d_{A2}'$, etc.). These equalities are valid assuming that, in the mixed sample, A and B can each be treated as a single, chemically independent component.

Equations (7.10.1) and (7.10.2) represent material balances that express the equilibrium compositions[2] of the A-B system as two equations in two unknowns. Since the bulk quantity (m_A, m_A', etc.) and the equilibrium volumes (V_1, V_1', etc.) are *measurable* quantities, the simultaneous solution of (7.10.1) and (7.10.2) allows the four concentrations d_{A1}, d_{A2}, d_{B1}, and d_{B2} to be determined.

If more than two equilibrium samples of A and B, each exhibiting two phases at the same temperature and pressure, are studied, the densities d_{A1}', d_{A2}', . . . are overdetermined, and a statistical treatment of their values can be carried out. In addition, this method can be applied to a given pair (or set) of samples at different (but common) temperatures, allowing the temperature-composition (T-X) phase diagram to be measured. This approach is used in determining the coexistence curve of a fluid (vapor and liquid constituents in a one-component system) near the critical point (see Experiment 12).

The simplicity of the method is very appealing. However, it should be clear (and it is mathematically demonstrable) that the approach begins to break down as m_A approaches m_A'. In the extreme case where two *identical* samples are considered, no unique information about the system composition can be obtained. Moreover, as the ratio of m_A'/m_A becomes very small (or large), the quality of the information deteriorates because the determined concentrations are sensitive to the error associated with reading the positions of the meniscuses. Thus, there exists an optimal mass ratio that provides minimum sensitivity, i.e., maximum precision in the determination of the equilibrium compositions using this methodology (for a given volumetric error). Hill was aware of this situation, and in a paper with W. M. Malisoff (1926)[3] presented the analysis of this system optimization. The opportunity for simple methodological optimization of experimental conditions is relatively unusual in experimental physical chemistry. Because the mathematical approach is straightforward, the Hill and Malisoff treatment will be described here in some detail.

Consider again two different samples of a two-phase liquid system consisting of components A and B at the same temperature and pressure. These samples are

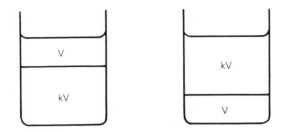

FIGURE 7.10.2   A set of complementary solutions of a two-phase, two-component system at the same temperature and pressure. The volumes of the two phases in the samples are exactly reversed.

designed so that the volumes are *exactly reversed* for each of the two phases. See Figure 7.10.2.

Thus if $V_1' = V$ and $V_2' = kV$, then $V_1 = kV$ and $V_2 = V$. k is the equilibrium volume ratio in the complementary samples and is > 1 in the example illustrated in Figure 7.10.2. Moreover, the subscripts 1 and 2 denote, respectively, the lower and upper phases. The material balance for component A in *each* sample is

$$m_A = d_{A1}kV + d_{A2}V, \tag{7.10.3a}$$

$$m_A' = d_{A1}V + d_{A2}kV. \tag{7.10.3b}$$

A similar set of equations can be written for component B. The composition of A in phase 1 can be conveniently expressed using determinants:

$$d_{A1} = \frac{\begin{vmatrix} m_A & V \\ m_A' & kV \end{vmatrix}}{\begin{vmatrix} kV & V \\ V & kV \end{vmatrix}} = \frac{km_A - m_A'}{(k^2 - 1)V}. \tag{7.10.4}$$

Equation (7.10.4) contains the equilibrium mass-based concentration, d_{A1}, as a function of the experimental measurables, m_A, m_A', and V, and the methodological variable, k. Using the propagation of errors expression (see pp. 6–10), the total uncertainty in d_{A1} can be written

$$\epsilon(d_{A1}) = \left(\frac{\partial d_{A1}}{\partial k}\right)_{m_A, m_A', V} \delta k + \left(\frac{\partial d_{A1}}{\partial m_A}\right)_{m_A', V, k} \delta m_A + \cdots, \tag{7.10.5}$$

where the remaining terms deal with the errors associated with m_A' and V. In equation (7.10.5), all terms but the first arise from intrinsic *experimental* uncertainties beyond our control. Since the first term involves k, the arbitrarily chosen volume ratio, its impact on $\epsilon(d_{A1})$ can be modified, i.e., minimized, by careful experimental design.

The value of k will be optimal when it yields the *minimum* error in d_{A1}. Thus its optimal value may be determined by setting the coefficient $(\partial d_{A1}/\partial k)$ equal to zero and solving the resulting equation. In this way, the value of k for which the measured d_{A1} is *least* sensitive (k_{opt}) is obtained; this corresponds to a minimization of the reading error in the meniscuses. Differentiating equation (7.10.4) with respect to k and equating to zero gives, after simplification,

$$k^2 - 2k\frac{m_A'}{m_A} + 1 = 0. \tag{7.10.6}$$

The solution of equation (7.10.6) is

$$k_{opt} = \frac{m_A'}{m_A} \pm \left[\left(\frac{m_A'}{m_A} \right)^2 - 1 \right]^{1/2}, \qquad (7.10.7)$$

or, rearranging equation (7.10.6),

$$\left(\frac{m_A'}{m_A} \right)_{opt} = \frac{k^2 + 1}{2k}. \qquad (7.10.6a)$$

Equation (7.10.7) provides a value of the optimal *volume* ratio in terms of the mass ratio of one component, A, in the two complementary samples. [Note that according to the initial premise, $k > 1$ and thus equation (7.10.7) requires that $m_A' > m_A$; moreover, the positive root in that equation must be used. If, in reality, however, $m_A' < m_A$, equations (7.10.6) and (7.10.7) are physically meaningless. The optimization relates, instead, to d_{A2} [equations (7.10.4) and (7.10.5)].

The result in (7.10.6a) would be more useful, clearly, if $(m_A'/m_A)_{opt}$ could be expressed in more specific terms. To obtain this information, we divide equation (7.10.3b) by (7.10.3a):

$$\frac{m_A'}{m_A} = \frac{d_{A1} + kd_{A2}}{kd_{A1} + d_{A2}}. \qquad (7.10.8)$$

Now, equating the expressions for m_A'/m_A in (7.10.6a) and (7.10.8), we have, after clearing terms, the cubic equation

$$d_{A1}k^3 - d_{A2}k^2 - d_{A1}k + d_{A2} = 0. \qquad (7.10.9)$$

This can be simplified by dividing by d_{A2} and defining r as d_{A1}/d_{A2}:

$$rk^3 - k^2 - rk + 1 = 0. \qquad (7.10.10)$$

The solution of equation (7.10.10) yields one real and two conjugate, imaginary roots. The physically meaningful result from (7.10.10) is

$$k_{opt} = \frac{1}{r} = \frac{d_{A2}}{d_{A1}}, \qquad (7.10.11)$$

r (or $1/r$) is called the *distribution ratio* (or distribution coefficient) of component A in the two phases and is a thermodynamic quantity. The optimal mass ratio in the two samples is [from equation (7.10.6a)]

$$\left(\frac{m_A'}{m_A} \right)_{opt} = \frac{r^2 + 1}{2r}. \qquad (7.10.12)$$

Consider this result. It is reasonable to expect the optimal volume ratio, k, (or mass ratio) to be equal to some *characteristic* (i.e., thermodynamic) property of the system, and this turns out to be its distribution ratio, r. (Note that k is assumed to be > 1; thus $1/r = d_{A2}/d_{A1}$ is also presumed to be > 1. If, in reality, however, $d_{A2} < d_{A1}$, then the denotation of phases must be reversed, i.e., the upper phase in Figure 7.10.2 is 1 and the lower one is 2.)

An analogous result can be obtained for the optimal volume and mass ratios for component B. In general, these optimal conditions are not the same for the two components because the mutual solubilities of A and B are different; thus, if the optimization condition derived above is $k_A = d_{A2}/d_{A1}$, the constraint pertinent to component B, denoted as k_B, is

$$k_B = \frac{d_{B1}}{d_{B2}}. \qquad (7.10.13)$$

Nevertheless, k_A and k_B are often close enough so that a satisfactory (although not absolute) optimization of the experiment can simultaneously accommodate both components.[4] From a practical point of view, it is desirable to choose a complementary system in which k is not too large because, in this case, relative volumetric errors become important.

Since the distribution ratio is not known before the experiment is performed (finding d_{A1} and d_{A2} is, after all, the objective of the experiment), a preliminary experiment can be performed to obtain approximate values of the equilibrium compositions. Using this information, the optimal value of k_A (or k_B), and hence m_A and m_A' (or m_B and m_B'), can be determined directly, or by iteration. The experiment can then be performed on an optimized (or nearly optimized) system. If optimized composition studies are to be performed as a function of temperature, a compromise value of k over the temperature range can be used. This is necessary because the system composition, hence k, generally changes with temperature.

From a practical point of view, it is desirable to prepare a predetermined *total* volume, V_t, of the samples from known masses of the two components rather than from the equilibrium volumes observed *after mixing*. We see from Figure 7.10.2 that $V_t = V(k + 1)$; hence $V = V_t/(k + 1)$. Assuming that the equilibrium compositions, d_{A1}, d_{A2}, . . . , are known (or can be estimated from an initial experiment), the expressions for the bulk masses in the two samples are obtained as follows (refer to Figure 7.10.2):

$$m_A = d_{A1}kV + d_{A2}V = \frac{V_t(kd_{A1} + d_{A2})}{k + 1}, \qquad (7.10.14)$$

$$m_B = d_{B1}kV + d_{B2}V = \frac{V_t(kd_{B1} + d_{B2})}{k + 1}, \qquad (7.10.15)$$

$$m_A' = d_{A1}V + d_{A2}kV = \frac{V_t(d_{A1} + kd_{A2})}{k + 1}, \qquad (7.10.16)$$

$$m_B' = d_{B1}V + d_{B2}kV = \frac{V_t(d_{B1} + kd_{B2})}{k + 1}. \qquad (7.10.17)$$

Equations (7.10.14) through (7.10.17) do not necessarily pertain to optimized samples; they indicate merely how to prepare *complementary* samples having a volume ratio k.

The bulk masses required for a pair of *optimized*, complementary samples can be readily obtained from (7.10.14) to (7.10.17) by using $k = k_A = d_{A2}/d_{A1}$, or $k = k_B = d_{B1}/d_{B2}$, whichever is more desirable. (Generally, the value of k closer to 1 is more desirable.) The following equations summarize the results:

Bulk mass	$k = k_A$	$k = k_B$	
m_A	$2C_A d_{A1} d_{A2}$	$C_B(d_{A1}d_{B1} + d_{A2}d_{B2})$	(7.10.18a,b)
m_B	$C_A(d_{B1}d_{A2} + d_{B2}d_{A1})$	$C_B(d_{B1}^2 + d_{B2}^2)$	(7.10.19a,b)
m_A'	$C_A(d_{A1}^2 + d_{A2}^2)$	$C_B(d_{A1}d_{B2} + d_{A2}d_{B1})$	(7.10.20a,b)
m_B'	$C_A(d_{B1}d_{A1} + d_{B2}d_{A2})$	$2C_B d_{B1} d_{B2}$	(7.10.21a,b)

where $C_A = V_t/(d_{A1} + d_{A2})$ and $C_B = V_t/(d_{B1} + d_{B2})$.

Finally, then, equations (7.10.18), (7.10.19), (7.10.20), and (7.10.21) can be used to prepare the optimized pair of complementary samples of any total volume if values of the equilibrium densities are known.

Safety Precautions

☐ Wear safety glasses or goggles.

☐ Be particularly careful in removing and manipulating the hot sample cylinders.

☐ Small amounts of 1-butanol (or other organic) vapors are released. Work in an open, ventilated laboratory. If the fumes are objectionable, assemble the apparatus in a fume hood.

☐ Do not allow the cylinders to build up pressure as they are heated; vent periodically by briefly removing the stoppers.

Procedure

Several different binary liquid systems can be studied using the Hill-Malisoff method. The choice depends on such factors as convenience, expense, and safety considerations. If mutual solubilities are determined at different temperatures, the miscibility diagram of the system can be constructed. In this case, it is desirable to study a system that has an upper consolute temperature that is in an experimentally convenient range (i.e., less than 100°C). The upper consolute temperature is the point above which the two liquids are miscible in all proportions. It is analogous to the critical temperature of a pure substance (see Experiment 12).

Unfortunately, many of the systems that manifest easily accessible upper consolute temperatures contain a noxious component and must therefore be handled in a fume hood. For example, the cyclohexane/aniline system has an upper consolute temperature of ca. 60°C; that of phenol/water is ca. 66°C. Other possible systems are methanol/cyclohexane and 1-butanol/water (or other butanol isomers and water).

In this experiment, the 1-butanol/water system, or the methanol/cyclohexane system, is studied between 0 and ca. 70°C. Since the upper consolute temperature of the former is ca. 125°C, only a portion of its miscibility curve can be constructed. The upper consolute temperature of the latter system, however, can be determined. Two different approaches are possible in this experiment. These will be described below. Your instructor will tell you in advance which you are to follow.

Preliminary Procedure A

Obtain from the literature the value of the distribution ratios of the 1-butanol/water (or methanol/cyclohexane) system. (Alternatively, your instructor will provide you with this information beforehand.) Use the k_{opt} value that is nearer 1:1 and equations (7.10.18), (7.10.19), (7.10.20), and (7.10.21) to find the four masses of the two components needed to prepare a certain volume (e.g., 8 to 9 mL) of the complementary solutions before you begin the experiment.

Preliminary Procedure B

(This method requires more laboratory time.) Prepare two 1-butanol/water samples that manifest two phases at room temperature. The volumes of the two phases need not be complementary but should be distinctly unequal. The densities of the phases will be determined for this unoptimized system from equations (7.10.1) and (7.10.2), see Figure 7.10.1, and these values will be used to construct samples according to equations (7.10.14) to (7.10.17), or (7.10.18) to (7.10.21).

$$d_{A1} = \frac{m_A V_2' - m_A' V_2}{D} \qquad d_{A2} = \frac{m_A' V_1 - m_A V_1'}{D},$$

$$d_{B1} = \frac{m_B V_2' - m_B' V_2}{D} \qquad d_{B2} = \frac{m_B' V_1 - m_B V_1'}{D},$$

where

$$D = V_1V_2' - V_1V_2.$$

General Procedure

1. Weigh directly into a tared 10-mL graduated cylinder (graduations of 0.1 mL) appropriate amounts ($\pm$ 10 mg) of 1-butanol and distilled water (or methanol and cyclohexane) as determined from Preliminary Procedure A or B. Use a Pasteur pipet to transfer the liquids. Likewise, add the calculated amounts of the components to the other graduated cylinder to prepare the complementary, two-phase mixture. Each cylinder and its stopper should be labeled or otherwise identified. Stopper the cylinders and gently invert each several times, venting the cylinder periodically.

2. Remove the stoppers and place each of the cylinders in the bath at the highest temperature (this will be different for the two systems). After a minute, firmly replace each stopper on the respective cylinder. After another minute or so, remove the cylinders one at a time using a test tube holder (or other appropriate device) and carefully invert several times. Replace the cylinder in the bath as quickly as possible. If possible, invert the cylinders directly in the bath. If at this point, or in subsequent stages, the two phases do not separate cleanly, remove the cylinder and *very gently* tap it on a firm surface.

3. Repeat the inversions after 1 to 2 min. Wait for a few minutes until the positions of the meniscuses of each sample have become established. Record these positions to the nearest 0.03 mL.

4. Remove the cylinders and place them in the next lower temperature bath (or add ice to the bath to lower the temperature by 5 to 10°C). Following the above procedure, invert the cylinders and read the meniscus positions after equilibrium is reached.

5. Continue the procedure until the lowest temperature (ca. 10°C) is reached.

6. For the methanol/cyclohexane system, the upper consolute temperature (or critical solution temperature), T_c, can be determined. Fill a clean cylinder with a mixture (total mass ~ 6 to 8g) that is about 77 percent cyclohexane. Mix thoroughly and place in a 2-L beaker that contains sufficient hot water (ca. 55 to 60°C) to completely cover the liquid in the cylinder. Invert the cylinder several times; vent periodically to release vapor. The liquid should appear as a homogeneous, one-phase system. Add sufficient quantities of crushed ice to the bath water so that the temperature drops 1 to 2°C per minute. Record the temperature at the first sign of a pale blue, hazy appearance to the cooling methanol/cyclohexane mixture. This phenomenon, called *critical opalescence,* appears just above T_c. It is caused by the strong light scattering that accompanies large fluctuations in the density within the sample as the two phases begin to separate. See p. 185 for a discussion of critical opalescence. If you overshoot T_c, the system will appear distinctly milky or cloudy. Slowly reheat the system by a few degrees until it homogenizes and repeat the cooling process (if you have time).

The detection of the onset of critical opalescence is best done by having a light source [a bright window, light bulb, or better yet, a low-power helium-neon laser *(wear safety goggles)*] illuminate the sample *at right angles* to the viewing axis.

Calculations

1. Tabulate the equilibrium volumes of the two phases in each cylinder at the different temperatures studied.

2. Using these data, along with the bulk masses of the two components, determine the equilibrium compositions (molarities or densities)[2] of the two components at each temperature.

3. Construct the solubility curve over the temperature range studied. If necessary consult your physical chemistry text or one of the texts cited below.

4. Determine the mutual solubilities of the two components, as well as the distribution coefficients at 20°C, and compare with the literature values.

Questions and Further Thoughts

1. Suppose the distribution coefficient of component *A* in a binary system is 20. According to the Hill-Malisoff method, the mass ratio, m_A/m_B, for optimal solubility precision is 20:1. If a 10-mL graduated cylinder is used in the experiment, the smaller volume contained in each tube is less than 0.5 mL. Does this correspond to maximal volumetric precision? How does one decide how to achieve the maximal *overall* experimental accuracy in an experiment such as this?

2. What other types of analytical methods can be used to determine the compositions of a two-phase liquid system? Discuss approaches that can be used (a) without removing samples, and (b) by withdrawing small aliquots.

3. What is the advantage of applying the Hill-Malisoff method to a binary system contained in a series of *N* samples (*N*>2)?

4. Can the methodology be extended to a ternary, two-phase system? Set up the initial equations analogous to (7.10.1) and (7.10.2). Ternary liquid equilibrium is conveniently studied by "titrating" a homogeneous binary phase consisting of components *A* and *B* with the other component, e.g., pure liquid *A*, until the mixed ternary system manifests a two-phase appearance. This condition is easily detected as an emulsion.

Notes

1. A. E. Hill, *J. Am. Chem. Soc., 45,* 1143 (1923).
2. For a multicomponent system, the more conventional composition variable is molarity (or mole fraction), in which case the treatment would be presented on a basis of the mole rather than mass. We choose to use density as the compositional variable because it is more experimentally direct, more easily visualized in the mathematical treatment, and it also allows one to confirm that the lower phase (i.e., 1 in Figure 7.10.2) is the denser phase. Note that the phase density is the sum of the component concentrations, i.e., $d_1 = d_{A1} + d_{B1}$, etc.
3. A. E. Hill and W. M. Malisoff, *J. Am. Chem. Soc., 48,* 918 (1926).
4. It should be realized that the extraction of k_{opt} from equation (7.10.9), or (7.10.10) identifies the value of *k* at which the function has a maximum (or minimum). The function itself is smooth near this extremum; thus the effect of the value of *k* on the error in d_{A1} [equation (7.10.5)] is not severe. Thus it is merely *helpful* to conduct the experiment on a system for which $k \sim k_{opt}$, not *crucial*.

Further Readings

A. W. Adamson, "A Textbook of Physical Chemistry," 3rd ed., pp. 341–343, Academic Press (Orlando, Fla.), 1986.

P. W. Atkins, "Physical Chemistry," 3rd ed., pp. 197–199, W. H. Freeman (New York), 1987.

G. W. Castellan, "Physical Chemistry," 3rd ed., pp. 319–322, Addison-Wesley (Reading, Mass.), 1983.

A. W. Francis, "Liquid-Liquid Equilibriums," Interscience (New York), 1963.

S. Glastone, "Textbook of Physical Chemistry," 2nd ed., pp. 723–726, Macmillan (London), 1960.

I. N. Levine, "Physical Chemistry," 2nd ed., pp. 320–322, McGraw-Hill (New York), 1983.

J. H. Noggle, "Physical Chemistry," pp. 322–325, Little, Brown (Boston), 1985.

J. E. Ricci, "The Phase Rule and Heterogeneous Equilibrium," Dover (New York), 1966.

Experiment 10

Mutual Solubilities of Liquids

NAME _____   DATE _____

Bath temperature _____

Meniscus positions

Sample A   Sample B

_____   _____

_____   _____

_____   _____

Bath temperature _____

Meniscus positions

Sample A   Sample B

_____   _____

_____   _____

_____   _____

Bath temperature _____

Meniscus positions

Sample A   Sample B

_____   _____

_____   _____

_____   _____

Bath temperature _____

Meniscus positions

Sample A   Sample B

_____   _____

_____   _____

_____   _____

Bath temperature _____

Meniscus positions

Sample A   Sample B

_____   _____

_____   _____

_____   _____

Bath temperature _____

Meniscus positions

Sample A   Sample B

_____   _____

_____   _____

_____   _____

Bath temperature _____

Meniscus positions

Sample A   Sample B

_____   _____

_____   _____

_____   _____

Bath temperature _____

Meniscus positions

Sample A   Sample B

_____   _____

_____   _____

_____   _____

Experiment 10

Mutual Solubilities of Liquids

NAME _____ DATE _____

Bath temperature _____

Meniscus positions

Sample A   Sample B

_____   _____

_____   _____

_____   _____

Bath temperature _____

Meniscus positions

Sample A   Sample B

_____   _____

_____   _____

_____   _____

Bath temperature _____

Meniscus positions

Sample A   Sample B

_____   _____

_____   _____

_____   _____

Bath temperature _____

Meniscus positions

Sample A   Sample B

_____   _____

_____   _____

_____   _____

Bath temperature _____

Meniscus positions

Sample A   Sample B

_____   _____

_____   _____

_____   _____

Bath temperature _____

Meniscus positions

Sample A   Sample B

_____   _____

_____   _____

_____   _____

Bath temperature _____

Meniscus positions

Sample A   Sample B

_____   _____

_____   _____

_____   _____

Bath temperature _____

Meniscus positions

Sample A   Sample B

_____   _____

_____   _____

_____   _____

☐ Experiment 11

The Vapor Pressure and Heat of Vaporization of Liquids

Objective

To measure the vapor pressure of a liquid (e.g., toluene, ethanol, methylethylketone) as a function of temperature and to determine its heat of vaporization.

Introduction

Consider a system composed of a pure liquid that is in coexistence with its vapor at a fixed temperature and (total) pressure. We will assume that the pressure of the system is caused by the substance itself, and thus that there are no other gases present in the vapor phase. The pressure in this case is referred to as the *orthobaric pressure*. Because liquid and vapor coexist, the rate of vaporization of the liquid is equal to the rate of condensation of the vapor, and this dynamic equivalence thus corresponds to *liquid-vapor equilibrium* for the substance. Although this system can be treated using kinetic molecular theory, we will consider the process from a macroscopic perspective based on the equivalence of the chemical potentials (molar free energies) of two phases in coexistence.

Suppose a pure substance, A, is present in both the liquid and gaseous phases, $A(l)$ and $A(g)$, respectively, at a temperature T and pressure P. We have already assumed that the only vapor present in the system is that produced by the substance A, i.e., $P = P_A$. P_A is called the saturation *vapor pressure* of A at the temperature T. The constraint of equilibrium between the two phases of A can be stated as

$$\mu_A(l)(T,P) = \mu_A(g)(T,P), \tag{7.11.1}$$

where $\mu_A(l)(T,P)$ and $\mu_A(g)(T,P)$ are the chemical potentials of liquid and vapor A at T and P, respectively. If the temperature is changed infinitesimally, to $T + dT$, and the pressure produced by A changes accordingly to $P + dP$ in order to *maintain equilibrium* between the two phases, then at the new T and P values, we also have

$$\mu_A(l)(T + dT, P + dP) = \mu_A(g)(T + dT, P + dP). \tag{7.11.2}$$

This can be rewritten in terms of the differentials of the chemical potentials:

$$\mu_A(l)(T,P) + d\mu_A(l) = \mu_A(g)(T,P) + d\mu_A(g). \tag{7.11.3}$$

Since from (7.11.1), the chemical potentials of liquid and vapor A at T and P are equal to each other, (7.11.3) becomes

$$d\mu_A(l) = d\mu_A(g). \tag{7.11.4}$$

The differential of μ for any pure material is $d\mu_i = V_{m,i}\, dP - S_{m,i}\, dT$, where $V_{m,i}$ and $S_{m,i}$ are the respective molar volume and entropy of the substance, i. Substituting this into (7.11.4) for the two phases of A, one obtains

$$V_{m,A}(l)\, dP - S_{m,A}(l)\, dT = V_{m,A}(g)\, dP - S_{m,A}(g)\, dT. \tag{7.11.5}$$

Equation (7.11.5) is a fundamental equation in one-component phase equilibrium. It should be noted that this is a general result that does not specifically require that the two phases be liquid and vapor. We will now suppress the subscript A because only one pure substance is being considered here.

Since we wish to know how the vapor pressure changes consequent to a change in T such that phase equilibrium is maintained, we solve equation (7.11.5) for (dP/dT):

$$\frac{dP}{dT} = \frac{S_m(g) - S_m(l)}{V_m(g) - V_m(l)} = \frac{\Delta S_m}{\Delta V_m}, \qquad (7.11.6)$$

where ΔV_m and ΔS_m are the volume and entropy changes for the phase transition, respectively. If we now consider the isothermal phase transition $A(l) \rightleftharpoons A(g)$ and use the Gibbs-Helmholtz equation, $\Delta G = \Delta H - T\,\Delta S$ for this transformation, it follows from (7.11.1) that $\Delta G = 0$ and thus

$$\Delta S = \frac{\Delta H}{T}. \qquad (7.11.7)$$

Substituting this expression for ΔS into (7.11.6) results in

$$\frac{dP}{dT} = \frac{\Delta H_m}{T\,\Delta V_m}. \qquad (7.11.8)$$

Equation (7.11.8) is known as the Clapeyron equation, after the French engineer, B. P. E. Clapeyron (1834). It was further developed by R. Clausius (1850) and is called, appropriately, the Clapeyron-Clausius equation.

In equation (7.11.8), $\Delta V_m = V_m(g) - V_m(l)$. The molar volume of the vapor in equilibrium with liquid greatly exceeds that of the liquid (for conditions well below the critical point). Using the ideal gas law to approximate $V_m(g)$ and ignoring the smaller $V_m(l)$ term, ΔV becomes

$$\Delta V \simeq V_m(g) = \frac{RT}{P}. \qquad (7.11.9)$$

Substituting this expression into equation (7.11.8) and rearranging:

$$\frac{dP}{P} = \frac{\Delta H_{m,\text{vap}}}{R}\frac{dT}{T^2}. \qquad (7.11.10)$$

It should be noted that equation (7.11.10), which is known as the *Clausius-Clapeyron* equation, resembles the van't Hoff equation:

$$d \ln K_P^{\circ} = \frac{\Delta H^{\circ}}{R}\frac{dT}{T^2}, \qquad (7.11.11)$$

where K_P° and ΔH° are the standard state equilibrium constant and enthalpy of the process. Thus the vapor pressure (expressed in atmospheres) is a true equilibrium constant (in the ideal gas approximation), and $\Delta H_{m,\text{vap}}$ in equation (7.11.10) is the molar *enthalpy of vaporization*.

Equation (7.11.10) can be readily integrated if one assumes $\Delta H_{m,\text{vap}}$ to be independent of temperature over the temperature range corresponding to the vapor pressure measurements. Thus integrating between an arbitrary temperature T' at which the vapor pressure is P' and a temperature T, one has

$$\ln P = -\left(\frac{\Delta H_{m,\text{vap}}}{R}\right)\left(\frac{1}{T} - \frac{1}{T'}\right) + \ln P',$$

or

$$(7.11.12)$$

$$\ln P = \left(\ln P' + \frac{\Delta H_{m,\text{vap}}}{RT'} \right) - \frac{\Delta H_{m,\text{vap}}}{RT}.$$

Equation (7.11.12) predicts that a plot of $\ln P$ vs. $1/T$ should be linear, with the slope equal to $-\Delta H_{\text{vap}}/R$. This is experimentally observed for many liquid and solid systems as long as the temperature range is not too large (on the order of several tens of degrees). These plots can appear curved if: (1) very precise pressure data are obtained, (2) the solid whose vapor pressure is measured undergoes a phase change within the temperature range (hence ΔH_{sub} changes), and (3) the heat of vaporization is particularly temperature-dependent. The temperature dependence of vapor pressure is treated in more detail in the appendix of this experiment.

Frequently, tables of vapor pressure data provide relationships of the form

$$\log_{10} P = A - \frac{B}{T}, \tag{7.11.13}$$

in which A and B are constants. It can be seen by comparing equations (7.11.13) and (7.11.12) that B is related to the heat of vaporization. Another convenient vapor pressure relationship frequently found in tables is *Antoine's equation,* which contains three parameters and allows somewhat more accuracy over a larger temperature range:

$$\log_{10} P = A' - \frac{B'}{t - C},$$

in which t is the Celsius temperature and the constants A', B', and C are obtained empirically to best represent the vapor presure over the temperature range for which Antoine's equation is valid. It should be noted that C is not necessarily 273.15, the conversion between the Celsius and Kelvin scales, but reflects the characteristics of the particular liquid (or solid). Furthermore, B' is not simply related to the heat of vaporization.

Experimental Method

In this experiment, the vapor pressure of the liquid [which may be toluene, methylethylketone (2-butanone), ethyl acetate, or some other appropriate liquid having a boiling point near 100°C] will be measured using a *static* method: the procedure does not require that any of the vapor be removed from the system. A simple device called an *isoteniscope,* first developed by Smith and Menzies (1910), will be used. It is shown schematically in Figure 7.11.1.

There are three parts to this apparatus: (1) the bulb containing the liquid, (2) a U-tube manometer that contains the same liquid, and (3) an external manometer (or other pressure reading device). The sample bulb and manometer are immersed in a constant-temperature bath.

First, the dissolved air is removed from the liquid (i.e., the liquid is deaerated) by gentle pumping using an aspirator or a mechanical vacuum pump. Then an external pressure is applied to the system so that the liquid levels of the isoteniscope manometer (acting essentially as a null manometer) are made equal. Under this condition the external pressure is equal to the pressure in the sample bulb (i.e., the vapor pressure of the substance). This method is particularly sensitive because the density of the liquid is quite low (e.g., the density of toluene is less than 1 g/cm^3). Thus, if the levels were unequal by 1 mm, the error in the pressure

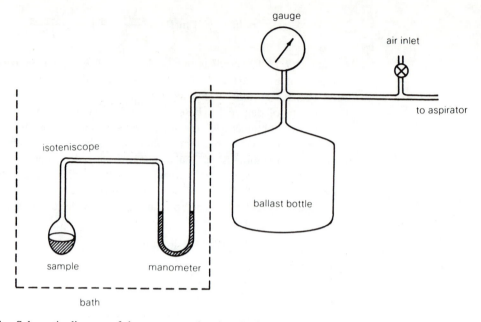

FIGURE 7.11.1   Schematic diagram of the apparatus showing the isoteniscope in a thermo-stated temperature bath, ballast tank, pressure-reading device, and air inlet valve.

(in torr) would be equal to the ratio of the density of toluene to that of mercury (i.e., 0.876/13.6), or 0.0644 torr.

Safety Precautions

- ☐ Safety goggles must be worn in the laboratory.
- ☐ Exercise caution in handling the liquid. If any liquid comes into contact with the skin, immediately wash the affected area with soap and water.
- ☐ The ballast vessel must be either wrapped with heavy tape or enclosed in a safety container.
- ☐ Exercise caution when working with the apparatus at the higher temperatures.
- ☐ Although there should be minimal evolution of organic vapors, consult your instructor if you smell fumes.
- ☐ The experiment should be performed in an open, ventilated laboratory.

Procedure

Refer to Figure 7.11.2, which is a more detailed diagram of the isoteniscope and associated apparatus used in this experiment.

1. First set up a water bath by filling a large beaker with water and placing it on a hot plate. Add a stirring bar and allow it to rotate *slowly* to ensure thermal equilibrium in the bath. Suspend a thermometer in the bath so that it is immersed deeply enough into the water to provide accurate readings but is kept away from the side of the bath.

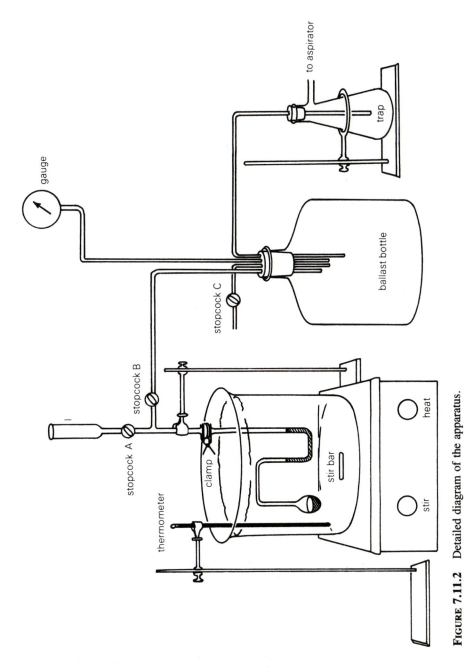

FIGURE 7.11.2 Detailed diagram of the apparatus.

2. Place the clean O-ring between the joints of the disconnected isoteniscope. Secure using the joint clamp; be sure *not to use* stopcock grease because it will contaminate the liquid. Clamp the apparatus to a ring stand; make sure that the isoteniscope is totally immersed in the water and that the manometer is secure and vertical. The joint and clamp, however, should be positioned above the surface of the water. Heat the bath water to just below the boiling point of the liquid being studied. (It may be helpful to use a 1000-W immersion heater periodically to save time.) While the water is being heated, evacuate the system by turning on the aspirator or mechanical pump. The pump should remain operating throughout the experiment. Only stopcock *B* should be open.

3. Pour about 5 mL of the liquid to be studied into the filling chamber *I*. After the bath has been heated to near 100°C, close stopcock *B,* and *gently* open *A* until the isoteniscope bulb is about half to three-quarters full and the liquid level in the U-tube is about 2 to 3 cm high. Close stopcock *A*.

4. Using the external pressure-controlling stopcock, *C*, raise the external pressure to about 600 torr. This will prevent the liquid from degassing too vigorously. Gradually open stopcock *B*. The liquid should start bubbling actively. Allow this degassing to continue for about 2 to 3 min.

5. Carefully adjust stopcock *C* until the levels of the liquid in the U-tube are nearly equal, or at least are stationary. It may be difficult to achieve exact pressure equilibration, but even a level difference of a few millimeters corresponds to good sensitivity. At this point read and record the external pressure as well as the bath temperature.

6. Turn off the hot plate and allow the bath to cool. Take pressure/temperature readings at about 5° increments as the bath temperature drops to ambient. Periodically add small amounts of crushed ice to hasten cooling, but do not allow cooling to proceed too rapidly so that the sample liquid and bath temperatures fail to correspond. Also, if the system cools too rapidly, the pressure is not compensated fast enough and air is drawn into the sample liquid.

7. If an open-end manometer or some other relative pressure reading device is used to measure the vapor pressure (such as a Bourdon gauge), you must determine the value of atmospheric pressure. Your laboratory instructor will show you how to read the laboratory barometer.

Data Analysis

1. Tabulate the measured vapor pressure and temperature data.

2. Convert the measured vapor pressures to absolute values by making the necessary correction using the actual atmospheric pressure. This is necessary if you are using an open-end manometer or certain Bourdon gauges. Consult your instructor.

3. Plot $\ln P_{vap}$ vs. $1/T$; see equation (7.11.12).

4. Perform weighted linear regression analysis (least squares) on the ($\ln P_{vap};1/T$) data and determine the heat of vaporization. By extrapolation, determine the normal boiling point of the liquid studied. Compare these results with the literature values.

5. Comment on the linearity of the $\ln P_{vap}$ vs. $1/T$ plot and the need to use the more complete relationships, equations (7.11.18) and (7.11.19). See the appendix below.

6. Estimate the errors in the experimental P_{vap} and T values and then determine the error in the heat of vaporization.

Questions and Further Thoughts

1. Does the liquid in the isoteniscope U-tube have to be the same as the liquid whose vapor pressure is being measured? If not, suggest some other liquids that might be used. What are the advantages and disadvantages of using another material? For what applications must some "inert" liquid be used in the U-tube?
2. What would the ln P_{vap} vs. $1/T$ plot of a *solid* look like, if a phase change (i.e., transition to another crystal structure) took place within the temperature range studied?
3. Why is the *entire* isoteniscope (sample bulb and U-tube) kept at the same temperature? What would happen if the sample bulb were cooler than the rest of the isoteniscope? Hotter?
4. What would the ln P_{vap} vs. $1/T$ plot look like if: (a) not all of the dissolved air had been removed in the beginning of the experiment, and (b) an amount of air entered the sample bulb as the system was cooling? What would be the effect of these problems on the value of the heat of vaporization obtained?
5. An indirect way of measuring vapor pressures is to determine the concentration of the vapor in equilibrium with its liquid. Propose a few methods by which the vapor concentration can be obtained *while it is in equilibrium* with the liquid phase.

Further Readings

A. W. Adamson, "A Textbook of Physical Chemistry," 3rd ed., pp. 265–284, Academic Press (Orlando, Fla.), 1986.

G. N. Castellan, "Physical Chemistry," 3rd ed., pp. 262–265, Addison-Wesley (Reading, Mass.), 1985.

I. N. Levine, "Physical Chemistry," 2nd ed., pp. 187–192, McGraw-Hill (New York), 1983.

J. N. Noggle, "Physical Chemistry," pp. 167–180, Little, Brown (Boston), 1985.

Appendix

If the temperature range is too large for equation (7.11.12) to hold, or if the vapor pressure measurements are very precise, the temperature dependence of ΔH_{vap} can be included. From

$$d \, \Delta H = \Delta C_p \, dT \qquad (7.11.14)$$

where $\Delta C_p = C_{P_V} - C_{P_L}$ (i.e., the difference in the respective heat capacities of the vapor and liquid phases), a more accurate form of equation (7.11.12) can be obtained. Integrating (7.11.14):

$$\Delta H = \Delta H' + \int_{T'}^{T} \Delta C_P \, dT, \qquad (7.11.15)$$

where T' is an arbitrary reference temperature and $\Delta H'$ is the corresponding transition enthalpy. If the heat capacities are now taken to be independent of T, (7.11.15) becomes

$$\Delta H = \Delta H' + \Delta C_P(T - T'). \qquad (7.11.16)$$

If this expression is substituted into equation (7.11.10), we finally obtain

$$\frac{dP}{P} = \frac{1}{R} [\Delta H' + \Delta C_P(T - T')] \frac{dT}{T^2}. \qquad (7.11.17)$$

Recalling that T' is a constant, this equation may be integrated to provide P as a function of T:

$$\ln P = -\frac{\Delta H_{vap}'}{nRT} + \frac{\Delta C_P}{R} \ln T + \frac{\Delta C_P T'}{nRT} + \text{constant}, \qquad (7.11.18)$$

where the constant is found by establishing a boundary condition: the vapor pressure of the substance at an arbitrary temperature. Equation (7.11.18) has the general form

$$\ln P = -\frac{A}{T} + B \ln T + C, \qquad (7.11.19)$$

where $A = \Delta H_{vap}'/R + \Delta C_P T'/R$, $B = \Delta C_P/R$, and C is a constant.

Equations (7.11.11) and (7.11.18) and (7.11.19) can be experimentally distinguished only if (1) the temperature range spanned in vapor pressure measurements is large enough, (2) the precision in P is high, or (3) ΔC_P for the system is particularly large—or some combination of these conditions.

Experiment 11

Vapor Pressure

NAME _____ DATE _____

Barometric pressure _____

Temperature   Manometer (Gauge) reading

————————   ————————————

————————   ————————————

————————   ————————————

————————   ————————————

————————   ————————————

————————   ————————————

————————   ————————————

————————   ————————————

————————   ————————————

————————   ————————————

————————   ————————————

————————   ————————————

————————   ————————————

————————   ————————————

————————   ————————————

————————   ————————————

————————   ————————————

————————   ————————————

————————   ————————————

Vapor Pressure

NAME _____   DATE _____

Barometric pressure _____

Temperature	Manometer (Gauge) reading
_____	_____
_____	_____
_____	_____
_____	_____
_____	_____
_____	_____
_____	_____
_____	_____
_____	_____
_____	_____
_____	_____
_____	_____
_____	_____
_____	_____
_____	_____
_____	_____
_____	_____
_____	_____

□ **Experiment 12**

Liquid-Vapor Coexistence Curve of CO_2 or SF_6 Near the Critical Point

Objective

To construct the liquid-vapor phase diagram of CO_2 or SF_6 near the critical point and to determine the critical temperature and molar volume.

Introduction

The coexistence curve is a phase diagram that illustrates the relationship between thermodynamic properties of a substance in equilibrium between different phases. Typical two-dimensional, one-component phase diagrams relate such properties as the molar volume, V_m, and temperature, T; V_m and pressure, P; and P and T. In this experiment, the V_m-T phase diagram of CO_2 or SF_6 will be constructed at temperatures just below the critical point.

A diagram of a V_m-T coexistence curve (at constant pressure) is shown in Figure 7.12.1. The points *on* the upper part of the curve represent the molar volumes of the vapor which is in equilibrium with the liquid, while points *on* the lower portion of the curve correspond to the molar volumes of the liquid phase of the substance as a function of temperature. The region outside and above the diagram corresponds to the substance being in a single phase: vapor. Likewise, the region in V_m, T space outside and below the coexistence curve corresponds to there being only liquid phase present. The point on the V_m-T phase diagram where the liquid and vapor densities equal each other is called the *critical point* and is denoted as $V_{m,c}, T_c$. The region of V_m, T space inside the coexistence curve corresponds to there being both liquid and vapor in equilibrium: a two-phase system. The behavior of liquids at the critical point was first described in detail by de la Tour in 1822.

From a direct experimental point of view, it is more convenient to plot the density, D, of the system vs. T. Since the density is proportional to the reciprocal of the molar volume, the D vs. T plot is inverted with respect to the V_m vs. T diagram and is portrayed in Figure 7.12.2. The interpretation of this diagram is

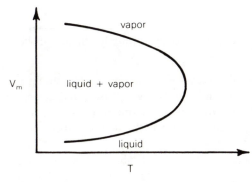

FIGURE 7.12.2 Phase diagram of a one-component system. Density, D, vs. T in the vicinity of the critical point, D_c, T_c.

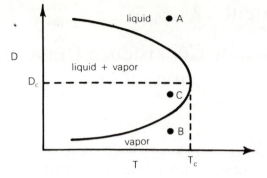

FIGURE 7.12.2   Phase diagram of a one-component system. Density, D, vs. T in the vicinity of the critical point, D_c, T_c).

analogous to that in Figure 7.12.1. The locus of points describing the upper part of the D-T curve represents the density of the liquid which is in equilibrium with vapor, and the lower curve indicates the corresponding equilibrium vapor densities. The one-phase regions are indicated: liquid at high densities above the curve and vapor at low densities below the curve. The critical point is located at D_c, T_c and corresponds to the position where the equilibrium liquid and vapor densities are equal to each other. Since the terms liquid and vapor are significant when they refer, respectively, to the high- and low-density phases of a substance, these terms lose distinction when the liquid and vapor densities are equal. Hence for $T > T_c$ (and $P > P_c$), the substance is called a *supercritical fluid,* without distinction as to liquid or vapor. A supercritical fluid is thus a high-density vapor or a low-density liquid. Supercritical CO_2, NH_3, and other materials receive wide attention as extraction agents. They can sometimes be used to remove biologically active substances from natural products under mild chemical and thermal conditions.

The coexistence curve can be interpreted in the following way:

1. When the overall, or bulk, density of the system (total mass/total volume) is greater than that of the liquid which is in equilibrium with vapor at a given T (point A in Figure 7.12.2), the system consists of only one phase: liquid (or, noncommittally, a high-density *fluid*).

2. When the bulk density is less than that of the vapor which would be in equilibrium with the liquid at a given T (point B in Figure 7.12.2), the system is also characterized by one phase: vapor (or low-density fluid). In both cases 1 and 2 the system lacks a phase boundary, or high-density/low-density interface.

3. If the bulk density is intermediate between the equilibrium liquid and vapor densities at the temperature T (point C in Figure 7.12.2), the system consists of liquid and vapor in coexistence. Accordingly, there will be a phase boundary: a *meniscus.*

The appearance of a meniscus is caused by the abrupt (nearly discontinuous) change in the refractive index of the two phases, and this arises from the disparity in the densities of the two phases. Accordingly, as T increases and the liquid and vapor densities approach each other, the meniscus becomes less well defined. At $T = T_c$ the meniscus ceases to exist.

In this experiment, a set of sealed tubes containing different amounts of CO_2 or SF_6 will be provided. The meniscus position in each of the tubes will be measured as a function of temperature. The equilibrium liquid and vapor densities of CO_2 or SF_6 will be determined from these data, and this information will be used

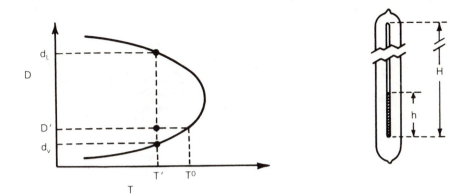

FIGURE 7.12.3   (a) *D-T* phase diagram. A sample at temperature T' has a bulk density, D'; the equilibrium vapor and liquid densities are d_V and d_L, respectively. (b) A sealed capillary tube containing the sample represented in (a); the overall height of the inner bore is H, and the height of the liquid phase is h.

to construct the coexistence curve. Figure 7.12.3 shows a *D-T* phase diagram and a sketch of a sample tube. The sample vessel is constructed from thick-walled capillary tubing of about 1 mm bore. This tube must withstand pressures up to about 73 atm.

In the phase diagram, D' refers to the bulk density of the sample in the sealed tube. The height of the liquid in this tube is denoted as h, while the overall height of the sealed inner capillary is H. Consider the system at a temperature, T'. In Figure 7.12.3, a tie line connects the liquid and vapor density curves at T'. Applying the lever rule to the *D-T* phase diagram, the segment a represents the *volume* of liquid present, while b represents the *volume* of vapor at equilibrium. The lever rule thus provides the relation

$$\frac{a}{b} = \frac{v_L}{v_V} = \frac{D' - d_V}{d_L - D'},$$

and $(7.12.1)$

$$\frac{v_L}{v_V} = \frac{h}{H - h},$$

where d_L and d_V are the *equilibrium* densities of liquid and vapor (called *orthobaric densities* when the system pressure is equal to the vapor pressure of the liquid only) at T'. It is assumed that the capillary bore is uniform. If z is defined as the ratio of the liquid height to the total height of the sample tube cavity, $z = h/H$, and equation (7.12.1) becomes

$$\frac{h}{H - h} = \frac{z}{1 - z} = \frac{D' - d_V}{d_L - D'}. \qquad (7.12.2)$$

This can be rearranged to express the bulk density:

$$D' = zd_L + (1 - z)d_V. \qquad (7.12.3)$$

Equation (7.12.3) provides a method for calculating the bulk density of a sealed sample tube if z is measured at a temperature at which both d_L and d_V are known. It is instructive to solve (7.12.3) for z:

$$z = \frac{D' - d_V}{d_L - d_V}. \qquad (7.12.4)$$

Examining equation (7.12.4) and Figure 7.12.3, one can predict the behavior of the meniscus height, i.e., z, in a sample tube as a function of temperature. As T increases, the ratio of lever arms (a/b) decreases and eventually approaches zero as $T \to T^0$. Since a/b represents the ratio of equilibrium liquid to vapor volumes in the sealed tube, the liquid phase vanishes at T^0 and therefore the bulk density, D', of the sample tube becomes equal to the vapor density, d_V, at T^0. From (7.12.4) it is apparent that as $D' \to d_V$, $z \to 0$; also, from (7.12.3), as $z \to 0$, $D' \to d_V$.

If the tube is filled in such a way that D' is *greater* than the mean (orthobaric) liquid and vapor densities, different behavior is seen. In this case as T increases, b/a decreases and approaches zero as $D' \to d_L$, thus, $z \to 1$ [equation (7.12.4)] in this case. Physically, the meniscus *rises* with increasing T and, in the limit, approaches the top of the sealed capillary tube.

Values of d_L and d_V are obtained in the following way: A set of N sealed tubes containing different amounts of CO_2 (or SF_6) is provided. Values of the bulk densities of these tubes, $\{D_i\}$, are provided, as are their overall inner bore lengths, $\{H_i\}$. For each sample tube, the meniscus height is accurately measured relative to the bottom of the inner bore (using a cathetometer) at a series of temperatures. Hence $\{z_i\}$ values are determined. At a given temperature d_L and d_V can be measured from the $\{D_i\}$ and $\{z_i\}$ data. Equation (7.12.3) is rearranged so that D is linear in z:

$$(D_i)_T = (d_L - d_V)_T (z_i)_T + (d_V)_T. \tag{7.12.5}$$

Since there are two unknowns at a given T (d_L and d_V), only two sample tubes containing both liquid and vapor are needed in applying equation (7.12.5). Since N is 6 to 8 (depending on the number of tubes provided and the particular temperature), d_L and d_V are overdetermined and their values can be obtained using linear regression. A plot of D vs. z should be linear according to equation (7.12.5), and d_L and d_V can be obtained from the slope and intercept.

As T is increased, the meniscuses of the tubes will vanish, one by one, and there will be progressively fewer $\{D_i, z_i\}$ data to analyze; hence, the accuracy of the d_L and d_V data decreases as T approaches T_c. For the last tube (i.e., the Nth) containing a meniscus, (7.12.5) cannot be used to obtain unique values of d_L and d_V. In this case, a plot of z vs. T for that tube can be made, and the temperature at which $z \to 0$ (or $z \to 1$), e.g., T_N, can be estimated. At T_N, $d_V = D_N$ (for $z \to 0$) and $d_L = D_N$ (for $z \to 1$). Once values of the orthobaric densities are obtained at the various temperatures, the coexistence curve is obtained by plotting these d_L and d_V data vs. T.

Theoretically, the temperature dependence of the orthobaric densities can be anticipated in what is called the *law of rectilinear diameters* (discovered by Cailletet and Mathias in 1886). This law states simply that the mean orthobaric density, $<d>$, is a linear function of T:

$$<d> = \frac{d_L + d_V}{2} = A + GT, \tag{7.12.6}$$

where A and G are constants for a particular substance. In principle, T_c could be obtained by determining the intersection of a plot of $<d>$ vs. T with the coexistence curve. Note that $<d> = d_c$ (i.e., $d_L = d_V = d_c$) at $T = T_c$. In practice, however, data points in the coexistence curve near T_c are very difficult to obtain, and thus the intersection point cannot be located accurately.

Another approach is to use the *law of critical exponents,* which states that for T not too far below T_c:

$$d_L - d_V = B\left(1 - \frac{T}{T_c}\right)^b, \tag{7.12.7}$$

where B and b are constants. It has been experimentally determined that $b \simeq 0.33$ for many substances. For CO_2 and SF_6, $b = 0.348$ and 0.342, respectively.[1] Equation (7.12.7) can be used to obtain the T_c from values of d_L and d_V. This is most easily done by assuming $b = 1/3$ and linearizing equation (7.12.7): $(d_L - d_V)^3 = (B^3/T_c)(T_c - T)$.

It is immediately apparent from equation (7.12.7) that as $T \rightarrow T_c$, $d_L \rightarrow d_V$. Another implication of (7.12.7) (and the apparent similarity of b for many substances) is that the shape of the coexistence curve is the same for all substances. Interestingly, other properties such as surface tension[1] and certain magnetic phenomena[2] follow a critical exponents law. T/T_c is called the reduced temperature, and (7.12.7) is an example of the law of corresponding states.

The law of rectilinear diameters can be rewritten in terms of the same temperature variable as is used in the critical exponents law. Accordingly, if t is defined as $(1 - T/T_c)$, equation (7.12.6) becomes

$$\frac{d_L + d_V}{2} = d_c + Ct, \tag{7.12.8}$$

where d_c is the critical density and C is a constant. Note that for $T = T_c$, $t = 0$ and $d_L = d_V \equiv d_c$. If equations (7.12.7) and (7.12.8) are solved simultaneously, d_L and d_V can each be obtained explicitly as a function of t (hence T):

$$d_{L,V}(t) = d_c + Ct \pm \frac{B}{2}t^b. \tag{7.12.9}$$

Equation (7.12.9) analytically represents the coexistence curve for a substance that obeys the rectilinear diameter and critical exponent laws.

Finally, we consider an interesting phenomenon concerning transitions between one-phase and two-phase systems. For $T < T_c$, a sample tube will contain two distinct phases at equilibrium (if $d_V < D < d_L$), while for $T > T_c$ only one phase exists. It might be said that below T_c, the system possesses order: a distinct phase boundary separates two different regions, while above T_c the system is totally disordered and is homogeneous throughout. As T increases and approaches T_c reversibly, the system responds in a continuous manner, the meniscus rising (or falling) and approaching the bottom (or top) of the tube. On the other hand, if a homogeneous sample, initially at a temperature greater than T_c, is cooled reversibly, a curious thing happens at $T = T_c$. An ordered system is produced from a chaotic one. One might wonder by what mechanism the system reacts to establish a meniscus in a *definite position* where one did not previously exist. If a sealed sample tube is actually observed as $T = T_c$, one sees wild fluctuations in density gradients and the sample appears turbid because of the light scattering caused by the abrupt increase in particle size as high-density (liquid)–low-density (vapor) clusters are formed. This phenomenon, called *critical opalescence,* can be observed whenever one-phase–two-phase transitions occur. This can be quite dramatically demonstrated in a two-component system at ambient pressures by slowly cooling a mixture of aniline and cyclohexane, or methanol/cyclohexane, below its critical (or upper consolute) temperature (ca. 60°C). See p. 163.

Safety Precautions

☐ Safety glasses must be worn during this experiment.

☐ The sealed capillary tubes are under high pressure. Do not remove any of the tubes from the temperature bath. If you need assistance in placing one of the tubes, etc., consult the instructor.

☐ Avoid eyestrain when using the cathetometer. Make sure it is properly focused. If there are several users, one person should use the instrument for a continuous period during which the focusing is properly set.

Procedure

1. A set of sealed capillary tubes containing known bulk densities of CO_2 or SF_6 is held in a rotatable rack that is immersed in a water bath. The bulk densities and the lengths, H, of the sample tubes will be provided. The position of the meniscus of a given tube will be measured (to ca. 0.01 cm) using a cathetometer, which is a telescope mounted on a vertical calibrated shaft. The proper use of the cathetometer is crucial to the successful completion of the experiment and will be demonstrated by the instructor. A diagram of the apparatus is shown in Figure 7.12.4.

2. Begin at a temperature of about 24°C for CO_2 (38° for SF_6). Record the temperature indicated by the sensitive thermometer before and after all the data are taken at that temperature. Accuracy to 0.1° (or better) is necessary. For each of the tubes, measure the position of the *bottom* of the inner capillary bore and record these values. Then read and record the positions of the meniscuses of each of the tubes. The crosspiece of the telescope should always be aligned with the bottom of the curved meniscus.

3. Increase the bath temperature by about 1°. After equilibrium is reached, record the new temperature. Carefully invert the capillary tube holder, using the knob attached to the side of the rack so that liquid runs down to the other ends of the tubes. Restore the rack to the original upright position and wait for the liquid

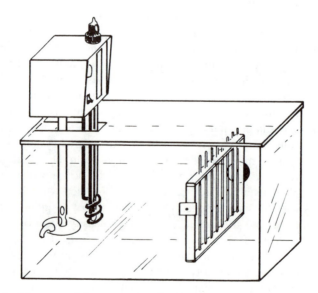

FIGURE 7.12.4   Diagram of the apparatus showing the arrangement of sealed capillary tubes on a rotatable rack.

to drain. If necessary, repeat the sample tube inversion. You can follow the progress of the liquid draining by viewing the meniscus through the cathetometer.

4. After the meniscus positions have equilibrated, read and record these values. Record the bath temperature after the meniscus readings are completed.

5. Increase the bath temperature by 1° increments and repeat the tube inversion and meniscus readings until a temperature of 28 to 29°C for CO_2 (42 to 43° for SF_6) is reached. Thereafter, increase the temperature in steps of about 0.5° until the liquid levels of all the tubes have vanished.

Calculations and Data Analysis

1. From your data, compute values of z for each tube at the different temperatures used. Tabulate the data as $z(T)$; T.

2. Plot D,z isotherms from the data assembled in step 1 and compute d_L and d_V using linear regression [see equation (7.12.5)]. For each plot make sure each of the points is appropriate in the linear regression analysis, i.e., that its deviation from the linear regression line is "normal."

3. From the orthobaric densities obtained from step 2, construct the D,T coexistence curve, indicating appropriate error bars.

4. Use linear regression to determine T_c and B by plotting $(d_L - d_V)^3$ vs. T [equation (7.12.7)]. Comment on the error in T_c obtained in this way. How does the value of T_c change if a value of $b = 0.348$ for CO_2 (or 0.3422 for SF_6) is used[1] in the critical exponents expression?

5. Once T_c is found, test the validity of the law of rectilinear diameters by plotting $d_L + d_V$ vs. t. Obtain values of d_c and C; see equation (7.12.8). Report also the critical molar volume, V_c.

6. (Optional) Using the measured values of d_c, B, and C, as well as the values of b that are given above, calculate the coexistence curve for CO_2 (or SF_6) using equation (7.12.9). Superimpose your d_L and d_V data on this calculated phase diagram. You can use a hand-held calculator (point by point, or by program); alternatively, you can write a computer program to generate $d_{L,V}(T)$ points and have these points plotted. Your d_L and d_V data should be superimposed on this plot.

7. Determine P_c using the T_c and V_c data by using various equations of state. For a van der Waals gas, the critical compressibility, $Z_c = P_c V_c/RT_c$, is 3/8. For a gas obeying the Redlich-Kwong equation,[3] $Z_c = 1/3$; and for the Dieterici equation, $Z_c = 2/e^2$. The Berthelot and modified Berthelot equations predict Z_c to be 3/8 and 9/32, respectively.[4]

8. Compare your critical data with literature values.

Questions and Further Thoughts

1. What would a system in liquid-vapor equilibrium look like in a "zero gravity" environment, for example, in a space laboratory? What would be the appearance of the meniscus?
2. What happens to the value of the liquid-vapor surface tension as the critical point is reached? Why? What is the value of the heat of vaporization of a liquid near the critical point?
3. The sealed capillary tubes had to be inverted to accelerate the attainment of phase equilibrium of the system after the temperature is changed. Why does this process take so long if two phases are not physically, or mechanically, mixed?
4. What is the mean free path of CO_2 (or SF_6) near the critical point? Compare this with the (inner) diameter and length of the capillary.
5. What is the value of the slope of the D,T phase diagram at the critical point? Use equation (7.12.9).

6. Consider a sample for which D' was equal to the mean of the orthobaric liquid and vapor densities. Describe what would happen to the position of the meniscus if the temperature of sample is raised to, and then above, the critical temperature.

7. Using the cathetometer is very tedious, and painstaking attention is required for good results. Because the distinction of the liquid and vapor phases can be made on the basis of their different densities, hence refractive indices, n_L and n_V, one could obtain the requisite data by using an optical probe to measure n_L and n_V. A relationship between the refractive index and density for a homogeneous nonpolar medium is expressed through the Lorentz-Lorenz function:

$$d = K\frac{n^2 - 1}{n^2 + 2}$$

where K is a constant, $(4/3\pi\alpha m)^{-1}$. α is the polarizability of the medium at the wavelength used to measure n, and m is the molecular mass. Consult a physics text or a reference on optics, and devise a method by which n could be measured for liquid and vapor in coexistence using a convenient optical source, for example, a low-power He-Ne laser.

See, for example, reference 1.

Notes

1. W. Rathjen and J. Straub, in E. Hahne and U. Grigull, eds., ''Heat Transfer in Boiling,'' pp. 425–451, Academic Press (New York), 1977.
2. H. E. Stanley, ''Introduction to Phase Transitions and Critical Phenomena,'' pp. 9–12, 39–49, Oxford (London), 1971.
3. I. N. Levine, ''Physical Chemistry,'' 2nd ed., pp. 199–208, McGraw-Hill (New York), 1983.
4. G. W. Castellan, ''Physical Chemistry,'' 3rd ed., pp. 33–47, Addison-Wesley (Reading, Mass.), 1983.

Further Readings

A. W. Adamson, ''A Textbook of Physical Chemistry,'' 3rd ed., pp. 11–24, Academic Press (Orlando, Fla.), 1986.

P. W. Atkins, ''Physical Chemistry,'' 3rd ed., pp. 28–31, W. H. Freeman (New York), 1986.

J. H. Noggle, ''Physical Chemistry,'' pp. 14–26, 180–181, Little, Brown (Boston), 1985.

Experiment 12

Liquid-Vapor Coexistence Curve

NAME _____ DATE _____

Tube #	#1	#2	#3	#4	#5	#6	#7
Bottom	_____	_____	_____	_____	_____	_____	_____

Temperature   *Meniscus position*

_____	_____	_____	_____	_____	_____	_____
_____	_____	_____	_____	_____	_____	_____
_____	_____	_____	_____	_____	_____	_____
_____	_____	_____	_____	_____	_____	_____
_____	_____	_____	_____	_____	_____	_____
_____	_____	_____	_____	_____	_____	_____
_____	_____	_____	_____	_____	_____	_____
_____	_____	_____	_____	_____	_____	_____
_____	_____	_____	_____	_____	_____	_____
_____	_____	_____	_____	_____	_____	_____

Experiment 12

Liquid-Vapor Coexistence Curve

NAME _____ DATE _____

Tube #	#1	#2	#3	#4	#5	#6	#7
Bottom	_____	_____	_____	_____	_____	_____	_____

Temperature   *Meniscus position*

_____	_____	_____	_____	_____	_____	_____
_____	_____	_____	_____	_____	_____	_____
_____	_____	_____	_____	_____	_____	_____
_____	_____	_____	_____	_____	_____	_____
_____	_____	_____	_____	_____	_____	_____
_____	_____	_____	_____	_____	_____	_____
_____	_____	_____	_____	_____	_____	_____
_____	_____	_____	_____	_____	_____	_____
_____	_____	_____	_____	_____	_____	_____
_____	_____	_____	_____	_____	_____	_____

□ Experiment 13

Solid-Liquid Equilibrium in a Binary System

Objective

To construct the solid-liquid temperature composition (*T-X*) phase diagram for the naphthalene-biphenyl system using thermal analysis, to estimate the heats of fusion of the components, and to compare the phase diagram with the ideal solubility curves of the components.

Introduction

The equilibrium between the solid phase of one component, *A*, and the liquid solution consisting of it and another component, *B* (see Figure 7.13.1) may be described by using the expression equating the chemical potential of *A* in the two phases:

$$\mu_A(s)(T,P) = \mu_A(l)(T,P) \tag{7.13.1}$$

where $\mu_A(s)$ is the chemical potential of solid *A* and $\mu_A(l)$ is the chemical potential of *A* that exists in the liquid phase containing *A* and *B* at a given temperature and pressure, *T* and *P*. Because *both* solid and liquid *A* coexist (at the particular applied pressure), the temperature of the system at equilibrium can be referred to as the *freezing point* of *A*. Moreover, because solid *A* is considered to be *pure* (although it is in contact with a liquid mixture of *A* and *B*),

$$\mu_A(s)(T_f,P) = \mu_A^{\bullet}(s)(T_f,P), \tag{7.13.2}$$

where $\mu_A^{\bullet}(s)(T,P)$ is the chemical potential of pure solid *A* at the equilibrium temperature T_f (which is the freezing point of *A*).

The chemical potential of *A* in the liquid phase can be expressed in terms of its activity, a_A,

$$\mu_A(l)(T_f,P) = \mu_A^{\bullet}(l)(T_f,P) + RT_f \ln a_A, \tag{7.13.3}$$

in which $\mu_A^{\bullet}(l)$ is the chemical potential of liquid *A* at unit activity, the pure liquid state at T_f and *P*. Replacing the chemical potentials in the fundamental equilibrium

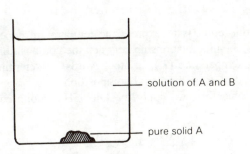

FIGURE 7.13.1 Pure solid *A* is in equilibrium with a saturated solution of *A* and *B* at constant temperature and pressure.

equation (7.13.1) with those expressed in (7.13.2) and (7.13.3) for the solid and liquid,

$$\mu_A^{\bullet}(s)(T_f) = \mu_A^{\bullet}(l)(T_f) + RT_f \ln a_A. \tag{7.13.4}$$

The explicit pressure dependence is now suppressed because P is presumed to be constant (e.g., 1 atm). To make (7.13.4) more useful, the activity of A in the A-B liquid mixture is expressed as the product of a composition variable, the mole fraction (x_A), and the activity coefficient, γ_A: $a_A = x_A \gamma_A$. The equilibrium condition (7.13.4) is now

$$\mu_A^{\bullet}(s)(T_f) = \mu_A^{\bullet}(l)(T_f) + RT_f \ln x_A + RT_f \ln \gamma_A. \tag{7.13.5}$$

In order to use this equation quantitatively, some means must be found to express γ_A as a function of composition and temperature. While there are many approaches that can be taken to achieve this end, they are often complicated and usually require specific knowledge about the nature of the system under consideration.

This experiment examines a mixture of two organic components that appear similar in structure, namely, naphthalene (N) and biphenyl (B). In the liquid phase containing both N and B, interactions between identical molecules (N-N and B-B) are very similar in magnitude to the interactions between different ones (N-B).

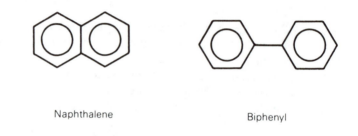

Naphthalene   Biphenyl

The consequence of this near equality in molecular interactions is that the activity coefficients of both components in the liquid mixture are very close to unity for all compositions. A system for which the foregoing assumptions are justified is called an *ideal solution*. This condition ($\gamma_A \simeq 1$) greatly simplifies the equilibrium expression (7.13.5), which now reads

$$\mu_A^{\bullet}(s)(T_f) = \mu_A^{\bullet}(l)(T_f) + RT_f \ln x_A. \tag{7.13.6}$$

The difference in the molar free energies of a pure liquid and solid is the molar free energy of fusion, i.e., $\mu_A^{\bullet}(l) - \mu_A^{\bullet}(s) \equiv \Delta G_{\text{fus},A}$. Thus, equation (7.13.6) becomes

$$\Delta G_{\text{fus},A}(T_f) = -RT_f \ln x_A. \tag{7.13.7}$$

The objective at this point is to express the temperature dependence of $\Delta G_{\text{fus},A}$ explicitly and thus obtain a relationship between the composition of A in the liquid phase (x_A) and the temperature (T_f), in other words, to determine the dependence of the freezing point of A on the liquid composition.

To pursue this objective, we apply the Gibbs-Helmholtz equation to the left-hand side of (7.13.7):

$$d \frac{\Delta G_{\text{fus},A}}{T} = -\frac{\Delta H_{\text{fus},A}}{dT/T^2}, \tag{7.13.8}$$

where T replaces T_f for generality (although it is still assumed that solid-liquid equilibrium exists). $\Delta H_{\text{fus},A}$ is the molar heat of fusion of solid A at the temperature T. In order to obtain $\Delta G_{\text{fus},A}$ as an explicit function of T, equation (7.13.8) must be integrated, and this requires knowledge of the temperature dependence of $\Delta H_{\text{fus},A}$. Over the limited temperature range pertinent to this experiment (ca. $10 - 90°C$), $\Delta H_{\text{fus},A}$ may be considered constant.

If (7.13.7) is rearranged to read $\Delta G_{\text{fus},A}/T = -R \ln x_A$ and this is substituted in equation (7.13.8), one obtains

$$d\,(-R \ln x_A) = -\frac{\Delta H_{\text{fus},A}}{dT/T^2}. \qquad (7.13.9)$$

Integration of equation (7.13.9) between the limits $\{x_A = 1;\ T = T_{f,A}^{\bullet}\}$ and $\{x_A; T\}$ results in (after simplification)

$$R \ln x_A = \Delta H_{\text{fus},A}\left(\frac{1}{T_{f,A}^{\bullet}} - \frac{1}{T}\right), \qquad (7.13.10)$$

where $T_{f,A}^{\bullet}$ is the melting point of pure A (at the constant pressure P).

Equation (7.13.10) is of fundamental importance in this experiment. It relates the solid-liquid transition temperature (i.e., the melting point) of a component (A) to its composition in the liquid phase. Equation (7.13.10) can be interpreted in two ways: (1) it represents a *freezing point depression* curve because it allows the calculation of the depressed melting point, T [note that in equation (7.13.10) $T < T_{f,A}^{\bullet}$] from the composition of A in the liquid phase, x_A; (2) it describes the *solubility curve* of A in the solvent B (where B is the other component present in the liquid phase) because x_A represents the *saturation* composition (or solubility) of A in the liquid phase at a temperature T.

Note that the solubility of A predicted from equation (7.13.10) depends only on the temperature and is thus independent of the nature of the other liquid component. This follows from the definition of an ideal solution in which the identity of the other component (which is supposed to resemble A) is irrelevant; for this reason, equation (7.13.10) is called the *ideal solubility equation*.

If we now turn to the other component in this system, i.e., B, and consider the equilibrium between solid and liquid B in a liquid solution of B and A, an expression analogous to equation (7.13.10) can be written

$$R \ln x_B = \Delta H_{\text{fus},B}\left(\frac{1}{T_{f,B}^{\bullet}} - \frac{1}{T}\right), \qquad (7.13.11)$$

where x_B is the mole fraction of B in the liquid phase, $\Delta H_{\text{fus},B}$ is the molar heat of fusion of pure B (also presumed to be temperature-independent), and $T_{f,B}^{\bullet}$ is the transition temperature of pure B.

In a binary system, $x_A + x_B = 1$, and equations (7.13.10) and (7.13.11) can be merged to describe the complete isobaric solid-liquid phase diagram. This is shown schematically in Figure 7.13.2. The composition variable is x_A. Since both solubility (or depressed melting point) functions (7.13.10) and (7.13.11) curve downward with increasing x_A and x_B, there will be an intersection point in the phase diagram. This point is indicated in Figure 7.13.2 at (x_A', T') and is called the *eutectic point*. Physically, it corresponds to a unique A-B composition in which the liquid is in equilibrium with *both* pure solid A and pure solid B.

To appreciate the significance of the eutectic point, consider the phase rule. The net number of degrees of freedom ($f = c - p + 2$) for a two-component system at constant pressure with three phases in equilibrium is zero; the system is invariant. Thus there is *only* one temperature and *only* one composition at which the eutectic can occur. The physical characteristics of the system at the eutectic composition will be described on the next page.

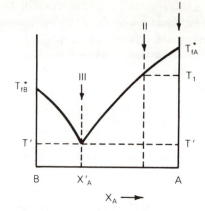

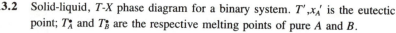

FIGURE 7.13.2 Solid-liquid, *T-X* phase diagram for a binary system. T', x_A' is the eutectic point; $T_A^{\bullet}$ and $T_B^{\bullet}$ are the respective melting points of pure A and B.

It is useful to examine the limiting slopes of the ideal solubility curves near unit mole fraction. Consider equation (7.13.10) and the phase diagram in Figure 7.13.2. For compositions very rich in component A, $x_A \to 1$ ($x_B \to 0$), and $\ln x_A = \ln(1 - x_B) \simeq -x_B$. Equation (7.13.10) now reads

$$-Rx_B \simeq \Delta H_{\text{fus},A} \left(\frac{1}{T_{f,A}^{\bullet}} - \frac{1}{T} \right) \qquad (x_A \to 1). \qquad (7.13.12)$$

This limiting expression can be used with $x_A + x_B = 1$ and rearranged to yield

$$x_A \simeq \frac{\Delta H_{\text{fus},A}}{R} \left(\frac{1}{T_{f,A}^{\bullet}} - \frac{1}{T} \right) + 1 \qquad (x_A \to 1). \qquad (7.13.13)$$

From (7.13.13), the derivative of x_A with respect to T is $\Delta H_{\text{fus},A}/RT^2$, and this term can be used with the observed limiting slope of the A-B phase diagram near $x_A = 1$ ($T \simeq T_{f,A}^{\bullet}$) to obtain the heat of fusion of A from the phase diagram. In an analogous way, $\Delta H_{\text{fus},B}$ can be obtained from the limiting slope of the A-B phase diagram near $x_A = 0$.

Theoretically, the eutectic point (T', x_A') for an ideal binary system can be determined by solving equations (7.13.10) and (7.13.11) simultaneously (assuming the heats of fusion and melting points of A and B are known). Because of the transcendental nature of equations (7.13.10) and (7.13.11), however, (T', x_A') cannot be obtained in closed form, and an iterative procedure, or numerical analysis, is required.

Experimental Method

The phase diagram will be determined using a technique called *thermal analysis*. In this approach the temperature of a cooling sample of known A-B composition is measured as a function of time as the system is brought from a molten state to a totally solidified one. (Alternatively, the temperature of a solid mixture of A and B can be monitored as the sample is heated until the system is entirely molten.) Thus a cooling curve is obtained. The shape of this curve is used to obtain both the transition temperature of the sample and the eutectic temperature for the A-B system. In thermal analysis, it is assumed that heat is withdrawn from the sample in a controlled way, either at a constant rate or in some other systematic manner.

For example, if the sample is immersed in a large cold (or hot) bath, the rate of heating (or cooling) will be proportional to the temperature difference between the sample and the bath (Fourier's law of heat flow).

Some representative cooling curves, shown in Figure 7.13.3, are discussed below.

1. The cooling curve of pure A (point I in Figure 7.13.3): Starting at a temperature above $T^{\bullet}_{f,A}$, the cooling of the liquid proceeds relatively rapidly. As $T \rightarrow T^{\bullet}_{f,A}$, solid A begins to form, and the temperature stabilizes because a two-phase, one-component system is invariant (at constant pressure). As the sample begins to solidify at $T = T^{\bullet}_{f,A}$ heat is released to the surroundings and thus the cooling curve halts, or levels off. When all the liquid has solidified, cooling resumes and the temperature drops.

Actually, the temperature of the melt often first drops *below* $T^{\bullet}_{f,A}$ until a small crystal has formed. The incipient formation of solid (a disorder-order transition) is a discontinuous process and requires the intervention of a stimulus such as the addition of a seed crystal, the mechanical agitation of the system, or the presence of a surface defect in the sample tube. The temporary dip in the cooling shown in Figure 7.13.3 (I) is called *supercooling* and is a direct manifestation of the departure from equilibrium.

The position of the halt in cooling curve I can be used to assign the value of $T^{\bullet}_{f,A}$. Similar behavior is expected for the cooling curve of pure B. It should be emphasized that the presence of supercooling indicates that a cooling curve does not continually monitor the *equilibrium* temperature of the system.

2. The cooling curve of an A-B mixture having a composition x_A (point II in Figure 7.13.3): Starting at a temperature above T_1, the depressed melting point of A, the cooling of the melt proceeds relatively rapidly (similar to 1). As the depressed melting point, T_1, is reached, solid A begins to form, and at this point heat is liberated, decreasing the rate of cooling and causing a break in the cooling curve. As cooling continues and more solid A forms, the rate of cooling continues to slow down.

The composition of the liquid phase changes and follows the A solubility curve; i.e., the liquid becomes increasingly rich in B. The relative amounts of liquid and solid present as the system cools are given by the application of the lever rule in the two-phase region of the phase diagram.

When the eutectic temperature T' is reached, solid B also starts to form. This

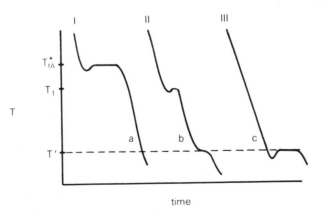

FIGURE 7.13.3   Cooling curves of three samples of A-B mixtures. Curves I, II, and III pertain to compositions shown in Figure 7.13.2.

solid mixture will have the eutectic composition (x'_A). As in the case of a pure substance, cooling slows even further, and the cooling curve halts, becoming nearly horizontal. Depending on the actual A composition, an inflection in the cooling curve may be observed at $T \simeq T'$ (the initial supercooling of A at $T^{\bullet}_{f,A}$ may mask the halt in the cooling curve). When all of the A and B have solidified, the cooling rate again increases, as for a pure substance.

3. The cooling curve of the A-B system that has the eutectic composition, x'_A (point III in Figure 7.13.3): The cooling curve of a liquid mixture of A and B having the eutectic composition appears to be that of a pure compound (having an apparent melting point T'). Cooling of such a sample never passes through the two-phase region of the phase diagram.

Cooling curves will be obtained for seven naphthalene/biphenyl compositions in addition to the two pure components. The temperature of a molten sample, contained in a small test tube, will be monitored as it cools using a small thermocouple (iron-constantan) sheathed in a closed 1/16-in. stainless-steel tube. The thermocouple probe is coupled to an electronic device that produces an analog signal (a voltage) proportional to the Celsius temperature of the probe. Thus the output is referenced to the ice point, 0°C. This device is called a temperature-to-analog converter (TAC). The TAC, which produces 1 mV/°C, is connected either to a strip chart recorder or to the input channel of a laboratory computer. Thus the cooling curves will be either directly displayed using the strip chart recorder, or acquired and analyzed using software (computer instructions) provided in the laboratory.

A weighed mixture of A and B (each to ± 0.01 g) is placed in a small test tube and is liquefied by immersion in boiling water. After the molten sample is homogenized, the thermocouple probe is inserted in the melt. A small teflon disc prevents the probe from wobbling and also keeps it centered in the sample tube. The sample tube is then placed in a jacketing test tube that prevents the sample from cooling too rapidly. The assembled apparatus is then quickly immersed in an ice bath and the temperature is monitored as the sample cools. It should be emphasized that the cooling curve obtained in this way is that pertaining to a *quiescent* system. It is assumed that the temperature sensed and transmitted by the TAC is the *equilibrium* temperature of the cooling A-B melt. Preferably, the system can be mixed or stirred during the cooling process to ensure both phase and thermal equilibrium.

For each of the seven A-B mixtures (as well as the two pure components), cooling curves will be analyzed with the objective of determining the depressed (or pure) melting point, and the eutectic temperature. In principle, T' can be obtained from any cooling curve of an A-B mixture. In practice, the accuracy with which T' can be determined varies with the composition of the system.

Safety Precautions

☐ Safety goggles must be worn at all times when performing this experiment.
☐ Naphthalene and biphenyl samples should be weighed in (or near) a fume hood.
☐ Avoid getting these chemicals on the skin; gloves may be worn for extra precaution. If these substances come in contact with the skin, wash the affected area with soap and water.
☐ Exercise particular caution when handling the sample tube in the boiling water bath. If hot water splashes on the skin, immediately inform your instructor.
☐ This experiment must be performed in a well-ventilated laboratory. If the presence of organic fumes is objectionable, inform your instructor. In this event, the apparatus should be set up in a fume hood.

Procedure

A sketch of the assembled cooling curve apparatus is shown in Figure 7.13.4. It is convenient to place the assembled apparatus in an ice bath in which a magnetic stirrer gently circulates the water.

Option A: Strip Chart Recorder

A strip chart recorder can be used to obtain the temperature vs. time plot. This graph directly corresponds to the cooling curve. A recorder having a 0- to 100-mV full-scale pen deflection and a chart speed of about 2 cm (or 1 in.) per minute should be used.

   If you are using the computer-assisted data acquisition and analysis method, see Option B.

   1. Using an analytical or suitable trip balance, weigh out a (nominal) 2-g sample of naphthalene on weighing paper and quantitatively transfer the material to the small inner test tube. An accuracy of 0.01 g is adequate.

   2. *Prepare the sample.* Lightly attach a rubber stopper (size 00) to the sample tube and immerse in a beaker of boiling water until the sample has melted. (A heat gun may also be used.) Remove the tube from the water and wipe it dry. Remove the stopper from the tube and insert the probe. The probe stopper should fit snugly and the teflon spacer should keep the probe in the center of the tube. The tip of the probe should be in the lower third of the melt and must not touch any part of the inside surface. See Figure 7.13.4.

   3. *Assemble the apparatus.* Make sure the TAC leads are properly connected. Place the outer jacket over the sample tube so that the rubber stopper is seated in the inner sample tube and *also* grasps the rim of the outer jacket (see Figure 7.13.4). This stopper should be tight enough to allow the apparatus to be picked up as a whole but not so tight that the inner tube cannot be easily removed.

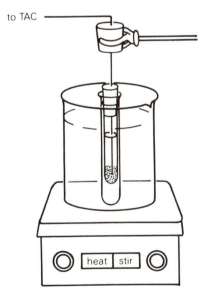

to TAC

heat | stir

FIGURE 7.13.4   Cooling curve apparatus. The jacketed sample tube is immersed in hot water to melt the mixture. The sample tube is then placed in an ice bath to initiate cooling. The temperature probe is connected to the temperature-to-analog converter.

4. Immerse the assembled apparatus in the boiling water until the sample is fully liquefied and the temperature (which can be monitored on the strip chart recorder or the computer) is above the melting point.

5. *Obtain the cooling curve.* Switch on the TAC and the strip chart recorder (25 mm/min or 1 in./min is appropriate). Quickly place the apparatus in the ice bath and secure with a pinch clamp. There must be a copious amount of crushed ice in the bath. Gentle mixing of the water with a magnetic stirrer is advisable unless there is interference with the TAC output. The cooling curve is complete when the temperature of the system falls well below T' (about 10°C is satisfactory).

6. After the cooling curve has been traced out on the chart paper, remove the apparatus from the ice bath and gently remove the outer jacket. Place the sample tube in boiling water until the sample melts and carefully remove the thermocouple probe.

7. *Prepare the next sample.* Let the molten sample cool and add about 0.20 g (0.01-g accuracy) of biphenyl to the sample tube. Try to avoid having the added material stick to the inside of the tube. Melt the mixture and gently swirl to homogenize the liquid. Make sure any solid still remaining on the inside walls has been brought down and mixed.

8. Reassemble the apparatus (step 3) and obtain the cooling curve of this sample (steps 4 and 5).

8a. If time permits, obtain a *heating curve.* The cooled sample and apparatus can be directly placed in a bath of gently boiling water. The system temperature is then monitored on the chart paper as the sample heats up to the completely molten state.

9. Remove the apparatus from the ice bath, heat the solid, remove the probe, and add 0.40 g of biphenyl to the sample tube. Melt and mix the sample, and assemble the apparatus. Obtain the cooling curve.

10. Add another 0.40-g quantity of biphenyl to the sample tube and acquire the cooling curve.

11. Add a final 1.0-g amount of biphenyl to the sample mixture and obtain the cooling curve.

At this point, you will have obtained *five* cooling curves: one of pure naphthalene and four corresponding to *cumulative* additions of 0.2, 0.6, 1.0, and 2.0 g of biphenyl.

The second part of the experiment consists of preparing samples and acquiring cooling curves by following the above procedures, but starting with pure biphenyl.

12. Heat the frozen mixture from the last cooling curve and remove the probe. Dispose of the liquid sample in the waste container provided. Clean the thermocouple probe and the sample tube with toluene and dry *thoroughly*. Use a heat gun, if available.

13. Starting with a 2.0-g sample of biphenyl, obtain the cooling curve.

14. Subsequent additions of naphthalene will be made to this sample, and the cooling curve of each (and also, if time allows, the heating curve) will be acquired. These additions are 0.20 g, 0.40 g, and another 0.40 g.

At the end of the experiment, you will have acquired a total of nine cooling curves, and possibly several heating curves.

Calculations and Data Analysis

1. Calculate the mole fractions of N and B for each of the samples prepared. Tabulate.

2. Examine the cooling curves and determine the transition temperatures for pure N and B ($T^\bullet$) as well as the depressed melting points of the binary systems. The objective is to find (or estimate) the positions of the halts in the cooling curves (see cooling curves I and II in Figure 7.13.3). Tabulate.

3. From the cooling curves of the binary mixtures, estimate the position of the eutectic temperature, T'. The eutectic point will appear as an inflection in the cooling curve (see curve II, Figure 7.13.3). This feature is rather subtle for compositions low in N (or B) but appears more distinct as the eutectic composition is approached. Tabulate the values of T' obtained for each sample, indicating the reliability (error limits) of each determination. Decide which value best represents the true eutectic temperature. Compare $T^\bullet$ and T' values with those in the literature for N and B. (Consult the International Critical Tables or some other suitable reference.) With respect to a given systematic procedure for determining the eutectic temperatures of the various N-B compositions, does the value of T' appear to be constant? Comment on this in your report.

4. Construct the phase diagram by plotting the depressed melting points vs. x_N (or x_B). Show the measured eutectic position. From the intersection of the solubility curves, determine the eutectic composition. Label the regions of the phase diagram.

5. Plot the mole fraction of N vs. the depressed freezing point. Do likewise for mole fraction B vs. freezing point. Include only the first two compositions in addition to the pure compounds. From each of the plots, estimate the limiting tangent as x_N (or x_B) $\rightarrow$ 1, and from these values determine estimates of the heats of fusion of N and B. See equation (7.13.13).

6. Using the literature values for the melting points and heats of fusion of N and B, determine the eutectic temperature and composition assuming the ideal solubility law, equations (7.13.11) and (7.13.12). This can be done either by plotting the ideal solubility curves, T vs. x_N (or n_B), on the same paper and graphically determining the intersection point, or by solving (7.13.11) and (7.13.12) simultaneously using an iterative process. Compare x' and T' obtained in this way with your experimental value as well as that from the literature.

Option B: Computer-Assisted Data Acquisition and Analysis

The TAC output is coupled to a laboratory microcomputer. First the computer will be "brought up" and the program loaded into the computer memory. You will be told how to do this by the instructor. The software (computer instructions) is menu-driven; this means that statements will appear on the CRT (computer screen) indicating the procedures to be followed or indicating the options for subsequent steps. In the latter case, you will have to decide how the experiment is to be carried out. In order to use the computer program properly, you have to thoroughly understand methodology of the experiment.

First, the TAC-computer interface is calibrated by immersing the TAC probe into an ice bath and then into boiling water. The ambient atmospheric pressure is entered to allow the actual water boiling point to be used as the upper calibration point.

After the proper commands are initiated, the cooling (or heating) curve is acquired and set up in a data file. The T vs. time data are displayed on the CRT *as* the cooling curve is acquired. In order to locate the transition temperatures more accurately, a derivative analysis of the cooling curve is performed. The rationale is that the slope (first derivative, FD) of the cooling curve becomes very small (and approaches zero) at the halt. Likewise, the FD will show a *minimum* at the inflection point that is often the manifestation of the eutectic transition in the cooling curve.

After the cooling (or heating) curve is acquired, a command can be initiated to compute and display the FD plot. In order to enhance the sensitivity of the analysis, the range of the vertical scale, for example $\pm 0.5°C$, can be selected. When an appropriate derivative sensitivity is identified by visual inspection, the FD plot can be obtained in "hard copy," and as well, the listing of these data can be obtained. These computer output documents are used for data analysis. The depressed melting points and eutectic positions are determined from an examination of the minima in the FD plots.

The construction of the *T-X* phase diagram follows the same sample preparation and apparatus setup as described above. At least one complete cooling curve must be obtained for each sample. It is desirable to determine a heating curve for as many samples as possible. Data analysis is likewise carried out as presented above. The computer output should be included as an appendix in your report.

Questions and Further Thoughts

1. Suppose that ΔH_{fus} were not treated as independent of T but were given by the relationship

$$\Delta H_{fus} = \Delta H^\bullet_{fus} + \Delta C_P(T - T^\bullet)$$

   where $\Delta H^\bullet_{fus}$ is the heat of fusion at the temperature $T^\bullet$ and ΔC_P is the difference in heat capacities of the liquid and solid. Starting with equation (7.13.8), derive the integrated expression similar to (7.13.10) which would apply in this case.

2. In deriving the ideal solubility law and in considering the use of the cooling curves to obtain data pertinent to the phase diagram, the assumption of *equilibrium* between the liquid and solid phases is pervasive and crucial. What procedural steps in this experiment are (or should be) taken to ensure this condition?

3. What difficulties would arise if one tried to stir or agitate the molten sample during cooling?

4. How could one obtain activity coefficients in a thermal analysis experiment?

Further Readings

R. A. Alberty, "Physical Chemistry," 7th ed., pp. 229–231, Wiley (New York), 1987.

P. W. Atkins, "Physical Chemistry," 3rd ed., pp. 195–196, 200–201, W. H. Freeman (New York), 1986.

G. W. Castellan, "Physical Chemistry," 3rd ed., pp. 324–328, Addison-Wesley (Reading, Mass.), 1983.

I. N. Levine, "Physical Chemistry," 2nd ed., pp. 322–324, McGraw-Hill (New York), 1983.

E. M. Washburn and J. W. Read, *Proc. Natl. Acad. Sci., 1*, 191 (1915).

Experiment 13

Solid-Liquid Equilibrium

NAME _____ DATE _____

Notes

Solid-Liquid Equilibrium

NAME _____   DATE _____

Notes

☐ Experiment 14

Liquid-Vapor Equilibrium in a Binary System

Objective

To obtain the liquid-vapor temperature-composition (T-X) phase diagram for a binary system which shows nonideal behavior (e.g., 1-propanol/water, ethanol/cyclohexane, or acetone/chloroform).

Introduction

We consider in this experiment the phase equilibrium between the liquid and vapor phases of a two-component system. To retain generality, these components are denoted as A and B. A and B are understood to be liquids at ambient temperature and pressure. We start by writing the fundamental equation of phase equilibrium: the equality of the chemical potential of a given component in the two phases. For component A in the liquid mixture of A and B, the chemical potential is

$$\mu_A = \mu_A^{\bullet}(l)(T,P) + RT \ln a_A = \mu_A^{\bullet}(l)(T,P) + RT \ln \gamma_A x_A, \quad (7.14.1)$$

where $\mu_A^{\bullet}(l)$ is the chemical potential of *pure* liquid A at the temperature and (total) pressure of the system, a_A is the activity of A in solution, and γ_A and x_A are its activity coefficient and mole fraction. A similar expression can be written for B:

$$\mu_B = \mu_B^{\bullet}(l)(T,P) + RT \ln a_B = \mu_B^{\bullet}(l)(T,P) + RT \ln \gamma_B x_B. \quad (7.14.2)$$

A mixture of liquid A and B for which $\gamma_A = \gamma_B = 1$ is called an *ideal solution*. The physical significance of an ideal solution is that the intermolecular interactions between the different component molecules A and B are considered to be identical to those between the pure constituents A or B. In other words A-B association is neither larger nor smaller than the A-A or B-B association.

Consider now the vapor phase consisting of A and B in equilibrium with the liquid A-B mixture. The chemical potential of A is expressed as

$$\mu_A(g) = \mu_A^{\circ}(g)(T) + RT \ln P_A, \quad (7.14.3)$$

in which $\mu_A^{\circ}(g)$ is the chemical potential of pure gaseous A at a temperature T and at the ideal gas, standard state, 1 atm pressure, and P_A is the partial pressure (in atmospheres) of A in the vapor above the A-B liquid mixture. Equation (7.13.3) assumes that the vapor of A behaves ideally, i.e., the fugacity is equal to the partial pressure. A similar equation describes the chemical potential of B vapor.

Phase equilibrium demands that $\mu_A(l) = \mu_A(g)$. Setting equations (7.14.1) and (7.14.3) equal to each other gives

$$\mu_A^{\bullet}(l)(T,P) + RT \ln \gamma_A x_A = \mu_A^{\circ}(g)(T) + RT \ln P_A. \quad (7.14.4)$$

If we consider the limit of $x_A \rightarrow 1$, i.e., a system of pure liquid A in equilibrium with its vapor (and hence $a_A = 1$), we have from equation (7.14.4)

$$\mu_A^{\bullet}(l)(T,P_A) = \mu_A^{\circ}(g) + RT \ln P_A^{\bullet}. \quad (7.14.5)$$

$P_A^\bullet$ is the vapor pressure of pure liquid A at the temperature T. Although the pressure implicit in the first term in equation (7.14.4) pertains to the *total* pressure above the A-B mixture and the pressure implied in the first term of equation (7.14.5) is the *vapor* pressure of A, the two pure liquid terms in these two equations can be considered equal (because the Gibbs free energy of a liquid varies very slowly with the applied pressure). Therefore, if equation (7.14.5) is subtracted from equation (7.14.4),

$$RT \ln \gamma_A x_A = RT \ln P_A - RT \ln P_A^\bullet = RT \ln \left(\frac{P_A}{P_A^\bullet} \right) \qquad (7.14.6)$$

From equation (7.14.6) it follows that $\gamma_A x_A = P_A/P_A^\bullet$, or

$$P_A = x_A P_A^\bullet, \qquad (7.14.7)$$

if $\gamma_A = 1$ (ideal solution). This result, known as Raoult's law, expresses the partial pressure of the vapor (P_A) in terms of the liquid composition (x_A). In an ideal solution, Raoult's law holds over the entire range of compositions, $0 < x_A < 1$. Because in an ideal solution, A and B molecules are indistinguishable in terms of their mutual intermolecular forces, equation (7.14.7) can be extended to the other component, B (or other components in a multicomponent ideal solution). Hence,

$$P_B = x_B P_B^\bullet. \qquad (7.14.8)$$

Since $x_A + x_B = 1$ in a binary system, it follows that both P_A and P_B can be expressed as a linear function of a single composition variable, x_A (or x_B). Likewise, the total pressure (P_T), which from Dalton's law is

$$P_T = P_A + P_B = (P_A^\bullet - P_B^\bullet)x_A + P_B^\bullet, \qquad (7.14.9)$$

is also a linear function of x_A.

Based on the mechanistic definition of an ideal solution presented above, a *nonideal* solution is one in which the molecular association between A molecules is disrupted in some way by the addition of B molecules (or vice versa). This means that A-B interactions are either *stronger,* or *weaker,* than A-A and B-B interactions. In the former case [in which mixing is exothermic (why?)], the vapor pressure of A (or B) is suppressed in an A-B mixture relative to that predicted from Raoult's law, equation (7.14.7); this is an example of a *negative* deviation from Raoult's law. A and B molecules can be said to form an association complex of some kind.

In the latter case [in which mixing is endothermic (why?)], the component vapor pressures are both larger than those predicted from Raoult's law, and the phenomenon is denoted as a *positive* deviation from Raoult's law.

Although we will not be using formal equations to represent vapor pressures in a nonideal solution in this experiment, it is instructive to consider an example of how the partial pressures of the components might be expressed. One simple approach introduced by Margules in 1895 is to express P_A and P_B by the equations

$$P_A = P_A^\bullet x_A \exp [W(x_B)^2], \qquad (7.14.10a)$$

and

$$P_B = P_B^\bullet x_A \exp [W(x_A)^2], \qquad (7.14.10b)$$

where the parameter W is the same for both components. Note that for an ideal solution, $W = 0$. For a system showing positive deviations from Raoult's law,

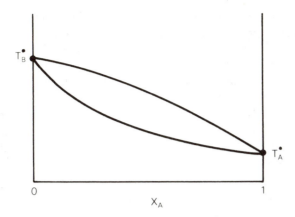

FIGURE 7.14.1 *T-X* phase diagram for a binary, ideal solution of two liquids. $T_A^\bullet$ and $T_B^\bullet$ are the respective boiling points of *A* and *B*. The system is at constant pressure.

$W > 0$. Equations (7.14.10a, 7.14.10b) are found to be satisfactory descriptions of partial pressures of liquid mixtures exhibiting moderate deviations from ideality.

The construction of a phase diagram which represents liquid-vapor equilibrium can be approached in several ways. One method is to measure the *vapor pressure* of the component(s) of the system as a function of liquid composition at constant temperature. A comparison could then be made with Raoult's law for an ideal solution, or with equations (7.14.10a, 7.14.10b) for a nonideal system.

From an experimental point of view, however, it is more convenient to measure the *boiling point* of a system as a function of composition at constant (applied) pressure. Boiling point–liquid composition data are then assembled to produce a *T-X* phase diagram, the objective of this experiment.

Even for an ideal solution for which Raoult's law is valid, the boiling point is not a linear function of *liquid* composition. Figure 7.14.1 shows a schematic diagram of a *T-X* phase diagram. The lower curve represents the temperature at which the *first* trace of vapor appears as a function of the bulk composition (defined as the composition of component *A* in the system as a whole, liquid *and* vapor phases). The upper curve indicates the temperature at which the *last* trace of liquid disappears (vaporizes) as a function of bulk composition. The lower region, labeled *L*, denotes a one-phase region (liquid), while the upper region, labeled *V*, represents the vapor one-phase region. The middle area between the two curves corresponds to the two-phase zone in which the system exhibits liquid-vapor equilibrium.

If we now consider a nonideal system having negative deviations from Raoult's law, the system will show a *maximum* in the boiling point–composition phase diagram. Moreover this maximum will be above the boiling points of either of the pure component boiling points. Such a phase diagram is shown in Figure 7.14.2.

This maximum appears at a unique *A-B* composition called the *azeotrope composition*. Azeotrope is derived from the Greek: to boil unchanged. The physical significance of the azeotrope is that a distillation of an *A-B* mixture at the azeotropic composition occurs at a sharp boiling point with the *A-B* vapor phase at the same composition as the liquid phase. There can be no separation into the *A-B* components. There is an analogy between an azeotrope and a eutectic (in a solid-liquid phase diagram) in that the phase change brought about on either system occurs sharply and thus it behaves as a pure substance. A eutectic, however, exists

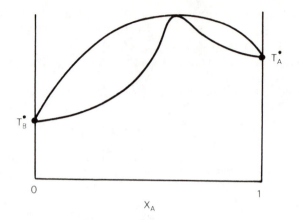

FIGURE 7.14.2   *T-X* phase diagram for a nonideal system showing negative deviation from Raoult's law, and thus a boiling point maximum.

even for ideal solid mixtures, whereas an azeotrope is a manifestation of a highly nonideal solution. Finally, it should be emphasized that the appearance of the phase diagram shown in Figure 7.14.2 depends on the applied pressure; hence, the azeotropic composition changes with the pressure of the system.

Experimental Method

Since the objective of this experiment is to construct the *T-X* phase diagram of a liquid-vapor system, the experimental approach requires removing a sample of vapor (condensed to a liquid) which is *in equilibrium* with the liquid. A sample of the liquid is also removed and the compositions of the two samples are determined. This is performed for *A-B* mixtures of various bulk compositions under constant (usually atmospheric) pressure.

The equipment to be used is shown in Figure 7.14.3. It is a distillation apparatus (called an Othmer still) that allows distillate to accumulate in a chamber (*D*) and to return to the pot (*P*), where it is rapidly mixed with the residue. A three-way T-bore stopcock (*S*) allows distillate to recirculate, or permits the collection

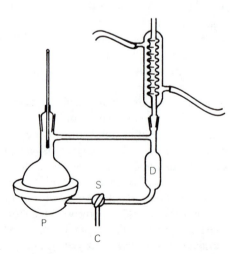

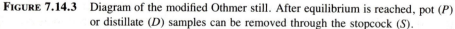

FIGURE 7.14.3   Diagram of the modified Othmer still. After equilibrium is reached, pot (*P*) or distillate (*D*) samples can be removed through the stopcock (*S*).

of either *equilibrated* distillate or residue through a collection port (*C*). The apparatus is constructed in such a way that when the stopcock is appropriately set, liquid can flow *only* from the distillate chamber back into the pot. This "recycling mode" allows pot and distillate samples to represent the respective *equilibrium* compositions once a constant vapor temperature is established.

The liquid is heated by means of a specially designed electric heating mantle. A water-cooled condenser that is open to the atmosphere prevents the loss of material from the system. A sensitive thermometer indicates the equilibrium liquid-vapor temperature.

There are three particularly important requirements in carrying out this experiment: (1) making sure that the distillate and residue samples are obtained under equilibrium conditions, (2) reading the equilibrium temperature accurately, and (3) analyzing the *A-B* compositions of these samples accurately. Equilibration of the distillate and residue is achieved by allowing the condensed liquid that accumulates in the distillation chamber to circulate back into the pot for a sufficient length of time until a stable vapor condensation temperature is reached. The temperature is monitored for at least 2 to 3 min after it appears to stabilize to make sure it indicates equilibrium conditions. The analytical method for determining the sample compositions depends on the nature of the components studied.

In this experiment different systems will be studied: 1-propanol/water, ethanol/cyclohexane, acetone/chloroform, or another binary mixture. Each laboratory group will be assigned one of these systems, or possibly another mixture. Different analytical techniques will be available, as appropriate: refractometry, gas chromatography (gc), or nuclear magnetic resonance (nmr) spectroscopy. The method used for the 1-propanol/water system is *refractometry*. This approach works well because the refractive indices of water and 1-propanol are sufficiently different from each other, and, moreover, water is more difficult to analyze using the other methods. The refractive indices of the samples are converted into the respective compositions through the use of a calibration curve. Ethanol/cyclohexane is amenable to gc analysis, while for the acetone/chloroform system, either gc or nmr can be employed. A general description of these techniques is presented in the appendix.

Safety Precautions

☐ Safety goggles must be worn in the laboratory at all times.
☐ Before adding liquid samples to the Othmer still, be sure that the refluxing has stopped. Stand back; organic vapors may be released when the thermometer fitting is removed from the pot.
☐ Chloroform must be stored and handled in a fume hood. As an additional precaution, gloves should be worn.
☐ Be sure you have been shown proper injection techniques in gc analysis. Exercise caution in handling the syringe.

Procedure

1. *Apparatus Setup*. If the apparatus is not assembled, first rinse the inside with a few milliliters of the alcohol (or acetone). Assemble the apparatus using pinch clamps (be sure not to overtighten); also see that the thermometer and condenser are vertical. Check to see that the water tubing is secure and properly arranged. Do not use stopcock grease *anywhere* on the apparatus. Connect the heating mantle line cord to a Variac (variable voltage transformer), which should, at first, be switched in the *off* position. The dial should be turned to zero.

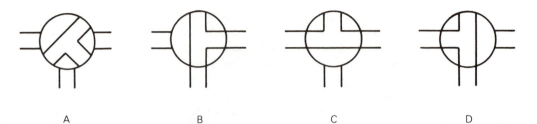

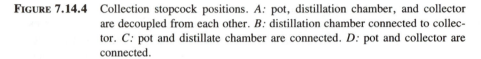

FIGURE 7.14.4 Collection stopcock positions. *A:* pot, distillation chamber, and collector are decoupled from each other. *B:* distillation chamber connected to collector. *C:* pot and distillate chamber are connected. *D:* pot and collector are connected.

The following procedure for the preparation of samples is given for the 1-propanol/water system. The same procedure should be followed for the ethanol/cyclohexane and acetone/chloroform systems.

Familiarize yourself with the three-way stopcock. Note that there are four useful positions: *A* decouples the pot, distillate reservoir, and sample port from each other; *B* connects the distillate reservoir with the sample port; *C* connects the distillate reservoir to the pot; and *D* connects the pot to the sample port. These positions are illustrated in Figure 7.14.4.

2. *Alcohol-Rich Mixtures.* Using the stopcock, isolate the pot, distillate reservoir, and sample port from each other (*A*) and remove the thermometer/adapter. Add about 100 mL of pure alcohol. Since volume accuracy is not required at this point, a graduated cylinder can be used. Subsequent additions of liquid are made directly through the thermometer port on the pot.

3. Make sure the cooling water is flowing through the condenser. Rotate the Variac dial to the 0 position and then switch it on. Increase the voltage to 60 to 100 V ac and wait until the liquid begins to boil. Adjust the voltage as necessary. Condensation on the thermometer and the upper regions of the pot should be brisk; the condensate should begin to form and collect in the distillate reservoir. Set the stopcock position to allow the distillate to flow back into the pot (*C*). After the temperature remains constant for a few minutes, record its value. Read and record the ambient atmospheric pressure.

4. Isolate the pot, distillation reservoir, and sample port. Turn the Variac down until the boiling ceases and add about 3 mL of deionized water (or cyclohexane) to the pot through the thermometer port. Heat the mixture to an active boil until enough distillate has collected in the reservoir so that its level is slightly above that in the pot; then position the stopcock so that distillate can flow into the pot. Continue the distillation until the temperature becomes constant. It may take several minutes for liquid-vapor equilibrium to be reached. Record this temperature.

5. *Collect Samples.* Lower the temperature until boiling abates. Connect the distillate reservoir to the sample port (*B*) and allow about 1 mL to drain into the collection vial. Discard. Add a fresh sample to the vial. Cap the vial and label it: DISTILLATE #1. Next, obtain a residue sample by connecting the pot to the sample port (*D*). Repeat the above procedure, and label the sample: RESIDUE #1. Withdraw the remaining liquid from the distillate reservoir and return it to the pot through the thermometer port.

6. Additional Systems. Repeat the distillation-equilibration and collection procedure after adding, successively, 5-, 8-, 10-, 15-, and 50-mL increments of water (or cyclohexane) to the pot. Collect and label the distillate and residue samples in vials.

7. *Water- (or Cyclohexane)-Rich Mixtures*. After the last set of samples has been collected, turn off the heater and allow the system to cool sufficiently. Disassemble and discard the liquid in the pot. Rinse with deionized water and reassemble. Add about 100 mL of water (or cyclohexane) and determine the boiling point as in the case of pure alcohol. Then add, in succession, 3-, 5-, 12-, and 20-mL increments of alcohol and, following the above procedure, determine the system boiling points, collecting in each case the distillate and residue samples in labeled vials.

8. *Sample Analysis*. 1-propanol/water. The compositions of the distillate and residue samples of the 1-propanol/water system will be determined by measuring the refractive indices of these liquids. If necessary, the calibration and sample data can be obtained by one member of a group while the other performs the distillation and sample acquisition. These activities should be switched about halfway through the experiment.

While the refractive indices are being measured, it is very important that the temperature of the sample plate of the refractometer be kept constant (to within 1°) because the refractive index of a liquid is a function of its density and hence its temperature.

The procedure for using the Abbe refractometer will be outlined and demonstrated by the instructor. Carefully read and record the refractive indices of all the distillate and residue samples. Do not discard the vials.

9. *Calibration*. In order to convert the refractive index to composition, a calibration curve of refractive index as a function of 1-propanol/water composition must be obtained. For this purpose, the following solutions will be prepared (or supplied): 0, 10, 20, 40, 60, 75, 85, 95, and 100 percent water (by *volume*). Place samples of these mixtures in small, individually labeled vials and then measure and record the refractive indices. Record the atmospheric pressure after the distillation experiment is completed.

8a. *Sample Analysis*. ethanol/cyclohexane. Your instructor will explain and demonstrate the use of the gas chromatograph. The apparatus must be completely equilibrated to the appropriate gas flow and temperature conditions before samples are to be analyzed. The objective is to determine the relative amount of each component from the area under the respective response curve. (In some cases, this information can be obtained satisfactorily from the maximum intensities of the curves.) Your instructor will advise you about the best method.

Small amounts (typically a few microliters) of the liquid samples are injected into the chromatograph. The loading of the syringe and the proper injection technique will be explained by your instructor. Be sure not to force the needle into the inlet septum.

9a. *Calibration*. Because the chromatographic response depends on the nature of the particular compound (the eluent), chromatograms of samples having known ethanol/cyclohexane composition must be obtained. Prepare (at least four) samples by pipeting the following volumes (mL) of ethanol[cyclohexane] into labeled sample vials: 1[5], 1[2], 2[1], and 5[1]. Obtain chromatograms of these mixtures. If there is insufficient time to acquire these calibration data, you will be provided with a calibration curve. Be sure that the chromatograms of the samples are obtained under identical conditions relative to the calibration data.

Data Analysis

1. Convert the volume compositions of the 1-propanol/water (or ethanol/cyclohexane) calibration samples to mole fractions using the densities of the pure liquids at the appropriate temperature.

2. Tabulate and plot refractive index vs. mole fraction 1-propanol. Draw a smooth curve through these points. A polynomial-fitting algorithm may be used to generate this curve.

2a. Tabulate and plot the ratio of the integrated response curves (or peak heights) vs. mole fraction. Draw a smooth curve through the points. The curve should pass though the origin. A polynomial-fitting routine may be used to generate the calibration curve.

3. Tabulate the refractive indices of the distillate and residue samples. From the calibration curve or its equation, convert these data to mole fraction 1-propanol.

3a. Tabulate the ratios of the integrated response curves for the distillate and residue samples. From the calibration curve, convert these data to mole fraction compositions.

4. Construct the *T-X* phase diagram by plotting the equilibrium boiling points of the systems studied vs. the composition of distillate (vapor curve) *and* residue (liquid curve). Label the phase diagram.

5. Plot the 1-propanol (or ethanol) composition in the vapor (distillate) vs. the liquid composition (residue).

6. From your results, determine the temperature and composition of the azeotrope. Report this along with the pressure of the system. Compare with the literature value.

Acetone/Chloroform System

The same procedure for liquid mixture preparation, equilibration, and sample withdrawal described above is to be followed for the acetone/chloroform system. The fundamental difference lies in the method of data analysis (nmr).

The following volume combinations are recommended. *Acetone-rich mixtures:* Start with 70 mL of acetone; other compositions are made by four successive additions of 14 mL of chloroform. *Chloroform-rich mixtures:* Start with 70 mL of chloroform. The other compositions are prepared by three successive additions of 3.5 mL of acetone. Record the temperature so that mole fractions can be determined from the pure liquid densities.

Because the nmr analysis of an acetone/chloroform sample provides the ratio of acetone protons (of which there are six per molecule) to chloroform protons (only one per molecule), mole fractions can be obtained directly. A calibration curve is therefore unnecessary. The instructor will show you how to obtain an nmr spectrum of a sample and how to perform an integration of the acetone and chloroform signals. Integrated nmr spectra should be obtained for each of the distillate and residue samples. For those samples which are low in chloroform, the integration must be performed on amplified signals. Convert the integration data for each spectrum into mole fraction acetone.

Following the data analysis above, construct the *T-X* phase diagram for the acetone/chloroform system. Also plot acetone composition in the vapor vs. the composition in the liquid. Determine the azeotrope composition and temperature at the ambient pressure and compare with the literature value.

Questions and Further Thoughts

1. Explain why it is important to have the distillate recycle back into the pot in order to ensure that the distillate and residue samples represent equilibrium compositions.
2. It is important that in achieving liquid-vapor equilibration, distillate actively recycles back into the pot; that is, residue liquid should not be allowed to mix with distillate. How does the control of power supplied to the heating mantle affect this condition?

3. Refractometry is a rapid and simple way of determining compositions. It is a technique often used in analyzing relatively concentrated aqueous sugar solutions. The refractive index of a substance is the ratio of the speed of light in vacuo to that in the substance. Why does light travel more slowly in a liquid (for water $n = 1.333$)?

4. Refractive indices of aromatic molecules are higher than those of saturated hydrocarbons; for example, for benzene and cyclohexane, $n = 1.50112$ and 1.4290 at $20°$, respectively. [This is the reason that, when viewed in a clear beaker, benzene (or toluene) appears to partially disperse white light into a ''rim'' of colors.] Can you suggest a reason why $n_{C_6H_6} > n_{C_6H_{12}}$?

Further Readings

Othmer Still

D. F. Othmer, *Anal. Chem., 20,* 763, 1948.

Phase Equilibrium

R. A. Alberty, ''Physical Chemistry,'' 7th ed., pp. 214–225, Wiley (New York), 1987.

P. W. Atkins, ''Physical Chemistry,'' 3rd ed., pp. 166–187, W. H. Freeman (New York), 1986.

G. W. Castellan, ''Physical Chemistry,'' 3rd ed., pp. 296–309, Addison-Wesley (Reading, Mass.), 1983.

I. N. Levine, ''Physical Chemistry,'' 2nd ed., pp. 313–320, McGraw-Hill (New York), 1983.

J. H. Noggle, ''Physical Chemistry,'' pp. 347–351, Little, Brown (Boston), 1985.

Nuclear Magnetic Resonance

R. A. Alberty, op. cit., pp. 563–571.

P. W. Atkins, op. cit., pp. 485–493.

G. W. Castellan, op. cit., pp. 603–609.

I. N. Levine, op. cit., pp. 714–721.

Appendix

Gas Chromatography (gc)

A block diagram of a gas chromatograph is shown in Figure 7.14.5. An inert carrier gas (usually He), sometimes called the mobile phase, constantly sweeps through the apparatus, passing through the sample injection chamber, column, and detector. The sample is introduced into the flowing gas after it is injected in the heated chamber. The temperature of this chamber is regulated by an oven and is kept above the normal boiling points of the sample(s); the liquid is thus flash evaporated. It is important that the evaporation takes place rapidly so that the sample vapor is kept in a small zone as it is swept along in the carrier gas.

The column is made of copper, stainless steel, or sometimes glass tubing of small diameter. The column, which is coiled or folded, is packed with a powdered solid that has been mixed with a small amount of an inert, high-boiling substance such as polyglycols, silicone oils, or phthalate esters. These materials are called the stationary phase. The ability of a gas chromatograph to separate a mixture into its components arises from the different interaction energies of the components with the stationary phase. This can be considered in terms of the different heats of absorption (or solution) of the sample components and the high molecular weight stationary phase. Because of the differences in adsorption (or solution)

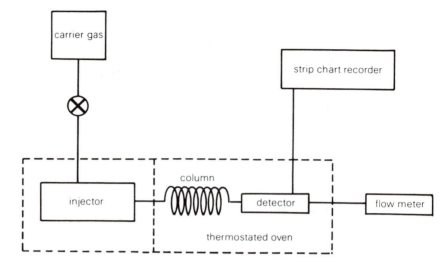

FIGURE 7.14.5   Schematic diagram of a gas chromatograph with thermal conductivity detection.

enthalpies, the *rates* of release of the components back into the carrier gas will vary. Thus a material that has a larger interaction enthalpy (because it may be more polar) will be swept through the column *after* one that has a smaller adsorption (or solution) enthalpy. The difference in the retention times of the components in the column depends on the nature of the stationary phase, the temperature, flow rate, and other factors.

After the components are removed from the column, they are passed into the detector, which produces an electrical response proportional to the amount of material in the carrier gas. This response is displayed on a strip chart recorder directly. The signal can also be electronically integrated for proper quantitative analysis.

Two methods of detection are commonly used in gas chromatography: thermal conductivity and flame ionization. The first technique takes advantage of the fact that the thermal conductivity of an organic vapor is smaller than that of He, the carrier gas. This is basically due to the larger mean free path of He. The gas flowing through the detector passes between a heat source (a heated filament) and a temperature sensor (either a thermocouple or a thermistor). The presence of even a minute amount of eluent material in the carrier gas is determined by monitoring the temperature of the sensor as a function of time. The resultant signal is amplified and displayed on a strip chart recorder.

Flame ionization detection is based on the fact that when a material is burned at a sufficiently high temperature ions are produced. In this application, the gas stream exiting the column is combusted in a H_2 flame (air is used as the oxidant). The presence of ions is detected by electrodes biased by a stable dc voltage. The current thus produced is amplified by an electrometer, and the signal is displayed on a recorder. Flame ionization detection is more sensitive than the thermal conductivity method.

Nuclear Magnetic Resonance (nmr) Spectroscopy

This is a very simple and general description of nmr spectroscopy designed to provide some background information for the experiment. Consult the references for more detailed discussions.

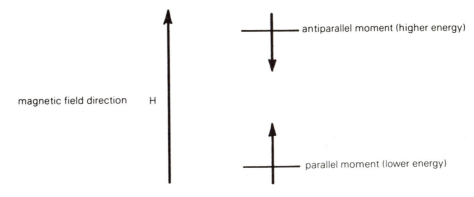

FIGURE 7.14.6   Nuclear spin state energies with orientations parallel and antiparallel to an external magnetic field.

Atomic nuclei possess magnetic moments. This means that when placed in an external magnetic field, the nuclei behave as microscopic magnets. The orientation of the nuclear magnetic moment (a vector) can adopt different discrete directions with respect to the lines of force of applied magnetic field. These orientations correspond to different potential energies. Hydrogen nuclei, or protons (^1H), display only two such orientations: parallel to the external magnetic field (lower energy) or antiparallel to the field (higher energy). See Figure 7.14.6.

The energy difference between these nuclear orientations depends linearly on the strength of the applied magnetic field. For a magnetic field strength of 1.00 tesla (a moderately strong magnet), this energy difference is 2.82×10^{-19} ergs (for ^1H). The absorption of this amount of energy in the form of electromagnetic radiation (which is in the microwave region of the spectrum, ca. 43 MHz) can cause the protons to undergo transitions from the lower to the upper energy levels represented above. Actually, the presence of a radiation field of the appropriate frequency causes both upward *and* downward transitions between the two energy levels. This phenomenon is called *resonance*.

The exact energy at which resonance can be brought about for a given hydrogen nucleus depends in a subtle, but important, way on the local environment of the proton. This, in turn, depends on the nature of the molecular structure to which the hydrogen atom is chemically bonded. Thus the different groups of protons in a molecule will have very small, but distinguishable, resonant frequencies (or energies). For example, in ethyl alcohol, CH_3CH_2OH, there are three different sets of protons, each having a different environment: the three methyl protons, the two methylene protons, and the hydroxy proton. The nmr spectrum of ethanol thus consists of three general features, having integrated areas in the ratio of 3:2:1.

In this experiment, acetone (CH_3COCH_3) and chloroform ($CHCl_3$) each have only one group of protons; hence, each molecule has only one resonance. Because the environments of these protons, however, are different, the acetone and chloroform resonances appear at different (and characteristic) frequencies. It should be noted (and appreciated) that this frequency difference is very small relative to the absolute resonance frequency. For example, this difference for acetone and chloroform is about 5 parts per million of the absolute resonance frequency, i.e., ca. 220 Hz, or 1.4×10^{-24} ergs. The relative integrated areas under these resonances have a ratio of 6:1 (molecule per molecule).

Experiment 14

Liquid-Vapor Equilibrium

NAME _____   DATE _____

Barometric pressure   Before: _____   After: _____

Alcohol-rich samples:

Volume X added	Boiling point	Refractive index distillate	Refractive index residue
_____	_____	_____	_____
_____	_____	_____	_____
_____	_____	_____	_____
_____	_____	_____	_____
_____	_____	_____	_____
_____	_____	_____	_____
_____	_____	_____	_____
_____	_____	_____	_____
_____	_____	_____	_____
_____	_____	_____	_____

X-rich samples:

Volume X added	Boiling point	Refractive index distillate	Refractive index residue
_____	_____	_____	_____
_____	_____	_____	_____
_____	_____	_____	_____
_____	_____	_____	_____
_____	_____	_____	_____
_____	_____	_____	_____
_____	_____	_____	_____
_____	_____	_____	_____
_____	_____	_____	_____
_____	_____	_____	_____

Experiment 14

Liquid-Vapor Equilibrium

NAME _____   DATE _____

Barometric pressure   Before: _____   After: _____

Alcohol-rich samples:

Volume X added	Boiling point	Refractive index distillate	Refractive index residue
_____	_____	_____	_____
_____	_____	_____	_____
_____	_____	_____	_____
_____	_____	_____	_____
_____	_____	_____	_____
_____	_____	_____	_____
_____	_____	_____	_____
_____	_____	_____	_____
_____	_____	_____	_____
_____	_____	_____	_____
_____	_____	_____	_____

X-rich samples:

Volume X added	Boiling point	Refractive index distillate	Refractive index residue
_____	_____	_____	_____
_____	_____	_____	_____
_____	_____	_____	_____
_____	_____	_____	_____
_____	_____	_____	_____
_____	_____	_____	_____
_____	_____	_____	_____
_____	_____	_____	_____
_____	_____	_____	_____
_____	_____	_____	_____

Composition/Refractive Index Data

NAME _____   DATE _____

Refractometer temperature: _____   _____   _____

Water, Composition, (v/v)   Refractive Index

_____   _____

_____   _____

_____   _____

_____   _____

_____   _____

_____   _____

_____   _____

_____   _____

Composition/Refractive Index Data

NAME _____ DATE _____

Refractometer temperature: _____ _____ _____

Water, Composition, (v/v)   Refractive Index

_____   _____

_____   _____

_____   _____

_____   _____

_____   _____

_____   _____

_____   _____

_____   _____

□ PART EIGHT

Transport Properties and Chemical Kinetics

□ Experiment 15

Viscosity of Liquids

PART I: LOW VISCOSITIES

Objective

To measure and analyze the viscosities of ideal (toluene/*p*-xylene) and nonideal (methanol/water) binary solutions and their components; to determine the activation energy to viscous flow.

Introduction

From a phenomenological point of view, the viscosity of a fluid is its resistance to flow. Viscosity measurements are often carried out for either of two main reasons. Viscosity is a quantitative property of a fluid, and although a particular sample might be highly complex, such as a blend of various resins or polymers, its viscosity serves to represent a physical property of that system. Viscosity, therefore, can be used as an empirical index in quality control applications concerning, for example, oils and resins, latex paints, or chocolate mousse. Another motivation for measuring viscosity is to determine a fundamental and intrinsic property of a liquid (as a solvent medium): the rate of mass transport, or diffusion, within the medium. In this application, for example, viscosity data can provide important information about chemical reaction kinetics. In this experiment, the fluid will be either a pure liquid or a mixture of liquids.

Considered macroscopically, viscosity is a frictional force that arises from the directed motion of molecules past each other in the liquid state. From a microscopic viewpoint, viscosity reflects the energetics of molecular association in the liquid state because in order for a liquid to flow, a force must be applied to overcome the attractive forces between the molecules. These forces are appreciable; they are manifest, for example, as latent heat of vaporization, surface tension, etc. The mathematical treatment of viscosity is best introduced by referring to Figure 8.15.1.

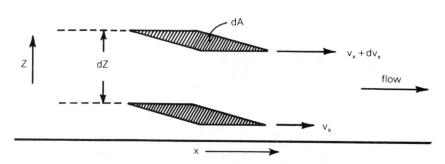

FIGURE 8.15.1 In viscous flow, a sheet of fluid having cross-sectional area, dA, is subjected to a force that causes it to move faster by an amount, dv, than an equivalent, adjacent sheet separated by a distance dz. The planes of the sheets are normal to the flow direction.

A liquid is presumed to be flowing smoothly in the x direction. Imagine that the liquid is composed of sheets of infinitesimal cross section, dA, which are oriented in the x-y plane, and that each sheet flows tangentially to its surface, in the positive x direction. If a given sheet is kept at a velocity, v_x, such that it exceeds the velocity of an adjacent sheet by an amount, dv_x, and the adjacent sheet is displaced by a distance, dz, the force required (per unit area) to maintain the motion of the given sheet, df_x, is given by

$$\frac{df_x}{dA} = \eta \left(\frac{\partial v_x}{\partial z}\right)_z. \tag{8.15.1}$$

The partial derivative, $(\partial v_x/\partial z)_z$, is the tangential velocity gradient, and η, the proportionality constant between f_x and this gradient, is defined as the *viscosity coefficient*. From equation (8.15.1), it can be seen that the cgs dimensions of the viscosity coefficient are g cm^{-1} s^{-1}. Equation (8.15.1) is called Newton's law of viscous flow, and fluids that behave according to (8.15.1) are called Newtonian fluids and are said to undergo *laminar* flow. Cases of nonlaminar, or non-Newtonian, flow (at ordinary temperatures and pressures) are unusual but not uncommon (e.g., "silly putty"). Materials whose viscosity decreases at high shear rates (e.g., paints that "thin out" as they are applied with a brush but then stiffen when quiescent) are examples of non-Newtonian fluids.

With respect to this experiment, a useful application of equation (8.15.1) to the case of mass transport through a circular tube of small internal diameter was derived by Poiseuille (1844):

$$\frac{dV}{dt} = \frac{\pi r^4 \Delta P}{8 \eta L}, \tag{8.15.2}$$

dV/dt is the volume flow rate of the liquid emerging from the tube; r and L are, respectively, the radius and length of the tube. ΔP is the pressure difference across the ends of the tube and is the driving force for the bulk flow. Equation (8.15.2) assumes that the flow rate is slow and uniform. The viscosity coefficient, η, is called the *poise* in recognition of Poiseuille. Thus one poise, P, is 1 g cm^{-1} s^{-1} (or dyne s). For many common liquids at room temperature, viscosities are about 0.002 to 0.04 P. For convenience, the centipoise (10^{-2} P), cP, is often used to report viscosity. In SI units, the viscosity coefficient is kg m^{-1} s^{-1} (or Pa s). 10^3 Pa s $= 1$ cP.

Mixtures

From the way viscosity is defined, it follows that a *mobile* liquid is one that has a relatively *low* viscosity. Another useful parameter that applies to fluid mobility is *fluidity*, F, which is simply the reciprocal of the viscosity coefficient:

$$F = \frac{1}{\eta}. \tag{8.15.3}$$

One particular advantage to the use of fluidity is that the fluidities of mixed solutions of nonassociating liquids are often found (empirically) to be additive. Thus for a binary solution of liquids A and B, each pure liquid having fluidities $F_A^\bullet$ and $F_B^\bullet$, respectively, the fluidity of a mixture containing mole fractions x_A and x_B may be approximated as

$$F \simeq x_A F_A^\bullet + x_B F_B^\bullet, \tag{8.15.4}$$

i.e., a mole fraction–weighted linear combination of the pure liquid fluidities. The viscosity of the mixture would then be

$$\eta = \frac{1}{x_A/\eta_A^\bullet + x_B/\eta_B^\bullet}, \tag{8.15.5}$$

and is obviously not linear in the composition variable, x_A (or x_B). Another approach for expressing the viscosity of a mixture is the following, proposed by Kendall (1913). For a binary solution,

$$\ln \eta = x_A \ln \eta_A^\bullet + x_B \ln \eta_B^\bullet. \tag{8.15.6}$$

In this context, an *ideal solution* can be defined as one in which the interaction energetics between the constituents is the same as between the pure components. More specifically, it is assumed that in such a mixture, the intermolecular interactions between identical molecules (e.g., *A-A* and *B-B*) are equal to those between different molecules (*A-B*). The failure of component fluidities to be additive in the mixed state would arise, then, either from the formation of association complexes between the components or from the destruction of such complexes that may be present in the pure component(s) after the pure components are mixed. Under these circumstances equations (8.15.5) and (8.15.6) would not be valid.

Temperature Dependence of Viscosity

It is found that over a reasonably wide temperature range, the viscosity of a pure liquid increases exponentially with the inverse absolute temperature. This was first expressed quantitatively by Arrhenius (1912):

$$\eta = A \exp\left(\frac{E_\eta}{RT}\right), \tag{8.15.7}$$

where A is a constant for a given liquid and E_η is sometimes called the activation energy to viscous flow of the liquid. Several theories have been proposed to rationalize equation (8.15.7). Simply viewed, however, an energy barrier must be surmounted in order for a molecule to "squeeze" by its neighbors if it is to undergo transport in the bulk medium. A plot of $\ln \eta$ *vs.* $1/T$ (called an Arrhenius plot) should, according to equation (8.15.7), be linear and have a slope equal to E_η/R.

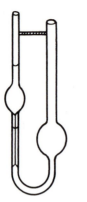

FIGURE 8.15.2   Ostwald viscometer.

Experimental Technique

The apparatus used in this experiment is called an Ostwald viscometer and is shown in Figure 8.15.2. Its design reflects the application of Poiseuille's law in that the viscosity of the liquid being measured flows through a uniform capillary tube. In principle, the viscosity of a fluid could be measured absolutely using equation (8.15.2) if its flow rate were determined and the physical dimensions of the viscometer were known. However, it is more practical to calibrate a given viscometer with a liquid of known viscosity. The wisdom of this empirical approach can be appreciated by noting that, from Poiseuille's law, the viscosity depends on r^4; thus the error in measuring the capillary radius enters fourfold in the measured viscosity. (See p. 10.)

Operationally, the experiment consists of measuring the time required for a given volume of liquid to flow through the viscometer capillary. The driving pressure that forces the liquid through the capillary is provided by gravity; hence, the difference in driving force between the measured and calibrating liquids is accounted for through the respective densities of these liquids. The Ostwald viscometer is designed to keep the separation of the upper and lower levels of the flowing liquid as constant as possible. This is accomplished by the spherical bulbs on the feed and receive ends of the apparatus. The volume of liquid that flows through the viscometer is determined by the positions of two lines that are inscribed on either side of the feed bulb. These lines are called fiducial (fixed reference) marks. The experiment, then, consists of measuring the time required for the liquid meniscus to pass between the upper and lower fiducial marks of the viscometer.

Poiseuille's law [equation (8.15.2)] can be integrated, and the viscosity can then be expressed as

$$\eta = \frac{\pi r^4 P t}{8VL},$$
(8.15.8)

where t is the elapsed time and V is the volume of liquid passing through the viscometer. The latter is constant for a given viscometer. Because the hydrostatic pressure, P, is proportional to the liquid density, D (the height is the same for both liquids), and the physical characteristics of the viscometer can be lumped into a constant, $k = \pi r^4/8VL$, the expression for viscosity becomes simply

$$\eta = kDt.$$
(8.15.9)

Thus, by measuring the flow time, t, for a liquid having a density, D, its viscosity can be determined relative to that of a reference liquid:

$$\eta = \frac{\eta_r D t}{D_r t_r},$$ (8.15.10)

where η_r, D_r, and t_r are the viscosity, density, and flow time of the reference liquid. It is important that the set of measurements considered above be carried out at a known and controlled temperature; hence, the η_r value pertains to this temperature.

Procedure

Ostwald Viscometer

1. Suspend the viscometer in one of the constant-temperature baths using a clamp around the rubber stopper attached to the viscometer. Make sure that the viscometer capillary is vertical and is below the surface of the water. The temperature of the bath should be adjusted to 25.0°C by adding, if necessary, small amounts of crushed ice. Pipet 5 mL of distilled water into the viscometer. (All subsequent measurements should be made using this volume of liquid and the same viscometer.) By placing the pipet bulb on the extended portion of the viscometer, and by gently squeezing the bulb, you should be able to push the liquid level up above the upper fiducial mark on the viscometer. Allow the water to run back down and start the electric timer exactly as the meniscus passes the upper fiducial mark. Stop the timer just as the meniscus passes the lower fiducial mark; record the elapsed time. Using the pipet bulb, bring the water back to the upper part of the viscometer and repeat the measurement. This should be done until two or three measurements agree to within about 0.2 s.

2. Remove the viscometer from the bath. Clean and dry the viscometer by running a few milliliters of clean acetone through it using a pipet. Drain the acetone from the viscometer and carefully attach a suction tube from the aspirator to the viscometer. Aspirate for about a minute until the acetone has completely evaporated.

3. Prepare and label solutions of water and methanol that are 20, 40, 60, 80 percent by volume methanol. Using the dried viscometer, determine the flow times of each of the methanol/water solutions at 25°C. Each measurement should be repeated until the flow times agree within about 0.2 s. After the measurement of one solution is completed, the viscometer should be rinsed with a few milliliters of the next solution to be studied. Complete by measuring the flow time for pure methanol.

4. Remove the viscometer from the bath. After draining the methanol, aspirate and reassemble the viscometer in the bath. Measure the flow times of a series of toluene/p-xylene solutions that are 0, 20, 40, 60, 80, and 100 percent by volume. Follow the procedure as in step 3.

5. Remove the viscometer from the bath, clean it with acetone, and aspirate. Suspend the viscometer in a water-filled 2-L beaker that is placed on a hot plate. Make sure the viscometer is fully immersed in the water. Add 5 mL of p-xylene and determine the flow time for a bath temperature of about 25°C. The exact temperature is not important as long as it is known to ±0.5°C. Measure the flow times at higher temperatures, roughly every 10° up to about 65°C. Make sure that the temperature is constant and that the viscometer has had time to equilibrate to a new temperature. A 1000-W immersion heater can be used to accelerate the

heating of the bath water; it must be disconnected well before the desired temperature is reached to avoid overshooting.

Data Analysis

1. Calculate and arrange in a table: (a) The viscosity of each pure liquid and liquid mixture using the viscosity of water at 25°C as a standard and equation (8.15.10). Relevant data are presented in the appendix. (b) The fluidity of each methanol/water and toluene/*p*-xylene mixture; compare with the fluidity calculated from (8.15.4).

2. Test the validity of equations (8.15.5) and (8.15.6) for the two binary systems studied.

3. Plot ln η *vs.* $1/T$ and determine the activation energy (and probable error) for viscous flow for *p*-xylene.

4. Comment on the "ideality" of the two solutions.

Questions and Further Thoughts

1. Using a propagation of errors technique as applied to Poiseuille's law [equation (8.15.2)], show that the relative error in the capillary radius, ϵ_r/r, enters fourfold in the relative error of the viscosity.

2. Is it possible to have a homogeneous, binary liquid solution that has a higher (or lower) viscosity relative to either of the components? What conclusions would you draw about the nature of intermolecular interactions in such a mixture?

3. In the case(s) above, comment on the expected appearance of the vapor pressure–composition (*P-X*) phase diagram.

Further Readings

R. A. Alberty, "Physical Chemistry," 7th ed., pp. 811–812, Wiley (New York), 1987.

P. W. Atkins, "Physical Chemistry," 3rd ed., pp. 10–11, 595, 615, W. H. Freeman (New York), 1986.

G. W. Castellan, "Physical Chemistry," 3rd ed., pp. 760–762, Addison-Wesley (Reading, Mass.), 1983.

S. Glasstone, "Physical Chemistry," 2nd ed., pp. 496–503, Macmillan (London), 1946.

I. N. Levine, "Physical Chemistry," 2nd ed., pp. 435–442, McGraw-Hill (New York), 1983.

J. H. Noggle, "Physical Chemistry," pp. 445–449, Little, Brown (Boston), 1985.

Appendix

Density of *p*-Xylene

T, °C	D, g/mL
20	0.879
25	0.857
30	0.852
35	0.848
40	0.843
45	0.839
50	0.834
55	0.830
60	0.825

Density of Methanol/Water Mixtures at 25°C

Methanol, volume %	Methanol, weight %	D, g/mL
0	0	0.997
20	16.54	0.971
40	34.57	0.944
60	54.33	0.909
80	76.02	0.859
100	100	0.788

Density of Toluene/*p*-Xylene Mixtures at 25°C

Toluene, volume %	D, g/mL
0	0.857
20	0.858
40	0.859
60	0.859
80	0.860
100	0.861

Experiment 15

Viscosity (Part I)

NAME _____   DATE _____

Percent methanol	Time			
0	___	___	___	___
20	___	___	___	___
40	___	___	___	___
60	___	___	___	___
80	___	___	___	___
100	___	___	___	___

Percent toluene	Time			
0	___	___	___	___
20	___	___	___	___
40	___	___	___	___
60	___	___	___	___
80	___	___	___	___
100	___	___	___	___

Temper-ature	Time			
___	___	___	___	___
___	___	___	___	___
___	___	___	___	___
___	___	___	___	___
___	___	___	___	___
___	___	___	___	___
___	___	___	___	___
___	___	___	___	___
___	___	___	___	___
___	___	___	___	___

Experiment 15

Viscosity (Part I)

NAME _____   DATE _____

Percent methanol	Time			
0	____	____	____	____
20	____	____	____	____
40	____	____	____	____
60	____	____	____	____
80	____	____	____	____
100	____	____	____	____

Percent toluene	Time			
0	____	____	____	____
20	____	____	____	____
40	____	____	____	____
60	____	____	____	____
80	____	____	____	____
100	____	____	____	____

Temperature	Time			
____	____	____	____	____
____	____	____	____	____
____	____	____	____	____
____	____	____	____	____
____	____	____	____	____
____	____	____	____	____
____	____	____	____	____
____	____	____	____	____
____	____	____	____	____
____	____	____	____	____

□ Experiment 16

Viscosity of Liquids

PART II: HIGH VISCOSITIES

Objective

To measure the viscosity and its temperature dependence on glycerol using the falling-sphere method.

Introduction

The Ostwald viscometer is obviously not very useful for measuring very viscous liquids because of the long flow times required. One simple and direct approach that can be used for very viscous liquids is called the falling-sphere method. This is a direct application of Stokes's law (1850), which deals with the movement of a smooth sphere through a quiescent, continuous medium. Because the moving object must separate layers of the intervening liquid, it experiences a "viscous drag." Stokes's law states that the force required to propel a small *sphere* of radius, R, through a medium of viscosity, η, with a velocity, v, is

$$F = 6\pi R \eta v. \qquad (8.16.1)$$

This is a fundamental law in mass transport and is also applied to the case of molecular diffusion (see Experiment 18). If the propelling force is caused by gravity, and the sphere falls at constant velocity, the acceleration is thus zero, and the force due to viscous drag, equation (8.16.1), equals the net gravitational force, i.e.,

$$6\pi R \eta v = 4/3\pi R^3 (D_S - D_L)g, \qquad (8.16.2)$$

where D_S and D_L are the respective densities of the sphere and the medium, and g is the acceleration due to gravity. D_L accounts for the buoyancy force (Archimedes's principle). If the time, t, required for the sphere to fall a distance, L, is measured, the viscosity of the medium can be computed from equation (8.16.2)

$$\eta = \frac{2gR^2(D_S - D_L)t}{9L}. \qquad (8.16.3)$$

The falling-sphere method can be used to obtain an *absolute* measurement of the viscosity because the characteristics of the sphere can be measured with high accuracy. Unlike the expression for η based on Poiseuille's law, equation (8.16.3) does not contain the troublesome r^4 factor. Alternatively, the falling-sphere method can be used with an apparatus that has been calibrated with a liquid of known viscosity.

Interestingly, an application of Stokes's law [equation (8.16.2)] was utilized by Millikan et al. (1911) in the famous oil-drop experiment in which the absolute measurement of the charge of the electron was measured.

It must be realized that Stokes's law [equation (8.16.1)] is valid only in the limit of a continuous medium, that is, a medium that extends infinitely far in all directions from the sphere. This condition is not always satisfied when a small sphere (a few millimeters) falls in a medium that extends for only a few centimeters. A correction is thus required. Gibson and Jacobs (1920), following an

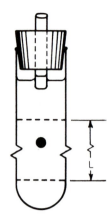

FIGURE 8.16.1   Falling-sphere apparatus. L is the length through which the fall time is measured. The small tube on top guides the sphere so that it falls through the center of the cylinder.

earlier treatment by Ladenburg (1907), describe two corrections to the Stokes's law equation. They applied Stokes's law to a small sphere falling through a cylindrical tube filled with a viscous liquid. One correction deals with the "wall effect" (lateral dimension) and the other with the "end effect," which accounts for the finite height of the fluid medium relative to the cylinder. If equation (8.16.3) is rearranged to express the velocity with which a sphere falls in a continuous medium, v_c, one obtains

$$v_c = \frac{2gR^2(D_S - D_L)}{9\eta}. \qquad (8.16.4)$$

If the sphere falls in a cylindrical container filled with the liquid to be measured (see Figure 8.16.1), the *measured velocity, v,* will be *smaller* than that corresponding to an infinite medium because of the slight "compression" of the liquid against the sides of the cylinder.

The velocity that the sphere would have in a continuous medium, v_c, is given by

$$v_c = v(1 + 2.4x), \qquad (8.16.5)$$

where x is the ratio of the diameter of the sphere to that of the cylinder. The "end effect" correction accounts for the finite distance of the falling sphere to the bottom of the cylinder:

$$v_c = v(1 + 1.65y), \qquad (8.16.6)$$

where y is the ratio of the sphere diameter to the total height of liquid in the cylinder. Thus the net expression for the continuous-medium velocity in terms of the measured velocity is

$$v_c = v(1 + 2.4x)(1 + 1.65y). \qquad (8.16.7)$$

Placing this in equation (8.13.3) allows one to calculate η from the modified Stokes equation:

$$\eta = \frac{2gR^2(D_S - D_L)t}{9L(1 + 2.4x)(1 + 1.65y)}. \qquad (8.16.8)$$

Experimental Design

A diagram of the apparatus is shown in Figure 8.16.1. It simply consists of a cylinder filled with the liquid to be measured and is immersed in a temperature bath. A small tube attached to the top of the cylinder extends below the surface of the liquid by about 3 cm. This releasing tube both guides the sphere to fall vertically through the center of the cylinder and also frees the sphere from any air bubbles that might adhere and cause anomalous behavior. It also acts to slow the initial descent of the sphere before it reaches the first fiducial mark. Fiducial marks are indicated on the cylinder and define the distance through which the falling ball is timed.

Procedure

The fall times of different spheres in glycerol will be measured at three temperatures, 0°, *ca*. 10°, and *ca*. 20°C. Glycerol-filled cylinders will be equilibrated in baths at these temperatures.

For the 0°C system, stainless-steel and bronze spheres, 1/16 in. diameter, will be used. The 10°C study will employ stainless steel and bronze (1/16 in. diameter), aluminum (3/32 in. diameter), and teflon (3/32 in.) spheres. You will use stainless-steel and teflon (3/32 in. diameter) spheres in the 20°C measurement.

At least three trials should be used for each type of sphere at each temperature. Because of the relaxation time problem, measurements should be done in a sequence of one temperature to the next, starting with the 0°C bath. Be sure to keep your data well organized as they are acquired.

First acquire the data needed for determining the sphere densities. Accurately weigh (to 0.1 mg) about 10 spheres (stainless steel, bronze, aluminum, and teflon). In calculating the densities, use the nominal (given) diameters.

Measure the overall height of glycerol in each of the cylinders as well as the distance between fiducial marks. (Use a cathetometer for optimal accuracy.)

Also measure the internal diameter of the cylinder. Use the calibrated stick or calipers provided.

Be sure to use tweezers in handling the spheres. Avoid distorting or scratching the spheres (especially teflon and aluminum).

Note: In performing the calculations, you will need the density of glycerol at the temperatures studied. For this determination, use the coefficient of expansion, which can be obtained from a handbook or the International Critical Tables.

Data Analysis

1. Calculate and report the sphere and cylinder characteristics. Using the measured glycerol densities, interpolate to obtain the densities at the same temperatures used in the viscosity measurements. Alternatively, use the literature value of the coefficient of expansion to obtain the necessary densities.

2. From these data, calculate the viscosity of glycerol at the different temperatures. Tabulate these results. Perform a propagation of error analysis and report the error in the measured viscosities.

3. Make an Arrhenius plot of the viscosities and determine the activation energy to viscous flow for glycerol. Report this along with the probable error.

Questions and Further Thoughts

1. The falling-sphere method for determining viscosities of highly viscous materials is very appealing because of its simplicity. Many substances that one might want to examine

with this technique are opaque (paint) or translucent (crude resins, etc.). Describe an experimental method by which the falling-sphere approach could be used in such cases. In your analysis, consider the desirability and possibility in computer interfacing the apparatus.

2. In equation (8.16.2), the factor $(D_S - D_L)$ accounts for the buoyancy of the sphere in the fluid medium. The same principle is encountered in the characterization, separation, and purification of macromolecules (e.g., proteins) in sedimentation and ultracentrifugation. Consult a text on biochemistry or biophysical chemistry and see how Stokes's law is applied to these techniques.

3. If spheres are dropped in rapid succession, the fall times will gradually decrease. Thus a falling sphere will be affected by the "wake" left by a preceding sphere (a relaxation effect). Describe a mechanism by which this effect works; i.e., how does the medium manifest the "memory" of a fallen sphere?

4. As is mentioned in the discussion following equation (8.16.4), the measured fall time of a sphere in a narrow cylinder is longer than that in a wide one. Propose a mechanism by which this "wall effect" operates. Describe an experiment that could be used to confirm equation (8.16.5).

Further Readings

See the discussion and References in Experiment 15.

W. H. Gibson and L. M. Jacobs, *J. Chem. Soc., 117*, 473 (1920).

A. Ladenburg, *Ann. Phys, 23,* 9 (1907).

I. Tinoco, K. Sauer, and J. C. Wang, "Physical Chemistry," 2nd ed., pp. 241–254, Prentice-Hall (Englewood Cliffs, N.J.), 1985.

Experiment 16

Viscosity (Part II)

NAME _____ DATE _____

Weight of spheres: _____   Number of spheres: _____   Type of sphere: _____

_____   _____   _____

_____   _____   _____

_____   _____   _____

Temperature	Fall distance	Sphere	Time
_____	_____	_____	_____
_____	_____	_____	_____
_____	_____	_____	_____
_____	_____	_____	_____
_____	_____	_____	_____
_____	_____	_____	_____
_____	_____	_____	_____
_____	_____	_____	_____
_____	_____	_____	_____
_____	_____	_____	_____
_____	_____	_____	_____
_____	_____	_____	_____
_____	_____	_____	_____
_____	_____	_____	_____

Viscosity (Part II)

NAME _____ DATE _____

Weight of spheres: _____   Number of spheres: _____   Type of sphere: _____

_____   _____   _____

_____   _____   _____

_____   _____   _____

Temperature	Fall distance	Sphere	Time
_____	_____	_____	_____
_____	_____	_____	_____
_____	_____	_____	_____
_____	_____	_____	_____
_____	_____	_____	_____
_____	_____	_____	_____
_____	_____	_____	_____
_____	_____	_____	_____
_____	_____	_____	_____
_____	_____	_____	_____
_____	_____	_____	_____
_____	_____	_____	_____
_____	_____	_____	_____
_____	_____	_____	_____

□ Experiment 17

Collision Diameters from Gas Viscosities Using the Evacuation Method

Objective

To measure the viscosities of several gases (e.g., He, Ne, Ar, SF_6) and to determine their gas-kinetic collision diameters.

Introduction

In studying gas phase reaction dynamics, it is important to know the effective sizes, e.g., collision diameters, of the molecules involved in the elementary steps of the reaction. One of the ways that this fundamental information is obtained is through measurements of gas viscosity. From the application of gas kinetic theory to "hard sphere" molecules, the square of the collision diameter, σ, may be expressed as follows:

$$\sigma^2 = \frac{5(MRT/\pi)^{1/2}}{16N_A\eta} \quad (\text{cm}^2), \quad (8.17.1)$$

in which M is the molecular weight, η is the viscosity, N_A is Avogadro's number, and the other symbols have the usual meanings. This equation is derived from the relationship between the viscosity of a (hard sphere) gas and its density, D, mean speed, $<c>$, and mean free path, λ:

$$\eta = \frac{5\pi}{32} D<c>\lambda \quad (\text{cm}). \quad (8.17.2)$$

Equation (8.17.1), which can be directly obtained from (8.17.2), may be rearranged to express the gas viscosity. Thus,

$$\eta = \frac{5(MRT/\pi)^{1/2}}{16N_A\sigma^2} \quad (\text{g s}^{-1}\text{ cm}^{-1}). \quad (8.17.3)$$

The cgs units of η are called the *poise*, P. The interesting point about equation (8.17.3) is that it predicts that η should be independent of gas pressure [this is because the pressure dependence of the gas density in equation (8.17.2) cancels that of the mean free path], and that η should vary with temperature as $T^{1/2}$ (from the temperature dependence of $<c>$). The first prediction seems counterintuitive, but it has been verified over a relatively wide range of pressures. The reasons that η becomes pressure-dependent at low and high pressures stem from departures from laminar flow; some of these complications will be discussed below. The $T^{1/2}$ dependence has been confirmed experimentally and is unusual because gases demonstrate the opposite type of temperature dependence from liquids; that is, liquid viscosities decrease with increasing temperature (see Experiments 15 and 16).

We will now develop the physical and mathematical basis on which the viscosity of a gas is experimentally determined. The mass transport of a fluid (in this case, a gas) arising from a pressure gradient is represented by the general equation

$$J_x = -C\frac{dP}{dx} \quad (\text{g s}^{-1}\text{ cm}^{-2}), \quad (8.17.4)$$

where J_x is the *flux* (mass transported per unit cross section per unit time) of gas in the x direction as a result of the pressure gradient, dP/dx; C is a positive constant that is inversely proportional to the viscosity. Equation (8.17.4) is known as Poiseuille's law (1840) (or in other applications as Fick's first law; see Experiments 18 and 20). The minus sign reflects the fact that the flow direction opposes the pressure gradient.

The flux can be expressed as $J_x = (1/A)(dm/dt)$, where A is the unit cross section (normal to the flow direction) and dm is the (infinitesimal) mass transported in a time dt. To apply equation (8.17.4) to the case of a fluid flowing through a straight cylinder of radius r, we must consider Newton's law of viscosity. It states that the shear force, F_x, required to move an infinitesimally thin sheet of fluid having unit cross section, A, in a flow direction parallel to the plane of the sheet is proportional to the velocity gradient of the sheet taken in a direction normal to the flow (dv_x/dy). See p. 228. The mathematical statement is

$$\frac{1}{A} F_x = -\eta \frac{dv_x}{dy} \qquad (\text{g s}^{-2} \text{ cm}^{-1}). \qquad (8.17.5)$$

Note that the proportionality constant in equation (8.17.5) is the *fluid viscosity*. Another way of stating equation (8.17.5) is that the drag force exerted on a unit plate of flowing fluid is proportional to the change in velocity of adjacent plates.

According to equation (8.17.5), a fluid flowing through a straight cylinder has the largest flow rate at the center of the cylinder (at the cylinder axis) and is *zero* at the walls. This particular fluid dynamic condition is known as *laminar* (or Newtonian) *flow* and is approached by many gases and (small molecule) liquids (called Newtonian fluids) as long as the flow rate (thus the pressure gradient) is not excessive. At very low pressures, gases deviate from Newtonian (or bulk) flow as the mean free path becomes comparable with the cylinder radius. In this situation, the gas flow is said to be in the *molecular*, or *Knudsen flow* regime, and viscosity, which is a bulk property, fails to have significance.

The application of Newton's law of viscosity to the flow of a fluid through a straight cylindrical tube of radius r leads to another statement of Poiseuille's law, i.e.:

$$\frac{dV}{dt} = -\frac{\pi r^4}{8\eta} \frac{dP}{dx} \qquad (\text{cm}^3 \text{ s}^{-1}). \qquad (8.17.6)$$

In this equation, which is used in viscosity measurements of liquids, dV is the volume of gas transported through the tube in time dt, and dP/dx is the pressure gradient in the tube along the direction of flow.

The infinitesimals dV and dm are related through the density, D, as $D = dm/dV$; thus equation (8.17.6) becomes

$$\frac{dm}{dt} = -\frac{\pi r^4 D}{8\eta} \frac{dP}{dx}. \qquad (8.17.7)$$

Because we are dealing with a gas at relatively low pressures (<1 atm), the density can be expressed in terms of its pressure, P, temperature, T, and molecular weight, M, through the ideal gas law, i.e., $D = PM/RT$. Applying this result to equation (8.17.7), and recalling that $n = m/M$, we can write

$$\frac{dn}{dt} = \frac{\pi r^4}{8\eta RT} \left(\frac{-P \, dP}{dx} \right). \qquad (8.17.8)$$

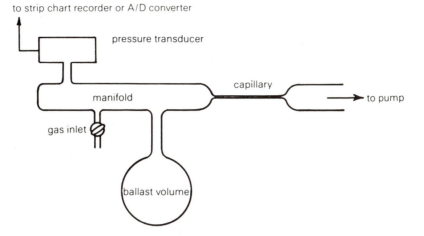

to strip chart recorder or A/D converter

pressure transducer

capillary

manifold

to pump

gas inlet

ballast volume

FIGURE 8.17.1   Schematic diagram of the evacuation apparatus.

Assuming the flow rate, dn/dt, to be independent of position (x) within the tube, the right-hand side of (8.17.8) containing the x dependence of P and the left-hand side of (8.17.8) can each be set equal to a constant. If we do this for the right-hand side of (8.17.8) and then integrate [between $P_1(x_1)$ and $P_2(x_2)$], the following result is obtained:

$$\frac{dn}{dt} = \frac{\pi r^4 (P_1^2 - P_2^2)}{16\eta RT(x_2 - x_1)}. \tag{8.17.9}$$

P_1 and P_2 are pressures at the beginning (x_1) and end (x_2) of the tube. Note that $P_1 > P_2$.

To clarify the rest of this development, a description of the experimental approach is presented. Refer to Figure 8.17.1. The gas is introduced to a manifold to which a large ballast volume (about 1 L), a pressure transducer (a device that converts the pressure to a voltage), and a capillary tube are connected. As the gas is evacuated through the capillary tube by a high-speed vacuum pump, the pressure of the system is continuously monitored by the transducer, whose output is displayed on a strip chart recorder. Alternatively, the transducer output is fed into the input port of a laboratory computer for direct data acquisition.

Because this experiment is carried out in such a way that $P_2 \ll P_1$, equation (8.17.9) simplifies to

$$\frac{dn}{dt} = \frac{\pi r^4 [P_1(t)]^2}{16\eta RTL}, \tag{8.17.10}$$

in which the time dependence of the pressure at the beginning of the tube, P_1, (the manifold pressure) is explicitly indicated. L is the total length of the capillary tube.

We conclude this part of the discussion by expressing the number of moles transported through the tube, dn, in terms of the manifold pressure, P_1. The ideal gas law furnishes this relationship: $dn = (V/RT)dP$. After rearrangement, equation (8.17.10) becomes

$$\frac{dP}{P^2} = \frac{\pi r^4 dt}{16\eta VL}. \tag{8.17.11}$$

V is the total volume of the manifold, ballast volume, etc. The integration of equation (8.17.11) with the boundary condition $t = 0$; $P = P_0$ furnishes

$$\frac{1}{P} = \frac{1}{P_0} + Kt, \tag{8.17.12}$$

in which the constant K is $\pi r^4/(16\eta VL)$. Thus, Poiseuille's law predicts that a plot of the reciprocal pressure of the system being evacuation vs. time should be linear. The slope can, in principle, provide an absolute value of the gas viscosity since r, V, and L can be determined. Note that the r^4 factor in K makes r a very sensitive characteristic. In practice, it is more satisfactory to obtain an empirical value of the constant K by calibrating the system using a gas of known viscosity, e.g., dry air. In this regard, the approach is similar to that followed in using the Ostwald viscometer in measuring liquid viscosities.

Corrections to Poiseuille's Law

Because equation (8.17.12) is based on the assumption of laminar, or streamline, flow, it is important to consider the conditions under which this condition is expected to apply. This understanding is, as we shall see, vital to proper experimental design. As mentioned above, laminar flow fails to occur at high pressure (where turbulence sets in) and low pressure (when molecular flow and other complications arise). These cases will be discussed more quantitatively. The goal is to identify the pressure regimes in which these two departures from Newtonian flow (hence Poiseuille's law) take place.

HIGH PRESSURE: Osborne Reynolds showed (1883) that if the average linear velocity of a fluid flowing through a straight tube *exceeds* a critical value, turbulent flow sets in; hence, Poiseuille's law becomes invalid. The linear velocity is the speed with which an infinitesimal cross section of the fluid moves through the tube. The value of the linear velocity that approximately marks the transition between streamline and turbulent flow, v', is given by

$$v' = \frac{1000\,\eta}{rD}, \tag{8.17.13}$$

where r is the tube radius and D and η are the density and viscosity of the fluid. Thus quantitative applications of Poiseuille's law should be made for fluids having flow velocities *below* v'. The quotient, $2rDv/\eta$, which is dimensionless, is called the *Reynolds number*. When the Reynolds number exceeds 2000, fluid flow may not be laminar.

In the case where the fluid is an ideal gas of molecular weight M, equation (8.17.13) becomes

$$v' = \frac{1000\,\eta RT}{rMP}. \tag{8.17.14}$$

The linear velocity of a gas flowing through a tube can be obtained from equation (8.17.6) by substituting $-P/L$ for dP/dx and $A\,dx/dt$ (where $A = \pi r^2$) for dV/dt. The linear velocity, v, is (dx/dt):

$$v = \frac{Pr^2}{8\eta L}. \tag{8.17.15}$$

The onset of turbulent flow occurs when $v > v'$, or when $v/v' > 1$. Using equations (8.17.14) and (8.17.15), the condition for *laminar flow* can be stated as

$$\frac{v}{v'} = \frac{Mr^3P^2}{8000\ RTL\eta^2} < 1. \tag{8.17.16}$$

Although equation (8.17.16) cannot be used to determine whether laminar flow occurs without knowing the gas viscosity (the object of the experiment), η can be estimated for this purpose. Gas viscosities are generally ~ 1 to 2×10^{-4} g cm^{-1} s^{-1}. In summary, the inequality in equation (8.17.16) places an upper limit on P for the application of Poiseuille's law.

LOW PRESSURE: If the pressure at which a gas being forced through a very narrow tube is decreased beyond a certain point, the flow rate of the fluid becomes *larger* than that predicted by Newtonian fluid mechanics. This is called the "slip condition," and it arises because the molecules adjacent to the tube walls do not move with zero velocity (as assumed in Newtonian flow); hence, the bulk gas "slips" through the tube. Poiseuille's law (8.17.6) can be written to include a slip correction (dP/dx is again replaced by $-P/L$):

$$\frac{dV}{dt} = \frac{\pi r^4 P}{8\eta L}\left(1 + \frac{4S}{r}\right), \tag{8.17.17}$$

S is known as the *slip coefficient* and has dimensions of length. Slip becomes important when the quantity $4S/r$ approaches unity. It can be seen that this deviation becomes more important in tubes of narrow bore. It turns out that the slip coefficient depends on the pressure (as stated above). In one approximation [due to Maxwell (1879)], $S \sim 2\lambda$, where λ is the mean free path of the flowing gas. From gas kinetic theory,

$$\lambda = \frac{RT}{\sqrt{2}\pi N_A\sigma^2 P}, \tag{8.17.18}$$

where σ is the collision diameter. The inverse pressure dependence of S can be seen in (8.17.18). Actually, σ is often determined from gas viscosity measurements. λ can be alternatively expressed as

$$\lambda = \frac{1.25\eta(RT/M)^{1/2}}{P}. \tag{8.17.19}$$

Thus to avoid slip conditions, which become significant when $4S/r \rightarrow 1$ [see equation (8.17.17)], the pressure must be *larger* than a value that can be obtained from equation (8.17.19) (using $S = 2\lambda$), i.e.:

$$P > \frac{10\eta(RT/M)^{1/2}}{r}. \tag{8.17.20}$$

Equations (8.17.16) and (8.17.20) place upper and lower limits, respectively, on the pressure in the evacuation experiment such that equation (8.17.12) is valid.

Collision Diameters

Once the gas viscosity is determined, its collision diameter can be obtained through equation (8.17.1). It is logical to compare σ obtained from gas viscosity with another experimental manifestation of molecular size, namely, the van der Waals b constant. b, having units of volume/mole, can be related to the gas collision diameter as follows. Figure 8.17.2 shows two identical spherical molecules in close contact during a collision. The center of molecule B cannot approach the

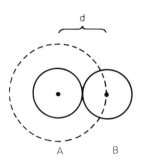

FIGURE 8.17.2   Two hard spheres in collisional contact. One particle contributes an excluded volume of $4\pi\sigma^3/3$, where σ is the collision diameter.

center of molecule A by an amount less than the distance equal to σ. Therefore, molecule A effectively removes from the total container volume, V, an *excluded volume*, V_{excl}. This excluded volume is just $(4/3)\pi\sigma^3$, and the net *available* volume to molecule B is

$$V_{\text{avail}} = V - (4/3)\pi\sigma^3. \qquad (8.17.21)$$

Now if successive molecules are added to the container, the net available volume accessible to the ith added molecule is

$$V_{\text{avail}} = V - \sum (i-1)V_{\text{excl}}, \qquad (8.17.22)$$

and the *average* volume available to a total of N added molecules is

$$\langle V_{\text{avail}}\rangle = V - \frac{1}{N}\sum_{i=1}^{N}(i-1)V_{\text{excl}}. \qquad (8.17.23)$$

Because V_{excl} is a constant, the summation is simply

$$\frac{1}{N}V_{\text{excl}}\sum_{i=1}^{N}(i-1).$$

The summation is equal to $N^2/2 - N/2$, and because N is large (on the order of N_A), this expression becomes $N^2/2$. Therefore, the average available volume is

$$\langle V_{\text{avail}}\rangle = V - \frac{N}{2}V_{\text{excl}}. \qquad (8.17.24)$$

But $\langle V_{\text{avail}}\rangle$ is just the ''net volume'' term in the van der Waals equation, $(V - nb)$, where n is the number moles of gas. We now have

$$V - nb = V - \frac{N}{2}V_{\text{excl}}. \qquad (8.17.25)$$

Dividing by n, we get

$$V_m - b = V_m - 1/2V_{\text{excl}}, \qquad (8.17.26)$$

where V_m is the molar volume and V_{excl} is the excluded volume of the gas. Finally, the desired result is presented as

$$b = 1/2V_{\text{excl}} = 2/3N_A\pi\sigma^3. \qquad (8.17.27)$$

This equation allows a comparison to be made between σ, the collision diameter, and b. Moreover, collision diameters can be estimated from van der Waals b constants.

Safety Precautions

☐ Safety goggles must be worn during this experiment.

☐ Make sure you have been instructed about proper vacuum manifold techniques. Use two hands to turn stopcocks; one hand is used to counterbalance the torque exerted by the other.

☐ The high-pressure gas cylinders must be firmly attached to a secure foundation. The reducing valves are adjusted to admit gas at a few psi above ambient pressure (< 5 psig). Do not adjust any of these controls.

☐ The vacuum pump vent must be connected to an exhaust system.

☐ A low-temperature trap should not be used with the vacuum pump unless reactive gases are used in the experiment.

Procedure

Option A: Strip Chart Recorder Analysis

1. The output of the pressure transducer (which produces, for example, 10 V per 1000 torr) is connected to a strip chart recorder. Make sure the transducer power supply is turned on and that the range switch (which indicates the maximum pressure for full-scale pen deflection) is in the proper position. Your instructor will advise you about its proper use. Also make sure the zero position is set.

2. Your instructor will demonstrate the design and use of the manifold. When using the stopcocks, be sure to follow the "two-hand rule": use one hand to counterbalance the torque exerted on the stopcock plug by the other. The procedure for filling and evacuating the rack will be reviewed. First, the system will be calibrated using dry air. First pump out the manifold, including the capillary tube and ballast bulb. As the pressure falls to a low value (ultimately to below 1 torr), the recorder pen should approach the zero position. Record the ambient temperature.

3. Evacuate the gas inlet line (to the gas-handling manifold) and then isolate the system by shutting off the main manifold stopcock. Gently crack open the dry air valve until the manifold pressure rises to an appropriate level (e.g., 100 torr).

4. The chart pen should respond by indicating the manifold pressure, and it should be stable. Isolate the gas inlet line from the manifold. Start the recorder chart drive (about 5 in./min is appropriate), and as the pen crosses a vertical line on the chart paper, open the capillary stopcock (the pump side). The pressure should start to fall immediately. After the pressure has dropped to about 5 to 10 percent of the initial value, evacuate the entire system by opening the main stopcock.

5. Pump out the gas inlet line and fill the system with another gas. Repeat the acquisition of evacuation curve. Follow the above procedure until all the gases to be studied have been measured. The instructor will indicate which gases are to be examined in your laboratory session.

Data Analysis

For each of the evacuation curves, prepare tables of the pressure. For each of the evacuation curves, prepare a table of (pressure, time) data. The points should be spread out relatively evenly between the maximum and minimum pressures tabulated. About 20 to 40 points should be tabulated for each evacuation curve.

Plot $1/P$ vs. time, using as much of the paper as practical while still retaining convenient P and t scales. Although these plots should be linear, some curvature might appear at early times. Using linear regression, determine the slopes of each of the curves. Be sure to note the pressure range used for the evacuation curves. Tabulate these values along with the probable errors.

Using the viscosity of dry air at the temperature of your experiment (a $T^{1/2}$ correction can be made if necessary), convert the evacuation curve slopes to viscosity values. Tabulate these results with their errors.

Finally, calculate the collision diameters of the gases from their viscosities and compare these values with the collision diameters obtained from the van der Waals b constants.

Procedure

Option B: Computer Acquisition of Data

The computer software (set of instructions) is menu-driven and as it runs, the program prompts you to perform certain steps. After showing you how to "load" the program into the computer memory, your instructor will guide you through one data-acquisition and -analysis cycle (for dry air). Read the procedure above regarding the operation of the manifold.

From the linear regression slopes obtained from the computer-aided analysis of the evacuation curves, tabulate the viscosities (and their probable errors) of the gases studied in the experiment. Determine and tabulate their collision diameters (and uncertainties). Compare these data with the collision diameters obtained from the van der Waals b constants.

See the last paragraph of the Data Analysis section above.

Questions and Further Thoughts

1. Many assumptions have gone into the development of equation (8.17.12). In your report summarize these assumptions and discuss the validity of each. If any of your evacuation curves have shown nonlinearity, speculate what the causes of this curvature might be.
2. The way in which the viscosity of gases depends on temperature and on molecular structure (mass) is opposite to that of the behavior of liquids. Rationalize the temperature and mass dependence of gas viscosity in terms of simple kinetic theory.
3. Use equations (8.17.14) and (8.17.16), a viscosity of 2×10^{-4} poise, $M = 40$, and $T = 300$ K, and r and L values consistent with the apparatus (or $r = 0.05$ cm, $L = 30$ cm), and determine the Reynolds number, hence the pressure above which turbulent flow should commence.
4. Use equation (8.17.20) and the above data to assess the pressure below which slip conditions should apply.

Further Readings

A. W. Adamson, "A Textbook of Physical Chemistry," 3rd ed., pp. 17–18, 55–62, Academic Press (Orlando, Fla.), 1986.

P. W. Atkins, "Physical Chemistry," 3rd ed., pp. 651–660, W. H. Freeman (New York), 1986.

G. W. Castellan, "Physical Chemistry," 3rd ed., pp. 671–672, 745–755, Addison-Wesley (Reading, Mass.), 1983.

W. Kauzmann, "Kinetic Theory of Gases," Benjamin, 1966.

I. N. Levine, "Physical Chemistry," 2nd ed., pp. 456–462, McGraw-Hill (New York), 1983.

J. H. Noggle, "Physical Chemistry," pp. 5–7, 445–451, Little, Brown (Boston), 1985.

J. R. Partington, "An Advanced Treatise on Physical Chemistry," vol. I, pp. 241, 881–887, Longmans (London), 1949.

Experiment 17

Gas Viscosity

NAME _____ DATE _____

Notes

Gas Viscosity

NAME _____   DATE _____

Notes

☐ Experiment 18

Measurement of the Diffusion Coefficient in Solution

Objective

To obtain the diffusion coefficient of sucrose in water and to determine its hydrodynamic radius.

Introduction

Studies of reaction kinetics and the bulk fluid properties of liquids often require fundamental knowledge of mass transport in the condensed phase. The movement of molecules (considered in this experiment as the solute in a dilute solution) arises from one of the most basic properties of fluids, namely, the constant random motion of the constituent molecules. Even when there is no directed driving force for mass transport, such as a concentration gradient or a gravitational field (i.e., an isotropic medium), molecules are observed to undergo constant random displacements. A well-known example of this phenomenon is Brownian motion in which a particle moves in continual zigzag patterns. Over a long time period there is no *net* displacement of the particle because in the absence of a concentration, chemical, or potential energy gradient the probability of motion in any one direction is equal to the probability of motion in exactly the opposite direction.

In this experiment, the directed displacement of molecules in solution will be considered. The driving force for this motion is a *concentration gradient;* thus solute will spontaneously flow from a region of higher concentration to one of lower concentration in a one-phase system. This process results in an entropy increase corresponding to the increase in the volume available to the solute as it moves into a more dilute environment; in this sense, we are dealing with an "entropy of mixing" phenomenon.

In the most general case, the driving force causing the change in local concentration of a solute is expressed in terms of the anisotropy (spatial asymmetry) of the chemical potential, μ. If we consider one-dimensional motion (along the x axis), the instantaneous velocity of a molecule, v_x, is proportional to the gradient of the chemical potential at that point, thus

$$v_x = -u\left(\frac{\partial \mu}{\partial x}\right)_x \qquad (\text{cm s}^{-1}), \qquad (8.18.1)$$

where the proportionality constant, u, is called the *mobility*. The negative sign indicates that the molecule moves in a direction from higher to lower chemical potential, i.e., opposite to the gradient. u can also be expressed in terms of another constant, D, called the *diffusion coefficient* as

$$u = \frac{D}{kT} \qquad (\text{s g}^{-1}), \qquad (8.18.2)$$

where k is the Boltzmann constant and T is the absolute temperature. Equation (8.18.2) is known as the *Nernst-Einstein relation,* from which it can be seen that D has dimensions of $\text{cm}^2\,\text{s}^{-1}$.

If the concentration of the solute, C, is the number of moles per dm^3, the *flux,*

J_x, is defined as the number of moles passing through a unit cross section, (cm^2) per second, hence $J_x = v_x C$, and equations (8.18.1) and (8.18.2) can be combined and written

$$J_x = v_x C = -\frac{D}{RT} C \left(\frac{\partial \mu}{\partial x}\right)_x \qquad (\text{moles cm}^{-2}\,\text{s}^{-1}). \qquad (8.18.3)$$

Now if the chemical potential is expressed in terms of a concentration standard state, $C° \equiv 1$ mol dm^{-3}, $\mu = \mu° + RT \ln C$, the concentration dependence of $d\mu$ becomes $d\mu = RTd(\ln C) = (RT/C)dC$, and if this relation is substituted into equation (8.18.3), we obtain

$$J_x = -D \left(\frac{\partial C}{\partial x}\right)_t. \qquad (8.18.4)$$

This is known as Fick's first law and is a more specific statement of mass transport than equation (8.18.3). See also the discussion of Fick's laws in Experiment 20. Another fundamental law of mass transport, known as *Fick's second law,* is obtained from equation (8.18.4) through an application of the conservation of mass known as the *law of continuity.* The change in the flow rate (flux) after a system has moved from one volume element to another is equal to the concentration difference of the volume elements (per unit time). Mathematically, this is expressed as

$$\left(\frac{\partial J}{\partial x}\right)_t = -\left(\frac{\partial C}{\partial t}\right)_x. \qquad (8.18.5)$$

By differentiating equation (8.18.4) with respect to x and equating the result with that in equation (8.18.5), one obtains

$$\left(\frac{\partial C}{\partial t}\right)_x = \frac{d}{dx}\left[D\left(\frac{\partial C}{\partial x}\right)_t\right]. \qquad (8.18.6)$$

If we take D to be independent of the diffusion distance x, which is to say the solute concentration, Fick's second law then reads

$$\left(\frac{\partial C}{\partial t}\right)_x = D\left(\frac{\partial^2 C}{\partial x^2}\right)_t. \qquad (8.18.7)$$

The solution of equation (8.18.7) will describe the space *and* time dependence of the solute concentration as it undergoes diffusion. In order to solve this differential equation and apply the solution to the determination of the diffusion coefficient, it is important to consider the physical methodology of the experiment. A homogeneous solution containing the solute at a concentration C is placed in a cylindrical vessel. This solution is suddenly but *quiescently* (i.e., without disturbance) placed in contact with pure solvent contained in a cylinder of the same diameter such that the circular surfaces of the two liquids come into direct contact. The solution and solvent columns, which are quiescent (i.e., no physical agitation), remain in contact for a measured period of time during which the solute spontaneously diffuses into the solvent. The two liquid columns are then separated at the same place as their original contact, and the amount of solute that has diffused into the solvent column is then determined. The procedure is shown schematically in Figure 8.18.1. The apparatus used in this experiment was devised by Polson[1,2] for use in determining the diffusion coefficient of a virus and is described in more detail below. The foregoing discussion of the experimental design

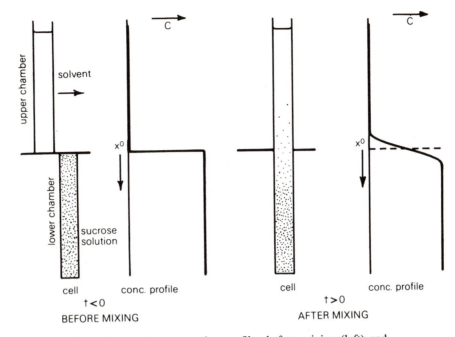

FIGURE 8.18.1   Polson cell positions and concentration profiles before mixing (left) and after mixing (right).

allows us to establish the boundary conditions appropriate to the solution of equation (8.18.7):

At $t \leq 0$, $C = C_0$ for $x > 0$ and $C = 0$ for $x < 0$.   (8.18.8a)

At $t > 0$, $C \to C_0$ as $x \to +\infty$ and $C \to 0$ as $x \to -\infty$,   (8.18.8b)

and

As $t \to \infty$, $C \to$ constant.

As is indicated in equation (8.18.8b) the columns of liquid are presumed at this point to be infinitely long.

Fick's second law, equation (8.18.7), subject to the above boundary conditions, has the following solution for $C(x,t)$.[3] It is important to realize that this solution is valid only for *finite* times, i.e., after diffusion has commenced ($t > 0$).

$$C(x,t) = \frac{C_0}{2}\left\{1 - \text{erf}\left[\frac{x}{2(Dt)^{1/2}}\right]\right\},$$ (8.18.9)

where erf denotes the error function,

$$\text{erf}(w) \equiv \frac{2}{\sqrt{\pi}}\int_0^w \exp(-q^2)\,dq.$$ (8.18.10)

Values of erf(w) can be obtained from standard mathematical tables. It should be noted that in this one-dimensional solution of Fick's second law, the space and time variables, x and t, are linked in the form ($xt^{-1/2}$); see equation (8.18.9).

In order to express the diffusion coefficient in terms of experimentally measurable quantities, equation (8.18.9) is first differentiated with respect to x:

$$\left(\frac{\partial C}{\partial x}\right)_t = \frac{-C_0}{2}\left[\frac{1}{(\pi Dt)^{1/2}}\right]\exp\left(\frac{-x^2}{4Dt}\right).$$ (8.18.11)

This is the concentration gradient expressed as a function of x and t. If we now evaluate this gradient at the original interface, $x = 0$ (see Figure 8.18.1), we have

$$\left(\frac{\partial C}{\partial x}\right)_{x=0}(t) = \frac{-C_0}{2(\pi D t)^{1/2}}. \tag{8.18.12}$$

The concentration gradient in equation (8.18.12) can now be substituted in Fick's first law, equation (8.18.4), and defining the flux as $J = (1/A)dn/dt$, we have

$$\frac{dn}{dt} = -AD \frac{-C_0}{2(\pi D t)^{1/2}} = \frac{AD^{1/2}C_0}{2(\pi t)^{1/2}}, \tag{8.18.13}$$

where dn is the number of moles transported through the cross-sectional area, A, (across the original boundary at $x = 0$) in the time dt. Equation (8.18.13) may now be integrated to provide the total number of moles of solute that diffuses across the $x = 0$ boundary during a time t:

$$\int_0^n dn = \frac{AD^{1/2}C_0}{2\sqrt{\pi}} \int_0^t \frac{dt}{t^{1/2}}, \tag{8.18.14}$$

or

$$n = A\left(\frac{D}{\pi}\right)^{1/2} C_0 t^{1/2}. \tag{8.18.15}$$

The length of the "solvent" cylinder is, in fact, finite, i.e., h cm high. The *bulk* (mixed) concentration of solute that has diffused during the time interval t is then $C = n/Ah$. Substituting n into equation (8.18.15) and solving for D provides the final result for the measured diffusion coefficient:

$$D = \frac{h^2(C/C_0)^2\pi}{t} \qquad (cm^2\ s^{-1}). \tag{8.18.16}$$

Interestingly, the diffusion coefficient can be used to determine not only quantitative transport properties of a solute but also certain structural characteristics of the solute in a given solvent environment. The relationship between the diffusion coefficient and the "size" of the solute is contained in the *Stokes-Einstein equation*. It expresses D_0, the diffusion coefficient at *infinite dilution*, in terms of the solvent viscosity, η, and the "effective" radius, R, of the solute:

$$D_0 = \frac{kT}{6\pi\eta R}, \tag{8.18.17}$$

k is the Boltzmann constant and T is the absolute temperature. There are several important assumptions implicit in equation (8.18.17). Two of these are that the solute is spherical, and considerably larger than the solvent molecules. Deviations from spherical geometry (such as oblate or prolate ellipsoids) could cause equation (8.18.17) to be in error up to about 30 to 40 percent.[3] For solutes of similar size to that of the solvent, the 6 in the denominator of (8.18.17) is sometimes replaced by 4 (see references 4 and 5). Stokes's law was actually developed to deal with macroscopic particles, and Einstein successfully applied it to the quantitative observation of the random displacements (Brownian motion) of colloidal particles. The falling-sphere method of measuring viscosities (see Experiment 16) is an explicit application of Stokes's law to a macroscopic body. Stokes's law was also invoked in the treatment of falling oil drops in the famous experiment by Millikan et al. in 1909.

It is remarkable that a fluid dynamic treatment of mass transport pertaining to

0.01- to 1-cm objects (which underlies the development of the Stokes-Einstein equation) can be successfully extended to the level of molecular dimensions, i.e., $\sim 10\text{Å}$. Einstein's application of Stokes's law to Brownian motion (1905) provided a theoretical basis that was used to support the then still unaccepted idea of atomicity—that molecules are specific, discrete entities.

The diffusion coefficient reflects the transport properties of the solute under *solvated* conditions and therefore the "effective radius," R, represents the radius of the solute that may contain bound (e.g., hydrogen-bonded) solvent molecules that on a time-averaged basis are part of the solute structure. The effective radius, which includes the associated solvation shell, is referred to as the *hydrodynamic radius*.

Using the Stokes-Einstein equation (8.18.17) to estimate the hydrodymamic radius of a solute requires knowledge of D_0. Some of the assumptions that underlie this equation would seem to make its application semiquantitative at best [after all, the measurement of *one number* (the diffusion coefficient) cannot provide much information about something as complex as molecular structure]. Nevertheless, some insight into the gross structure of solvated molecules (including biologically significant macromolecules) can be obtained. In this experiment, the diffusion coefficient of sucrose in aqueous solution will be determined at different concentrations, and D_0 will be obtained by extrapolation. The concentration dependence of D is observed to be very slight, and a linear extrapolation to zero concentration can be empirically justified.[6] The hydrodynamic radius will be estimated from D_0 and the Stokes-Einstein equation. Sucrose is a disaccharide in which two cyclic forms of simple sugars, glucose and fructose, are linked together. The structure is shown below.

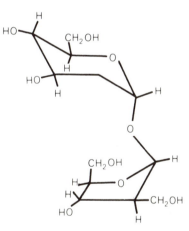

It is obvious that in aqueous solution there is appreciable hydrogen bonding between the solvent and the sucrose molecule.

Experimental Method

The Polson apparatus is shown in Figure 8.18.2.

It consists of two cylinders constructed from an inert material such as stainless steel, brass, or bronze. Six 1/4-in.-diameter cylindrical holes, or columns, precisely spaced 60° apart, are drilled into the two cylinders. The six holes completely pass through the upper cylinder and terminate in a blind end in the lower one. The two cylinders are in close, fluid-tight contact, the upper one being designed to rotate smoothly with respect to the lower one.

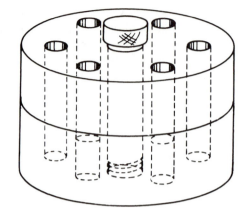

FIGURE 8.18.2   Diagram of the Polson cell.

The upper cylinder is first rotated to a position in which the two columns are in precise alignment. The solution (concentration C_0) is then added to three *alternate* chambers so that the liquid levels are slightly *above* the boundary plane between the two cylinders. The upper cylinder is next rotated by about 1/12 revolution so that the three empty chambers are situated above the boundary surface of the lower cylinder. These chambers are then filled with solvent, and the upper cylinder is *slowly* and steadily rotated *ahead* (i.e., clockwise, viewed from above) so that the solvent-filled chambers are precisely (and quiescently) placed above the solution-filled chambers. It is assumed that the engagement of the two liquids has not produced any eddy currents or turbulence. The upper cylinder contains notches allowing the user to determine when the upper and lower cylinder holes are in alignment.

At this point, a discontinuous concentration gradient is created between the upper and lower chambers (the concentration profile now represents a step function). A timer is then started. After an appropriate time (t) has elapsed, the upper cylinder is *slowly* rotated *back* by 1/12 revolution to separate the upper and lower chambers, and the contents of the three isolated upper chambers are combined and mixed. The solute concentration (C) of the sampled liquid is then determined, and the diffusion coefficient is calculated through equation (8.18.16). This procedure can be repeated using different diffusing times, or initial solute concentrations.

The success of the experiment rests not only on the ability to engage and disengage the liquids quiescently but also on the accuracy and sensitivity of the analytical procedure used to determine the diffused solute concentration. Diffusion coefficients of solutes in liquids generally range between 10^{-5} and 10^{-8} cm^2 s^{-1}. An examination of equation (8.18.16) shows that for a Polson apparatus having $h \sim 3$ cm and for a diffusion time of 600 s, the bulk concentration of solute, C, is only about 0.1 percent of the initial concentration, C_0. Since it is desirable to keep C_0 as low as possible so that D_0 can be obtained accurately, the analytical sensitivity of the solute is an important issue.

Analytical Procedure

The amount of sucrose that diffused into the solvent column is determined via a colorimetric assay developed by Dubois et al.[9] It is sensitive, fast, and if done properly, reproducible. A quantity of phenol (5 percent in water) is mixed with the sucrose-containing solution. Concentrated sulfuric acid is then added, turning the solution yellow. The reaction probably involves first the protonation of a hydroxyl group (or groups) of the sucrose followed by dehydration to form an unstable carbocation (carbonium ion). This species would undergo an electrophilic addition to the para position of the phenol to form a more stable carbocation that is the light-absorbing species. These steps are outlined below.

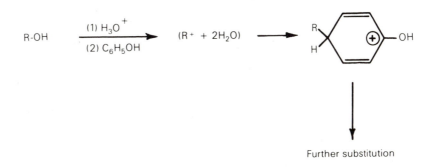

The absorption spectrum of this species is shown in Figure 8.18.3.

Safety Precautions

- ☐ Safety *goggles* MUST be worn.
- ☐ A laboratory coat that covers clothing and arms MUST be worn.
- ☐ Proper pipeting techniques must be followed. Never pipet by mouth. Consult your instructor if you need help.
- ☐ This experiment requires the use of concentrated sulfuric acid. This is an extremely corrosive and hazardous chemical. Exercise extreme caution in handling this material. If any acid comes in contact with the skin, immediately wash with copious amounts of water. Notify your instructor.
- ☐ Phenol odors may be produced. Work in a fume hood if possible.
- ☐ The experiment must not be carried out in a crowded laboratory.

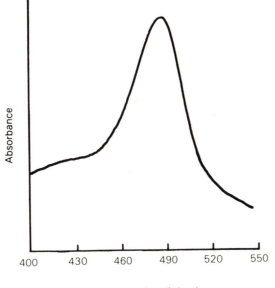

FIGURE 8.18.3 Absorption spectrum of the sucrose sample after reaction with phenol and concentrated sulfuric acid.

Procedure

Sucrose Assay Calibration

The spectrophotometer should be turned on and fully warmed up (at least 15 min).

To perform quantitative analysis of the sucrose solutions, a calibration plot must be obtained. Prepare a sucrose stock solution by dissolving 1.000 g of sucrose in enough distilled (or deionized) water to make 100 mL of solution.

Four calibrating solutions are made up. Dilutions ranging from ca. 1/5 to 3/2 percent of the stock solution are appropriate. For example: 1, 2, 1, and 4 mL of the stock solution are added to 500-, 250-, 100-, and 250-mL volumetric flasks, respectively, which are then filled with deionized water and mixed.

The sucrose assay is performed by pipeting 1 mL of the (dilute) sucrose solution to a clean, dry 25-mL Erlenmeyer flask. Pipet 1 mL of 5 percent aqueous phenol solution and mix. Concentrated sulfuric must then be added to this mixture. CAUTION! Be sure you know how to use the pipet-filling bulb properly to avoid spilling the acid.

CAUTION: This acid is extremely corrosive. Goggles *must* be worn. If any acid comes in contact with the skin or clothing, flood with copious amounts of cold water *immediately*.

To the sucrose-phenol solution, pipet 5 mL of concentrated sulfuric acid. The acid should be admitted directly to the *surface* of the solution to promote local heating. CAUTION! The flask must be pointed *away* from you and held near the top because of the heat released. The yellow color will appear immediately. Label the sample tube, set aside, and work up the other three sucrose samples. A spectrometric blank is prepared by mixing 1 mL of deionized water, 1 mL of 5 percent phenol, and 5 mL of concentrated sulfuric acid. Because this procedure produces a slight odor of phenol, the sample workup should be carried out in a fume hood. The pipet used to deliver the sulfuric acid should be placed in a clamped, 100-mL volumetric flask or other appropriate container that will catch the acid that drains from the pipet.

After the "fixed" sample has cooled to ambient temperature (the flasks can be immersed in cool water), *carefully* (remember: conc. H_2SO_4) transfer the solution to a spectrophotometer cuvette for analysis.

If you are unsure about the operation of the spectrophotometer, consult the instructor or read the description of this instrument in Experiment 19. With the instrument set at 495 nm, the 0 percent transmission is adjusted with the shutter closed; 100 percent is established using the clear spectrometric blank solution. After all of the "fixed" sucrose samples have equilibrated to ambient temperature, measure and record the optical densities at 495 nm.

Diffusion Measurements

Make up three sucrose solutions by dissolving 0.6, 1, and 2 g in enough deionized water to make 100 mL of solution.

Familiarize yourself with the Polson apparatus. Notice the markers that indicate whether or not the upper and lower chambers are aligned. Because the apparatus is not temperature-controlled, care should be taken to minimize contact with the hands. Read and record the ambient temperature immediately near the apparatus.

Adjust the upper cylinder so that the chambers are in alignment. Using a Pasteur pipet (or other suitable device) add about 1 mL of the dilute sucrose solution to three alternate chambers. The chambers must be filled so that the level of the solutions is about 1 cm *above* the boundary between the chambers. Rotate the upper cylinder forward (i.e., clockwise, viewed from above) about 1/12 revolution and add deionized water to the *unfilled* upper chambers. The water level should reach the top of the cylinder.

Now carefully and *slowly* rotate the upper cylinder *forward* so that the water-filled and solution-filled chambers are in precise alignment. Immediately start the timer or stopwatch. After about 3 min, rotate the upper cylinder *back* 1/12 revolution to separate the two chambers and stop the timer. Withdraw the liquid from the three filled upper chambers, add the samples together in a 25-mL Erlenmeyer flask, and label.

Remove all liquid from the apparatus and repeat the above procedure using nominal diffusing times of 5, then 7 min. After each run, the Polson cell should be dismantled, rinsed thoroughly with deionized water, and reassembled. It may be necessary to apply a *very* thin coating of silicone grease to the contact surfaces before reassembly; consult your instructor.

Repeat the procedure using the intermediate sucrose solution; diffusion times of 2, 3, and 5 min should be used. Finally, the concentrated sucrose solution is studied; somewhat shorter diffusion times should be used in this case. It may be necessary to adjust the contact times so that the absorbances of the samples do not exceed 0.8 to 0.9.

Using the same procedure described above, react the sucrose solutions with the phenol solution and sulfuric acid. After the samples have equilibrated to room temperature, measure the absorbances at 495 nm. Make sure the 0 and 100 percent levels of the spectrophotometer are checked after each measurement.

Finally, after the Polson cell is thoroughly rinsed, cleaned, and reassembled, measure the height of the upper cylinder, h, to the nearest 0.2 mm. Alternatively, your instructor will provide you with this information.

Data Analysis

1. Prepare a calibration curve from the sucrose solutions by plotting absorbance vs. sucrose composition (in g mL^{-1}). Use linear regression. The plot should pass through the origin.

2. Tabulate the diffusion data: initial concentration, diffusion time, absorbance.

3. Convert the sample absorbances to sucrose concentration using the calibration curve.

4. For each set of initial sucrose concentrations, plot C/C_0 vs. $t^{1/2}$, and using the measured (or provided) cylinder height, h, obtain the value of D [see equation (8.18.16)].

5. If your data permit, plot D vs. C_0 and smoothly extrapolate to $C_0 = 0$ in order to obtain D_0.

6. Estimate the hydrodynamic radius of sucrose using D_0 (or a mean value of D) and the solvent viscosity at the appropriate temperature; see equation (8.18.17).

7. Consult the literature (see the references), compare your result with reported values of D_0 for sucrose, and comment on the results.

Questions and Further Thoughts

1. Considering the nature of aqueous solvation of the sucrose molecule, i.e., the role of hydrogen bonding, how do you think D_0 (and hence the hydrodynamic radius) would change in an aqueous electrolyte, e.g., 0.1*M* KCl? Likewise, consider the above for a less ''polar'' solvent such as ethanol.

2. Suppose you were determining the diffusion coefficient of a molecule, M, which dimerizes in solution: $2M \overset{K}{\rightleftharpoons} D$ and you were not aware of this fact. Qualitatively, how would D depend on the total solute concentration? How about D_0? Assuming that the hydrodynamic radius of the dimer, R_D, were twice that of the monomer, R_M, and that the *observed* diffusion coefficient (at finite solute concentration), D_{obs}, reflected the mole fraction–weighted diffusion coefficients of the monomer and dimer, predict semiquantitatively how D_{obs} would depend on the solute concentration. Assume $K_D = 10^4$; $R_M = 1 \times 10^{-7}$ cm and that the solute concentration has a maximum value of $1 \times 10^{-4} M$.

Notes

1. A. Polson, *Nature, 154,* 823 (1944).
2. P. W. Linder, L. R. Nassimbeni, A. Polson, and A. L. Rodgers, *J. Chem. Educ., 53,* 330 (1976).
3. C. Tanford, ''Physical Chemistry of Macromolecules,'' p. 354, Wiley (New York), 1961.
4. J. T. Edward, *J. Chem. Educ., 47,* 261 (1970).
5. I. N. Levine, ''Physical Chemistry,'' 2nd ed., pp. 470–471, McGraw-Hill (New York), 1983.
6. Based on a treatment of charge mobility in electrolytes by Onsager,[7] Gordon[8] has presented an expression for the concentration dependence of D.
7. L. Onsager and K. Fuoss, *J. Phys. Chem., 36,* 2689 (1932).
8. F. Gordon, *J. Chem. Phys., 5,* 522 (1937); *7,* 89 (1939). See also L. J. Gosting and M. S. Morris, *J. Am. Chem. Soc., 71,* 1998 (1949).
9. M. Dubois, K. A. Gilles, J. K. Hamilton, P. A. Rebers, and F. Smith, *Anal. Chem., 28,* 350 (1956).

Further Readings

A. W. Adamson, ''A Textbook of Physical Chemistry,'' 3rd ed., p. 376, Academic Press (Orlando, Fla.), 1986.

R. A. Alberty, ''Physical Chemistry,'' 7th ed., pp. 821–827, Wiley (New York), 1987.

P. W. Atkins, ''Physical Chemistry,'' 3rd ed., pp. 612–613, 674–682, W. H. Freeman (New York), 1986.

A. Einstein, ''Investigations on the Theory of the Brownian Movement,'' R. Fürth, ed., Dover (New York), 1956.

I. N. Levine, ''Physical Chemistry,'' 2nd ed., pp. 464–471, McGraw-Hill (New York), 1983.

J. H. Noggle, ''Physical Chemistry,'' pp. 439–443, Little, Brown (Boston), 1985.

Experiment 18

Diffusion Coefficient in Solution

NAME _____ DATE _____

Sucrose Assay Calibration: _____

g sucrose in _____ mL

Sample dilution factor   Absorbance at 495 nm

_____   _____

_____   _____

_____   _____

_____   _____

_____   _____

Diffusion measurements Upper chamber height _____   Temperature _____

Solution concentration   Contact time   Absorbance at 495 nm

_____   _____   _____

_____   _____   _____

_____   _____   _____

_____   _____   _____

_____   _____   _____

_____   _____   _____

_____   _____   _____

_____   _____   _____

_____   _____   _____

_____   _____   _____

Experiment 18

Diffusion Coefficient in Solution

NAME _____   DATE _____

Sucrose Assay Calibration: _____   g sucrose in _____ mL

Sample dilution factor   Absorbance at 495 nm

_____   _____

_____   _____

_____   _____

_____   _____

_____   _____

Diffusion measurements Upper chamber height _____   Temperature _____

Solution concentration   Contact time   Absorbance at 495 nm

_____   _____   _____

_____   _____   _____

_____   _____   _____

_____   _____   _____

_____   _____   _____

_____   _____   _____

_____   _____   _____

_____   _____   _____

_____   _____   _____

_____   _____   _____

□ Experiment 19

The Kinetics of a Reaction in Solution

Objective

To measure the rate constant for the reaction between 2,4-dinitrochlorobenzene and piperidine in solution; to test the reaction mechanism; and to determine the Arrhenius parameters of the reaction.

Introduction

From both a practical and a theoretical point of view, quantitative information about the rate of a chemical reaction, including the way in which the rate depends on the concentrations of all relevant chemical species and on the temperature, is very valuable information. It can be used, for example, to predict the rate of reaction under a given set of conditions. Alternatively, the conditions affecting the reaction could be tailored to a specific application that requires a certain reaction rate. Moreover, the information obtained concerning the dependence of the reaction rate on the conditions can provide an understanding of how, on a molecular level, the reaction actually takes place, i.e., the *mechanism* of the reaction.

Knowledge of the reaction mechanism is very important and useful. The reaction mechanism describes in a detailed way how the process actually takes place, and includes such considerations as the orientation between the reacting molecules as they interact, and the number and nature of steps involved in the overall reaction. Because it is so fundamental, detailed knowledge of the reaction mechanism often allows one to deduce the role played by the solvent in a given reaction. For example, one might suspect that a change in solvent polarity would affect the reaction rate. Insights into the relationship between molecular structure and chemical reactivity are also gained from mechanistic studies. In addition, knowledge of the mechanism could lead to the design and use of a catalyst that would significantly accelerate the reaction rate; this would have significant practical consequences.

Some basic aspects of reaction kinetics will be introduced first. Consider a general chemical reaction in which the reactants *and* products are known. The balanced equation is written as

$$aA + bB \rightarrow cC + dD. \tag{8.19.1}$$

The speed of the reaction can be determined by measuring either the rate of disappearance of one (or more) of the reactants or the appearance of one (or more) of the products. The actual magnitude of the reaction rate will depend on which species is being measured because of the stoichiometric relationship that exists between these species. In order to avoid the need to specify which species is monitored in characterizing the rate, the reaction (8.19.1) can be expressed more conveniently as

$$cC + dD - aA - bB = 0, \tag{8.19.2}$$

or, in a more formal and compact way:

$$\sum v_i X_i = 0, \tag{8.19.3}$$

where X_i is the *i*th chemical species and v_i is its stoichiometric coefficient in the balanced equation. ν_i is positive for a product and negative for a reactant. Equation (8.19.3) follows the usual convention of expressing a change as final state minus initial state.

The rate of reaction, or reaction velocity, ν, is simply

$$\nu = \frac{1}{\nu_i}\frac{d[X_i]}{dt} \qquad \text{(moles dm}^{-3}\text{ s}^{-1}), \qquad (8.19.4)$$

where $[X]$ is the molar concentration of the ith species. ν is always positive because, for a reactant, both ν_i and $(d[X_i]/dt)$ are negative (the reactant concentration decreases with time); for a product, both ν_i and $d[X_i]/dt$ are positive quantities.

In general the reaction rate, ν, is a function of the concentration of one or more of the reactants and/or products, and can also depend on the concentration of species not represented in the *overall* reaction. ν is almost always found to depend on the temperature, T. Thus,

$$\nu = f([X_1], [X_2], [X_3], \dots [X_N], T). \qquad (8.19.5)$$

The equation represented in (8.19.5) is called a *differential rate law*. It should be emphasized that, in general, the concentration terms, $[X_i]$, in equation (8.19.5) are time-dependent. With respect to the general reaction illustrated in equation (8.19.2), one possible rate law is

$$\nu = k[A]^\alpha[B]^\beta[C]^\gamma[D]^\delta[Y]^\epsilon, \qquad (8.19.6)$$

where k is a constant (independent of time) called the *rate constant*, $[A]$, $[B]$, . . . are concentrations, and $[Y]$ is a species not represented in the balanced equation but whose concentration affects the rate of reaction. The point to be made about the expression in (8.19.6) is that the chemical species all appear as factors in the rate law, and they are all raised (in general) to different exponents.

It must be strongly emphasized that: (1) one does not know in advance of actually determining the rate law [equations (8.19.5) or (8.19.6)] which chemical species actually appear in the expression, (2) there is not necessarily any relationship between the stoichiometric coefficient of a species in the balanced equation (8.19.2) and the power to which that species (concentration) is raised in the rate law. These are important points. The rate law can only be determined by experiment, and it is the form of the rate law thus obtained that bears relevance to the mechanism of the reaction. Conversely, the mechanism deduced for a reaction leads to a prediction of what the rate law for the reaction should be. That mechanism could thus be tested by comparing the predicted and experimentally determined rate law.

Usually, the exponents in the rate law are integers or rational numbers (the ratio of integers). For reaction in solution, these exponents are commonly 1, 2, or -1. If the exponent of a given component is negative, increasing its presence in the system will slow the reaction rate down; such a species is called a *poison*. On the other hand, a material that does not appear in the overall reaction but whose presence speeds up the reaction rate will appear in the rate law with a positive exponent. This species is called a *catalyst*.

The exponent of a given species (e.g., α for the component A) is referred to as the *order* of the reaction with respect to that species. The sum of all the exponents in the rate law expressed in equation (8.19.6) is called the *overall* order of the reaction.

As mentioned above, the reaction rate is usually temperature-dependent. For many reactions, both in the gas phase and in solution, this dependence can be expressed in terms of the rate constant, $k(T)$:

$$k(T) = A \exp\left(\frac{-E_a}{RT}\right) \tag{8.19.7}$$

A rate constant whose temperature dependence is accounted for by equation (8.19.7) is referred to as an Arrhenius rate constant. A is the preexponential, or A factor, and E_a is the *activation energy*. E_a can be interpreted as the potential energy barrier that must be surmounted in converting reactants to products. If there are several such barriers in the reaction mechanism, E_a denotes the largest of these barriers. These barriers often involve the breaking or rearrangement of chemical bonds or mass transport through a fluid medium. In the latter case, E_a is close in value to E_η, the activation energy to bulk viscous flow. (See p. 229.)

The reaction studied in this experiment is an addition involving two organic molecules, 2,4-dinitrochlorobenzene (DNCB) and piperidine (Pip). The products of the reaction are known to be 2,4-dinitrophenylpiperidine (DNPP) and piperidine hydrochloride (Pip:HCl). The reaction is shown below.

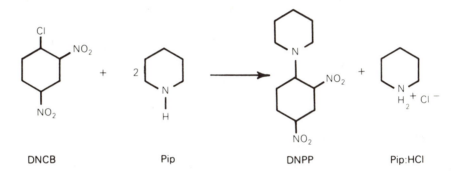

DNCB   Pip   DNPP   Pip:HCl

This reaction involves the formation of a nitroaniline derivative that is important in the manufacture of dyes. It has been studied in detail,[1,2] and the mechanism proposed contains two fundamental, or elementary, steps. The first of these involves the nucleophilic attack of the organic base, Pip, on the Cl-bearing C atom in DNCB. This is an example of an S_N2 mechanism in which the intermediate or transition state is stabilized by the electron-withdrawing property of the nitro groups, which are ortho and para to the C—Cl site. The C_1 atom is thus activated:

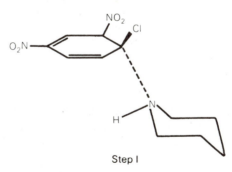

Step I

The formation of this intermediate is presumed to take place slowly because of the energy expended in the bond rearrangement.

The second elementary step proposed is also bimolecular and involves the participation of another piperidine molecule. In this step, the second piperidine functions simply as an organic base in removing the proton (H^+) from the piperidine that is being added to the dinitroaryl group. This step is an example of a base-catalyzed reaction because it speeds up the formation of the product from the intermediate while leaving the catalyst unchanged.

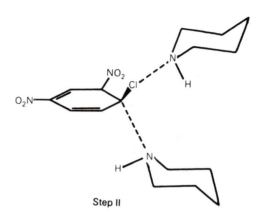

Step II

This two-step mechanism is summarized as follows:

1.
$$\text{DNCB} + \text{Pip} \underset{\text{fast, } k_{-1}}{\overset{\text{slow, } k_1}{\rightleftarrows}} \text{intermediate}$$

2.
$$\text{Intermediate} + \text{Pip} \xrightarrow{k_2} \text{DNPP} + \text{Pip:HCl}$$

Overall:

$$\text{DNCB} + 2\,\text{Pip} \longrightarrow \text{DNPP} + \text{Pip:HCl}.$$

Notice that the intermediate could also decompose unimolecularly to form DNPP and HCl, and HCl would rapidly react with Pip to form the hydrochloride salt. Thus, two additional elementary steps might be

3.
$$\text{Intermediate} \xrightarrow{\text{slow}} \text{DNPP} + \text{HCl}$$

4.
$$\text{HCl} + \text{Pip} \xrightarrow{\text{fast}} \text{Pip:HCl}.$$

It can be argued that because piperidine is already in solution in appreciable concentration (to allow step 1 to proceed at a reasonable rate), the rate of the bimolecular product formation step 2 is greater than that of the unimolecular process 3. If k_2 and k_3 denote the rate constants of these respective steps, the previous statement means that $k_2[\text{Pip}] > k_3$.

By applying the rate-determining step approximation to this mechanism, it can be proposed that since there is a slow step 1 followed by a fast step to the formation of product 2, the reaction rate is approximately equal to the rate of the slow elementary step. Hence,

$$v = \frac{d[\text{DNPP}]}{dt} \simeq k_1[\text{DNCB}][\text{Pip}]. \qquad (8.19.8)$$

If $[\text{DNCB}]_0$ and $[\text{Pip}]_0$ represent the initial concentrations of the reactants, and if $[\text{DNPP}]_0 = 0$, it follows from the stoichiometry of the reaction that at time t, the reactant concentrations are

$$[\text{DNCB}]_t = [\text{DNCB}]_0 - [\text{DNPP}]_t,$$

and $\qquad\qquad\qquad\qquad\qquad\qquad\qquad\qquad\qquad\qquad\qquad$ (8.19.9)

$$[\text{Pip}]_t = [\text{Pip}]_0 - 2[\text{DNPP}]_t.$$

Substituting the time-dependent expressions in equation (8.19.9) into the rate law (8.19.8) results in a differential equation in only *one* variable and the time:

$$\frac{d[\text{DNPP}]}{dt} = k_1\{[\text{DNCB}]_0 - [\text{DNPP}]\}\{[\text{Pip}]_0 - 2[\text{DNPP}]\}. \quad (8.19.10)$$

The variables can be separated, and the differential equation becomes

$$\frac{d[\text{DNPP}]}{\{[\text{DNCB}]_0 - [\text{DNPP}]\}\{[\text{Pip}]_0 - 2[\text{DNPP}]\}} = k_1 dt. \qquad (8.19.11)$$

The integral on the left-hand side of equation (8.19.11) is of standard form, and the differential equation can be integrated subject to the boundary condition: $t = 0$; $[\text{DNPP}] = 0$. The result is

$$\frac{1}{[\text{Pip}]_0 - 2[\text{DNCB}]_0} \ln\left\{\frac{[\text{DNCB}]_0([\text{Pip}]_0 - 2[\text{DNPP}])}{[\text{Pip}]_0([\text{DNCB}]_0 - [\text{DNPP}])}\right\} = k_1 t. \qquad (8.19.12)$$

This integrated rate law appears to be rather complex; that is, the time dependence of $[\text{DNPP}]$ is not of simple form. Although equation (8.19.12) could be tested by performing a numerical analysis of $[\text{DNPP}]$ vs. time data, it is far easier (and more instructive) to test the predicted rate law by applying two different experimental constraints:

Condition I

The reactants are stoichiometrically linked. This means that $[\text{Pip}]_0 = 2[\text{DNCB}]_0$. The significance of this ratio of reactants is that it is valid for *all time,* i.e., $[\text{PIP}]_t = 2[\text{DNCB}]_t$. This constraint poses a problem in equation (8.19.12) because of the singularity of the coefficient of the ln term. If this condition is applied to equation (8.19.10), however, the differential equation becomes much simpler:

$$\frac{d[\text{DNPP}]}{dt} = 2k_1\{[\text{DNCB}]_0 - [\text{DNPP}]\}^2. \qquad (8.19.13)$$

This is an example of second-order kinetics [rather than mixed second-order, as in equation (8.19.10)] and equation (8.19.13) is readily integrated to a more familiar form

$$\frac{1}{[\text{DNCB}]_0 - [\text{DNPP}]} = \frac{1}{[\text{DNCB}]_0} + 2k_1 t. \qquad (8.19.14)$$

Condition II

One of the reactants is present in great excess. In this case, we choose $[\text{Pip}]_0 \gg [\text{DNCB}]_0$; i.e., the system is flooded with piperidine. The constraint kinetically *isolates* the behavior of the other reactant, DNCB, since there is a negligible decrease in the piperidine concentration with time. Hence, $[\text{Pip}]_t \cong [\text{Pip}]_0$, and $[\text{Pip}]_0 \gg 2[\text{DNPP}]$. In this case equation (8.19.10) becomes

$$\frac{d[\text{DNPP}]}{dt} = k_1[\text{Pip}]_0\{[\text{DNCB}]_0 - [\text{DNPP}]\}. \qquad (8.19.15)$$

Integration of this differential equation provides, after rearrangement,

$$\ln\left\{1 - \frac{[\text{DNPP}]}{[\text{DNCB}]_0}\right\} = -\{[\text{Pip}]_0 k_1\}t. \qquad (8.19.16)$$

This is an example of a *pseudo-first-order* reaction because the system behaves just as if the formation of product was the result of a first-order process. The factor, $[\text{Pip}]_0 k_1$, is not a true constant since it depends on the arbitrary value of $[\text{Pip}]_0$.

Experimental Method

This experiment will be carried out over a 2-week period; one laboratory session is devoted to each of the limiting conditions described above. Under each of the reaction conditions, measurements will be carried out at two temperatures, 0 and 25°C. This will allow an estimate of the activation parameters to be made (two points 25° apart do not provide good reliability).

The course of the reaction could be followed in several ways. One would be to take advantage of the fact that one of the products (Pip:HCl) is an ionic salt and thus to use a conductometric procedure; another is to titrate the Cl⁻ ion that is formed. This experiment uses a photometric technique called spectrophotometry. The basic principles of this method are outlined in the appendix. A description of the apparatus, a spectrophotometer, is also included. This method is appropriate because while the reactants both absorb light principally in the ultraviolet (DNCB having an absorption maximum at ca. 250 nm), the aniline derivative has an intense absorption peak at 375 nm, much nearer to the visible. In ethanol solution, DNPP has a distinct yellow appearance, and its presence can be readily quantitated in the near ultraviolet–visible region. Absorption spectra of DNCB and DNPP are shown in Figure 8.19.1.

The reactants must be rapidly mixed and allowed to react at a given temperature. At measured time intervals, samples are withdrawn and analyzed for product. A method must be used to stop the reaction at the point of sample withdrawal; otherwise, the reaction time will be ambiguous. This is accomplished by immediately adding the reaction sample to a *quenching solution* of aqueous ethanol acidified with sulfuric acid. Once placed in this medium, the piperidine is very rapidly protonated, thus making it unreactive and stopping the reaction.

The DNPP analysis is based on Beer's law, which holds that a linear relationship exists between the *absorbance* of a species and its concentration (e.g., *M*). Absorbance, *A*, is defined as $-\log T$, where *T* is the transmittance, the ratio of light transmitted by the solution to that incident on the sample, $T = I/I_0$. Absorbance is sometimes called optical density, or OD. Experimentally, I_0 and I are obtained by determining the analyzing light intensity, first through pure solvent (I_0) and then through a DNPP-containing solution (I). Beer's law states that

$$A = \epsilon_\lambda [\text{DNPP}]\ell = -\log\left(\frac{I}{I_0}\right), \qquad (8.19.17)$$

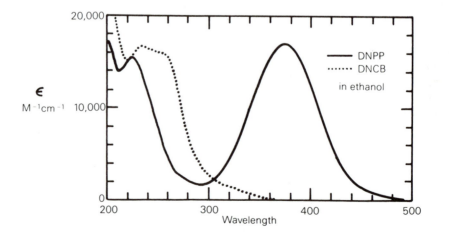

FIGURE 8.19.1   Absorption spectra of starting material, DNCP (...), and product, DNPP (____).

where ϵ_λ is the *molar extinction coefficient* ($dm^3\ mol^{-1}\ cm^{-1}$) at the wavelength λ, and ℓ is the optical path length in centimeters. [DNPP] is the molar concentration of the product.

The first step in this experiment is to obtain a calibration curve of absorbance vs. [DNPP] in order to get a value of $\epsilon_\lambda \ell$. Because the accuracy of a spectrometric measurement is optimal between A values of roughly 0.1 to 0.8, calibration curves for the two limiting conditions will be obtained at different wavelengths, 460 and 390 nm. These wavelengths are chosen to provide appropriate absorbances at the different (DNPP) values encountered in Parts I and II of the experiment.

During the laboratory period in which the reaction is studied under second-order conditions, the calibration curve appropriate for the DNPP concentrations encountered will be obtained at 460 nm. In this part, the [DNPP] will be relatively high, requiring an analysis wavelength on the tail of the [DNPP] absorption spectrum where ϵ is relatively small (ca. 870 $dm^3\ mol^{-1}\ cm^{-1}$).

Under the pseudo-first-order reaction conditions, established and studied in Part II, the DNPP concentration will be lower than in Part I. The analyzing wavelength to be used in this case is 390 nm where the DNPP absorbs more strongly; accordingly another [DNPP] calibration curve will be obtained at this wavelength.

Safety Precautions

- [ ] Safety *goggles* MUST be worn.
- [ ] A laboratory coat that covers clothing and arms should be worn.
- [ ] Proper pipeting techniques MUST be followed. Never, under any circumstances, pipet by mouth. Consult your instructor if you need help or information about the proper technique.
- [ ] This experiment involves the use of solutions of 1-chloro-2,4-dinitrobenzene (DNCB), which is a highly toxic irritant. When handling this material, gloves must be worn. If possible, prepare reaction samples in a fume hood. If any solution comes in contact with the skin, wash immediately with soap and water. Ask your instructor to show you the Material Safety Data Sheet (MSDS) for this and other chemicals encountered in the laboratory.
- [ ] If other reagents or chemicals come into contact with the skin, wash immediately with soap and water.
- [ ] Waste material should be deposited in a special, marked container.

Procedure

Part I, Simple Second-Order Kinetics

The following stock solutions will be provided: 0.100*M* DNCB and 0.400*M* Pip for the reaction, and 0.020*M* DNPP for the 460-nm calibration. If these are to be made up before the experiment, see the appendix for details. 50-mL volumetric flasks will serve as reaction vessels.

Initial reactant concentrations of $[DNCB]_0 = 0.020M$ and $[Pip]_0 = 0.040M$ are found to react sufficiently rapidly for a one-laboratory-period experiment.

SET UP AND EQUILIBRATE REACTANTS:

1. Add about 30 mL of 95 percent ethanol solvent and *exactly* 10 mL of the 0.100*M* DNCB solution to each of two 50-mL volumetric flasks. Place each volumetric flask in a different constant-temperature bath (e.g., 0 and 25°C). The other solutions that are to be added to these reactor flasks will also be kept in the respective temperature baths. The pipets used in the 0°C experiment should be immersed in 0° ethanol until needed.

2. Place approximately 10 to 20 mL of 0.400M Pip solution into each of two clean 50-mL volumetric flasks. Next add about 10 to 20 mL of solvent to each of two other clean 50-mL volumetric flasks. Place one set of these flasks in the 0° and the other in the 25° baths. There should then be three flasks in each of the baths. Allow them to equilibrate.

3. While these flasks are equilibrating, label eight 25-mL volumetric flasks (four for each temperature) and fill each with about 15 mL of the quenching solution that is provided. These flasks do not have to be temperature-equilibrated because the reaction stops as soon as the DNCB/Pip mixtures are added to the acidified ethanol.

460 NM DNPP CALIBRATION:

4. Obtain the data for the [DNPP] calibration at 460 nm. You will be shown how to use the spectrophotometer. Make sure you understand both the principle and the operation of the spectrophotometer because the success of the experiment rests on the proper acquisition of the calibration data, as well as on the photometric analysis of the reaction mixture.

Prepare a dilute DNPP solution by pipeting 1 mL of the DNPP stock solution (0.020M) into a 25-mL volumetric flask; fill to the mark with quenching solution, stopper, and mix. Determine the transmission (or absorbance) at 460 nm.

Set aside three 10-mL volumetric flasks. Pipet into the respective flask 2, 4, and 8 mL volumes of the diluted DNPP stock dilution; label each flask. Add quenching solution to each flask up to the mark; stopper and invert several times. Measure the transmission (or absorbance) of each of these solutions starting with the most dilute. These four readings will be used to construct the 460-nm calibration curve.

RUN THE REACTION:

5. The reaction vessels should be well equilibrated by now. Initiate the reaction by pipeting 5 mL of the temperature-equilibrated piperidine solution into the flask containing the DNCB in the 25°C bath. Start the timer (or record the time shown by your watch), immediately add the temperature-equilibrated solvent to the mark, stopper and shake vigorously, and quickly return the reactor flask to the bath. Repeat for the 0° bath and mark the time at which the Pip was added.

6. Starting with the 25°C sample, remove a 1-mL sample from the reaction mixture as soon as possible and deliver it to one of the 25-mL volumetric flasks containing the quenching solution. Read and record the elapsed time (if you are using a timer, do not shut it off). Do the same for the 0°C reaction. Continue to withdraw 1-mL samples from the reactor vessels, placing them in quenching solution; use (approximately) *five*-minute intervals for the samples in the 25°C bath and *fifteen*-minute intervals for the 0°C samples. In all cases, label the samples and record the elapsed times. At least four "timed" samples should be obtained for each reaction temperature.

7. As soon as possible after a given 1-mL sample has been added to the volumetric flask with the quenching solution, fill the flask to the mark with quenching solution and stopper and shake. Then determine the absorbance at 460 nm.

Procedure

Part II, Pseudo-First-Order Kinetics

In this part of the experiment, the reaction will be studied under pseudo-first-order conditions. The methodology is basically identical to that used in Part I. To keep [Pip] > [DNCB], the initial DNCB concentration will be much lower than that in

Part I, while the piperidine concentration will be the same. Initial concentrations of $[DNCB]_0 = 5.00 \times 10^{-4} M$ and $[Pip]_0 = 0.040M$ will be satisfactory.

The following stock solutions are needed: $[DNCB] = 0.010M$, $[Pip] = 0.400M$ for the reaction, and $[DNPP] = 0.010M$ for the 390-nm calibration.

SETUP AND EQUILIBRATION OF REACTANTS:

1. As before, DNCB, Pip, and solvent samples will be temperature-equilibrated. Pipet 2.5 mL of the 0.010M DNCB solution into each of two 50-mL volumetric reaction flasks; then add about 40 mL of solvent to each. Add about 15 mL of the 0.400M Pip solution to two other 50-mL flasks, and about 10 mL of solvent to two other flasks. Place a set of three of these flasks in each temperature bath. Consecutively number eight 10-mL volumetric flasks and add about 5 mL of quenching solution to each. These will receive the reaction samples.

390 NM DNPP CALIBRATION:

2. It is necessary to prepare a 0.4 percent solution of the 0.010M DNPP stock solution. This is conveniently done by first making 10 mL of a 10 percent solution and then by using this to make 25 mL of a further 25-fold dilution. Pipet into four clean 10-mL volumetric flasks 1, 2, 3, and 4 mL of the 0.4 percent DNPP stock solution. Fill each to the mark with quenching solution, stopper, and shake. Measure the absorbances of these four dilutions as well as the 0.4 percent stock solution at 390 nm. Make sure that the 100 percent transmission level is set using a clean spectrometer cell containing quenching solution.

RUN THE REACTION:

3. Pipet 5 mL of the Pip solution into the DNCB solution in the reaction flask. Start recording the time. Immediately fill with temperature-equilibrated solvent, stopper and shake thoroughly, and then replace it in the bath. Do likewise for the 0°C sample; be sure to use the 0°-equilibrated pipet for delivering the Pip solution.

4. Quickly withdraw a 1-mL sample from each reaction flask, starting with the 25° sample, and add to one of the 10-mL flasks containing quenching solution.

5. Continue to withdraw 1-mL samples from the reaction flasks, adding the samples to quenching solution and preparing 10-mL analysis samples. Samples should be withdrawn from the 25°C reaction vessel approximately every *2-1/2* minutes, and from the 0°C reaction every *fifteen* minutes. A total of at least four "timed" samples should be obtained for each reaction.

6. Measure the absorbance of the samples at 390 nm. Make sure that the spectrophotometer reads 0 percent T for the "dark" calibration, and 100 percent T for a sample cell filled with pure quenching solution.

Calculations and Data Analysis

1. Tabulate the [DNPP] calibration data obtained at 460 and 390 nm; this consists of absorbance and [DNPP] values. For each wavelength, plot absorbance vs. [DNPP] and obtain the least squares values of $\epsilon_\lambda \ell$. These plots should pass through the origin, and a linear regression program that fixes the y intercept at 0 should be used if available. Knowing the path length of the absorption cells, report the molar extinction coefficients at 460 and 390 nm. Note that the actual values measured depend on the spectrophotometer bandpass used and thus may not agree with the spectra shown in Figure 8.19.1.

2. Tabulate the {absorbance, time} data for the 0 and 25°C reactions for Parts I and II. Using the ϵ values obtained from the calibration curves, convert the absorbances into [DNPP] and enter these concentrations in the same table.

3. Using the appropriate dilution factor, transform the analyzed [DNPP] values into those pertinent to the reaction conditions. Then determine {[DNCB]$_0$ − [DNNP]} and tabulate with the reaction times.

4. For the Part I data, prepare plots according to equation (8.19.14) and obtain the linear regression slopes, and from these determine the specific rate constants, k_1 at 0 and 25°C. Likewise, analyze the data in Part II according to equation (8.19.16) and obtain the least squares values of the pseudo-first-order rate constants, and from them determine k_1 at 0 and 25°C.

5. From the two rate constants at the two temperatures, estimate the activation energy and frequency factor for the reaction.

6. Compare the values of k_1 obtained from Parts I and II (at each temperature). Comment on any discrepancies.

Questions and Further Thoughts

1. Another approach to the analysis of the mechanism suggested on p. 276 is to apply the steady-state approximation to the intermediate. Show that this approach gives equation (8.19.8) under conditions where k_2[Pip] $\gg k_{-1}$.

2. Indicate the expected reaction order if k_2[Pip] $\ll k_{-1}$. Derive the modified equation appropriate to conditions I and II under this circumstance and suggest which forms can be plotted to give a linear result.

3. Considering the structure of the proposed intermediate (see Step I), and its decomposition into the product (see Step II), what would the effect on the reaction rate be if a secondary amine such as di(*t*-butyl)amine were used? Explain.

4. The stability of the intermediate can be considered in terms of two simple factors: steric bulkiness of the amine, and its electron-donating ability. How does each of these factors affect the rate of reaction? Suggest amines that would allow these factors to be "varied" experimentally (as independently as possible).

5. Transition-state theory allows a rate constant to be expressed in terms of the thermodynamic parameters, activation entropy ($\Delta S^{\ddagger}$) and activation enthalpy ($\Delta H^{\ddagger}$). For a second-order reaction in solution, the rate constant can be expressed as

$$k = \frac{k_B T}{h} \exp\left(\frac{\Delta S^{\ddagger}}{R}\right) \exp\left(-\frac{\Delta H^{\ddagger}}{RT}\right),$$

where k_B is the Boltzmann constant and h is Planck's constant. T is the absolute temperature, and $\Delta H^{\ddagger}$ can be written in terms of E_a, the Arrhenius activation energy [see equation (8.19.7)]: $\Delta H^{\ddagger} = E_a - RT$. Using your values of the second-order rate constant at 298 K and the activation energy, determine the entropy and enthalpy of activation for the reaction. Express the rate constant in the standard state of 1 mole dm^{-3}. Do this also for the standard state of 1 mole cm^{-3}.

Notes

1. O. L. Brody and F. R. Cropper, *J. Chem. Soc., 1950*, 507 (1950).
2. J. F. Bunnett and H. D. Crockford, *J. Chem. Educ., 33*, 552 (1956).

Further Readings

R. A. Alberty, "Physical Chemistry," 7th ed., pp. 679–687, Wiley (New York), 1987.

P. W. Atkins, "Physical Chemistry," 3rd ed., pp. 687–705, W. H. Freeman (New York), 1986.

G. W. Castellan, "Physical Chemistry," 3rd ed., pp. 799–813, Addison-Wesley (Reading, Mass.), 1983.

I. N. Levine, "Physical Chemistry," 2nd ed., pp. 502–513, McGraw-Hill (New York), 1983.

J. H. Noggle, "Physical Chemistry," pp. 466–496, Little, Brown (Boston), 1985.

Appendix

A schematic diagram of the spectrophotometer used in the experiment (e.g., a Bausch & Lomb Spectronic 20) is shown in Figure 8.19.2. The analyzing light is produced by a high-intensity tungsten bulb (lamp) whose output is guided onto the surface of a diffraction grating. This grating disperses, or breaks up, this polychromatic ("white") light into a range of wavelengths. The selection of a monochromatic band of light is achieved by positioning the grating at a certain angle relative to the incident light. Thus the wavelength control (the wavelength cam) simply rotates the grating. An exit slit determines the wavelength range selected from the grating (bandpass), and allows a narrow beam of light to impinge on the sample.

The light passes through the sample and falls on the surface of the phototube which converts the incident light energy to an electric current. The output current is amplified and displayed on the meter on top of the instrument. The occluder is used to block the phototube from the analyzing light and allows the "zero" light level (0 percent transmission) to be set. The phototube and associated electronics are designed to produce an output signal that is *linear* with the incident light intensity. Thus an I_0 value is established with the sample tube filled with solvent, and this is arbitrarily used to set the meter at 100 percent transmission. I is then determined with a sample tube containing the solution to be analyzed. The magnitude of the transmitted light is directly displayed on the meter. It is important to periodically check (and reset if necessary) the 0 percent (dark) and 100 percent (solvent) levels during an experiment. It is also crucial to reset the 100 percent level when the wavelength is changed. This is because both the light intensity passed by the grating system and the intensity-to-current response of the phototube depend on the wavelength of the light.

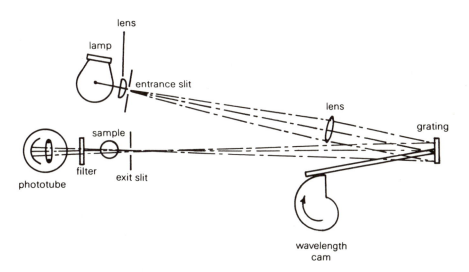

FIGURE 8.19.2   Schematic diagram of single-beam spectrophotometer.

Experiment 19

Kinetics of a Reaction in Solution

NAME _____ DATE _____

First laboratory period

Condition (I): Second-Order Kinetics:

Calibration curve at 460 nm:

Concentration	OD_{460}
_____	_____
_____	_____
_____	_____
_____	_____

Temperature = 0°C:

Time, min	OD_{460}
_____	_____
_____	_____
_____	_____
_____	_____
_____	_____

Temperature _____

Time, min	OD_{460}
_____	_____
_____	_____
_____	_____
_____	_____
_____	_____

Kinetics of a Reaction in Solution

NAME _____ DATE _____

Second laboratory period

Condition (II): Pseudo First Order

Calibration Curve at 390 nm:

Concentration	OD_{390}
_____	_____
_____	_____
_____	_____
_____	_____

t_∞ Data:

OD_{460} _____

OD_{390} _____

Temperature = 0°C:

Time, min	OD_{390}
_____	_____
_____	_____
_____	_____
_____	_____
_____	_____

Temperature _____

Time, min	OD_{390}
_____	_____
_____	_____
_____	_____
_____	_____
_____	_____

Experiment 19

Kinetics of a Reaction in Solution

NAME _____   DATE _____

First laboratory period

Condition (I): Second-Order Kinetics:

Calibration curve at 460 nm:

Concentration	OD_{460}
_____	_____
_____	_____
_____	_____
_____	_____

Temperature = 0°C:

Time, min	OD_{460}
_____	_____
_____	_____
_____	_____
_____	_____
_____	_____

Temperature _____

Time, min	OD_{460}
_____	_____
_____	_____
_____	_____
_____	_____
_____	_____

Kinetics of a Reaction in Solution

NAME _____   DATE _____

Second laboratory period

Condition (II): Pseudo First Order

Calibration Curve at 390 nm:

Concentration	OD_{390}
_____	_____
_____	_____
_____	_____
_____	_____

t_∞ Data:

OD_{460} _____

OD_{390} _____

Temperature = 0°C:

Time, min	OD_{390}
_____	_____
_____	_____
_____	_____
_____	_____
_____	_____

Temperature _____

Time, min	OD_{390}
_____	_____
_____	_____
_____	_____
_____	_____
_____	_____

☐ **Experiment 20**

Kinetics of a Diffusion-Controlled Reaction

Objective

To determine the rate constant and collision diameter of a diffusion-controlled reaction (photoexcited anthracene and carbon tetrabromide) using the technique of fluorescence quenching.

Introduction

This experiment examines a very fast bimolecular reaction between two different species in solution. We will assume that the reactants, A and B, which are electrically neutral, undergo independent, random motion. There is a certain probability that A and B will encounter each other at some close distance, R, where R is approximately equal to the sum of the molecular radii. This arrangement is called an *encounter complex*. Because there is a tendency for the solutes A and B to maintain constant random motion, it is inevitable that, once loosely held in this complex, they will subsequently separate from each other unless a chemical reaction (or other definitive process) first links them together. Since this random motion takes place in a "bath" of solvent molecules (assumed to be unreactive with respect to A and B), the separation of the A-B encounter complex will be impeded by the neighboring solvent molecules. This artificial holding together of the two molecules is called the "cage effect." The important point is that there is a kinetic competition between a net (thus measurable) reaction between A and B via the collision complex, and the release of A and B from the solvent cage back into the solvent medium where no reaction occurs. These processes can be represented by the following kinetic scheme:

$$A_{\text{solv}} + B_{\text{solv}} \underset{k_{-1}}{\overset{k_1}{\rightleftharpoons}} (A\text{-}B)_{\text{solv}},$$

$$(A\text{-}B)_{\text{solv}} \xrightarrow{k_2} \text{product(s)},$$

(8.20.1)

where A_{solv}, e.g., represents the solvated A molecule, and $(A\text{-}B)_{\text{solv}}$ is the encounter complex. The rate constants for these elementary steps are denoted as k_1, for the bimolecular formation of the encounter complex, or diffusion into the solvent cage; k_{-1}, for the unimolecular dissociation of the complex, or diffusion out of the solvent cage; and k_2 for the "unimolecular" reaction between A and B in the complex.

Since the encounter complex is present in very small concentration, it is amenable to the steady-state approximation, according to which the net formation rate of $(A\text{-}B)_{\text{solv}}$ is equal to zero. Thus,

$$\frac{d[A\text{-}B]}{dt} = k_1[A][B] - (k_{-1} + k_2)[A\text{-}B] = 0,$$

and the reaction rate can be expressed as

$$\text{Rate} = k_2[A\text{-}B] = \frac{k_1 k_2}{k_{-1} + k_2} [A][B].$$

For convenience, we define the second-order rate coefficient, k_{obs}, as

$$k_{obs} = \frac{k_1 k_2}{k_{-1} + k_2}.$$

(8.20.2)

This analysis leads to two limiting cases with respect to k_{obs}: one in which reaction between A and B is very slow compared with their departure (or separation) from the solvent cage, i.e., $k_{-1} \gg k_2$, and the other in which A-B reaction is much faster than their separation, or $k_2 \gg k_1$. The first case describes a reaction that is under *chemical control*, with $k_{obs} \sim k_1 k_2 / k_{-1}$, and the second pertains to a *diffusion-controlled* reaction for which $k_{obs} \sim k_1$. It is the latter situation that is considered in this experiment. More details of the reaction, which involves fluorescence quenching, will be described after a discussion of the necessary theory.

Diffusional Mass Transport

The basic issue in a diffusion-controlled reaction concerns the dynamics of mass transport in the condensed phase. The fundamental equations describing mass transport are embodied by Fick's laws of diffusion. These laws are encountered and discussed in Experiments 17 and 18. Fick's first law says that the number of molecules (dn) diffusing through a unit area, A, per unit time (dt) in the x direction (i.e., the flux, J_x) is proportional to the concentration gradient at that point, i.e.,

$$J_x \equiv \frac{1}{A} \frac{dn}{dt} = -D \frac{dC}{dx},$$

(8.20.3)

where D, the proportionality constant, is called the *diffusion coefficient*. The cgs units of D are $cm^2 \ s^{-1}$. The minus sign in equation (8.20.3) indicates that transport goes *against* the concentration gradient (i.e., from high to low concentration values).

Fick's second law states that the change in the concentration of molecules, dC, diffusing across an infinitesimally thin plane per unit time, dt, is proportional to gradient of the flux:

$$\left(\frac{dC}{dt} \right)_x = -\left(\frac{dJ_x}{dx} \right)_t.$$

(8.20.4)

Again, the minus sign ensures that concentration increases in time in response to a flux that decreases with increasing x. If D is assumed to be independent of x, the expression for J_x from Fick's first law may be substituted into the second to give

$$\left(\frac{dC}{dt} \right)_x = D \left(\frac{d^2C}{dx^2} \right)_t.$$

(8.20.5)

Since we are concerned with a three-dimensional and isotropic space, i.e., equal forces in all directions, Fick's second law can be written in more general form as

$$\left(\frac{dC}{dt} \right)_{x,y,z} = D(\nabla^2 C)_t,$$

(8.20.6)

where ∇^2 is the Laplacian operator, $\partial^2/\partial x^2 + \partial^2/\partial y^2 + \partial^2/\partial z^2$.

The solution of Fick's second law is an equation that expresses concentration as a function of *space*, i.e., distance (r), and *time* (t). We are interested in solving equation (8.20.6) with the boundary conditions

$$C(r,0) = C_0,$$

$$C(\infty,t) = C_0,$$

$$C(r = R,t) = 0,$$

where $r = [x^2 + y^2 + z^2]^{1/2}$ and C_0 denotes the bulk concentration.

The first boundary condition states that initially, the bulk concentration prevails throughout the system; the second says that even after the reaction starts ($t > 0$) the concentration very far away from the reactant is constant (C_0); and the third indicates that the reactant concentration is zero at a distance equal to the sum of the reactant collision radii. It should be realized that the diffusion coefficient, D, contained in equations (8.20.3), (8.20.5), and (8.20.6) is the *sum* of the individual diffusion coefficients of the reactants, $D_A + D_B$. This essentially allows the motion of one of the reactants to be considered *relative* to the other.

The solution of Fick's second law for this set of boundary conditions (first rendered by Smoluchowski in 1917) yields, for the space-time dependence of C,

$$C(r,t) = C_0\left[1 - \left(\frac{R}{r}\right)\right]\mathrm{erfc}\left[\frac{r - R}{2(Dt)^{1/2}}\right], \tag{8.20.7}$$

in which erfc is the *co-error function,*

$$\mathrm{erfc}(x) = 1 - \frac{2}{\sqrt{\pi}}\int_0^x \exp(-y^2)dy.$$

The time dependence of reactant at the reaction boundary $r = R$ becomes expressible in terms of the flux of reactant, J_R, or its rate of transport across a hypothetical spherical surface with radius R. Thus

$$J_R = \frac{4\pi RDN_A C_0}{1000}\left[1 + \frac{R}{(\pi Dt)^{1/2}}\right] \quad (\text{molecules s}^{-1})_1 \tag{8.20.8}$$

where N_A is Avogadro's number. Equation (8.20.8) indicates that the flux is actually time-dependent. This time dependence comes from $R/(\pi Dt)^{1/2}$, which is sometimes referred to as the *transient term*. Physically, the transient term accounts for the fact that initially nearby reactant molecules do not have to diffuse through the bulk medium in order to react. After a short time, however, these nearby reactant molecules are depleted, and the flux approaches a constant, or steady-state, value, $4\pi RDN_A C_0/1000$. This can be seen mathematically in that time is represented in equation (8.20.8) as $t^{-1/2}$.

Because J_R represents the rate of passage of one of the reactants (having a bulk concentration C_0 in moles dm^{-3}) through a spherical reaction surface with the other reactant at the center, the rate of the reaction represented by equation (8.20.1) is merely J_R itself. Hence the bimolecular, diffusion-controlled rate constant k_1 is J_R/C_0,

$$k_1 = \frac{4\pi RDN_A}{1000}\left[1 + \frac{R}{(\pi Dt)^{1/2}}\right] \quad (\text{dm}^3 \text{ mol}^{-1} \text{ s}^{-1}). \tag{8.20.9}$$

This rate constant is not a true *constant,* however, because of the transient term. As t becomes large enough,

$$k_1 => \frac{4\pi RDN_A}{1000} \quad (\text{dm}^3 \text{ mol}^{-1} \text{ s}^{-1}), \tag{8.20.10}$$

and equation (8.20.10) can be used to determine the diffusion-controlled rate constant if the mutual diffusion constant ($D_A + D_B$) and collision radii ($R_A + R_B$)

are known. It should be emphasized that the units of k_1 presented above are dm^3 $mol^{-1} s^{-1}$, if R and D are expressed in cm and $cm^2 s^{-1}$, respectively.

Often, however, R and D are not known, and an indirect method is used to estimate k_1. This approach, developed by Einstein using Stokes's law (see Experiment 19), allows the diffusion coefficient to be expressed in terms of a bulk property, the solvent viscosity, η. Thus

$$k_1 \sim \frac{8RT}{3000\eta} \quad (dm^3 \ mol^{-1} \ s^{-1}), \quad (8.20.11)$$

in which R is the gas constant. Equation (8.20.11), which is often referred to as the Stokes-Einstein-Smoluchowski (SES) equation, holds when the reactants A and B are different and their sizes are larger than that of the solvent molecules.

Fluorescence Quenching

The reaction studied in this experiment takes place between an *electronically excited* molecule and a *quencher,* a species that removes the electronic excitation. The advantages of using an electronically excited molecule as one of the reactants are: (1) the reaction begins only when the system is exposed to light; (2) the concentration of electronically excited molecules can be readily followed fluorometrically (via fluorescence detection); and (3) the reaction itself is intrinsically *fast:* it has no activation energy per se. Experiment 24 also utilizes fluorescence quenching as a "kinetic" technique. The overall reaction of fluorescence quenching can be represented as

$$A* + Q \rightarrow product(s)$$

where $A*$ represents the electronically excited (fluorescent) state of a molecule (e.g., anthracene), Q denotes the quencher (e.g., CBr_4) that extinguishes the $A*$ fluorescence, and "product(s)" indicate the species into which the fluorescent molecule is eventually transformed. This deexcitation process may occur through a charge-transfer intermediate $(A^+ - Q^-)$. For the system encountered in this experiment, electron transfer from $A*$ to the quencher occurs very rapidly. In general, it is possible that the consequence of fluorescence quenching is the return of A to its electronic ground state, i.e., $A + Q$.

The bimolecular quenching reaction above competes with intrinsic, first-order, fluorescence decay of $A*$. A more complete scheme is

$$A \longrightarrow A* \quad \text{(photoexcitation)},$$

$$A* \xrightarrow{k_r} A + h\nu \quad \text{(fluorescence)},$$

$$A* \xrightarrow{k_{nr}} A \quad \text{(nonradiative decay)},$$

$$A* + Q \xrightarrow{k_q} \text{electron transfer} + \cdots \quad \text{(quenching)}.$$

If a solution containing A and Q is irradiated with a *steady-state* light source (at a wavelength at which A absorbs), a very small time-independent concentration of $A*$ is produced. The probability that $A*$ will fluoresce, P_f, is the ratio of the fluorescence (or radiative) rate constant to the *sum* of *all* the rate constants that deplete $A*$, namely,

$$P_f = \frac{k_r}{k_r + k_{nr} + k_q[Q]}, \quad (8.20.12)$$

where k_r, k_{nr}, and $k_q[Q]$ are, respectively, the radiative, nonradiative, and pseudo-first-order quenching rate constants. Since $[Q] \gg [A^*]$, the diffusion-controlled, bimolecular quenching step becomes pseudo-first-order. If the amount of light absorbed by A is constant (i.e., independent of $[Q]$), the fluorescence *intensity, I,* measured by the detector (i.e., the response) is proportional to P_f. In the absence of quencher, $I = I_0 = k_r/(k_r + k_{nr})$, and the ratio of fluorescence intensities of unquenched to quenched samples of A becomes

$$\frac{I_0}{I} = 1 + \frac{k_q[Q]}{k_r + k_{nr}}. \qquad (8.20.13.)$$

Identifying $1/(k_r + k_{nr})$ as the reciprocal of the *fluorescence lifetime* of A^* in the *absence* of quencher (τ_0), equation (8.20.13) can be written as

$$\frac{I_0}{I} - 1 = k_q \tau_0 [Q]. \qquad (8.20.14)$$

This is known as the steady-state *Stern-Volmer* (S-V) relation as applied to fluorescence quenching. It predicts that a plot of $(I_0/I - 1)$ vs. $[Q]$ should be linear, having an intercept of zero and a slope equal to $k_q\tau_0$. (See also Experiment 24.) The latter is called the Stern-Volmer constant, K_{SV} (units of $dm^3 \ mol^{-1}$). If fluorescence quenching is diffusion-controlled (which is the case in the anthracene/CBr_4 system), k_q can be identified with k_1 [see equations (8.20.1), (8.20.2), and (8.20.9) to (8.20.11)].

In the absence of the transient effect, which is a type of "static" quenching in that some A^* molecules are quenched by nearby Q molecules that do not have to fully undergo transport through the solvent medium in order to approach A^* within a distance R, the S-V plot should be linear, and knowledge of τ_0 allows the determination of k_q. Alternatively, if the process is known to be diffusion-controlled, and if k_1 is determined from the SES (or some other) equation [e.g., Equation (8.20.10)], the fluorescence lifetime of A^* can be estimated from the measured K_{SV}. This is a case in which a dynamic property (a rate constant) can be obtained from a static experiment (I_0/I measurements).

The S-V relation can be expressed in terms of the diffusion-controlled rate constant that contains the transient term, e.g., equation (8.20.9). The time dependence was treated by Ware and Novros, who applied equation (8.20.9) to the condition pertinent to this experiment, namely, that the system is considered to be suddenly exposed to a steady-state excitation source, called a step function (see Figure 8.20.1).

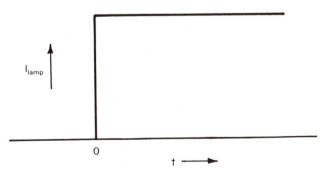

FIGURE 8.20.1 Step function excitation. At time = 0, absorbing radiation is suddenly "switched on."

In this case, the S-V equation becomes more complicated:

$$\frac{I^{\circ}}{I} = \frac{1 + 4\pi RDN_A/1000[Q]\tau_0}{Y},$$

where

$$Y = 1 - \frac{b}{a^{1/2}}\pi^{1/2}\exp\left(\frac{b^2}{a}\right)\mathrm{erfc}\left(\frac{b}{a^{1/2}}\right),$$

in which

(8.20.15)

$$a = \frac{1}{\tau_0} + \frac{4\pi RDN_A}{1000[Q]},$$

$$b = \frac{4R^2(\pi D)^{1/2}N_A}{1000[Q]}.$$

If $Y = 1$, equation (8.20.15) reduces to equation (8.20.14) in which $k_q = k_1$ from (8.20.10). (Verify this!)

The Stern-Volmer plot based on equation (8.20.15) will be curved upward. If the mutual diffusion coefficient of the reacting pair is known, the S-V plot of experimental data can be fit using equation (8.20.15) with R used as a "fitting" parameter. For a satisfactory match between experiment and theory, however, the value of R that provides a good fit of the data must be physically reasonable. That is, R should be approximately equal to the sum of the molecular diameters of A (actually A^*) and Q.

Safety Precautions

☐ Always wear safety glasses or goggles; these glasses should block ultraviolet light. Ordinary plastic safety goggles or glasses may not be effective in absorbing all the ultraviolet radiation. Check with your instructor.

☐ Do not allow solid anthracene to come in contact with the skin. Wear gloves when handling anthracence and CBr_4.

☐ Ozone is sometimes produced by ultraviolet light sources. If you detect this gas, which has a sharp, slightly acrid odor, notify your instructor and immediately shut off the source. Increase ventilation, and leave the room immediately.

☐ The compressed-gas cylinder used in deaerating should be securely strapped to a firm support. The delivery pressure should never exceed a few (e.g., 5) psig.

☐ The experiment should be performed in an open, well-ventilated laboratory.

Procedure

1. Using freshly sublimed or purified anthracene and CBr_4, prepare 250 mL of a ca. $1.00 \times 10^{-4}\ M$ solution of anthracene (AN) in spectrometric (or fluorometric) quality n-hexane. Using this solution as "solvent," prepare 25 mL of a ca. $1.50 \times 10^{-2}\ M$ "stock" solution of CBr_4.

2. Prepare at least eight dilutions of the AN/CBr_4 stock solution using the AN solution as solvent. The most dilute should be 5 percent, and the rest should be nearly evenly spaced up to the stock solution value. 10-mL volumes of each solution are appropriate. 0 and 100 percent samples should be included for uniformity. Label these samples.

3. Starting with the 0 percent sample (AN only), pour the solution into a fluorescence cell and deaerate with dry N_2 for about 2 min. Avoid using an excessive flow rate that will cause the solution to splatter from the cell or to otherwise result in undue solvent evaporation. Promptly stopper the cell.

4. Record the full fluorescence spectrum using the instrumental conditions previously outlined by the instructor. Label the spectrum while it is recorded.

5. Using exactly the same excitation wavelength as above, measure the fluorescence intensity of the most dilute and subsequently more concentrated AN/CBr$_4$ solutions. Each sample must be deaerated in a consistent procedure (i.e., identical bubbling time). Instead, however, of obtaining the entire fluorescence spectrum, only the AN maximum need be monitored (near 398 nm). Make sure you are *at* the maximum by first scanning that region slowly. If you have any doubt as to the steadiness of the exciting source, switch back to the AN sample and compare its maximum intensity with that of the first sample run.

6. As increasingly more concentrated samples are examined, the fluorescence intensity will diminish to the point that the instrument gain should be increased to ensure maximal reading sensitivity. The background, or "dark check," should be established when the instrument sensitivity is changed.

7. The following procedural points should be borne in mind: (a) the temperature of the sample should be held constant to 0.5 to 1°C; (b) the exposure times of all the samples (especially those rich in CBr$_4$) to the excitation source (and even room light) should be minimized; and (c) samples should be stored in the dark until they are used.

Data Analysis

1. Tabulate the data as I_f (AN fluorescence intensity), and $[Q]$ (CBr$_4$ concentration).

2. Make a Stern-Volmer plot of the data.

3. Superimpose on this plot the straight line expected if the transient term is ignored. The following information can be used: Ware and Novros reported a value of $D = 4.35 \times 10^{-5}$ cm^2 s^{-1} in *n*-heptane at 25°C. To obtain D for *n*-hexane at some other temperature, assume that D is proportional to T/η. They also reported a value of $\tau_0 = 1.81 \times 10^8$ s^{-1}. You can assume this to be temperature-independent.

4. Also indicate the straight line expected on the basis of the SES equation. Tabulate the two calculated k_1 values and their error limits.

5. Using equation (8.20.15), calculate several points for the S-V plot using a fixed value of R. If you are not using a computer to generate these points, choose them judiciously so you can discern the general shape of the calculated plot. Start with a value of 6 Å and increase R in steps of 0.5 Å until satisfactory agreement with the data is obtained.

Questions and Further Thoughts

1. Why is the anthracene fluorescence quenching diffusion-controlled, as opposed to chemical-controlled? Thus justify why the reaction between AN* and CBr$_4$ is very fast.
2. What experiment(s) could be performed to determine whether a stable photoproduct is formed as a result of the fluorescence quenching?
3. What could be done to obtain supporting evidence for the existence of a charge-transfer (or ion pair) intermediate, e.g.,

$$\text{AN*} + \text{CBr}_4 \rightarrow (\text{AN}^+)(\text{CBr}_4^-) \rightarrow \text{AN} + \text{CBr}_4,$$

in the quenching process?
4. Verify that if $Y = 1$ in equation (8.20.15), $k_q \Rightarrow k_1$; see equations (8.20.14) and (8.20.10).
5. One of the undesirable complications associated with this type of experiment is the presence of *ground-state* complexes that might exist between AN and CBr$_4$. If such a complex competes for light absorption with uncomplexed, free anthracene, it could

cause a decrease in fluorescence intensity without there having to be a diffusion between AN* and CBr$_4$. This is sometimes called static quenching. For this reason, it is desirable to work with quencher concentrations as low as possible. What experiment could be done to obtain evidence of a ground-state complex? How could the equilibrium constant of such a complex be determined?

6. What effect would a *polar* solvent, e.g., acetonitrile (CH$_3$CN), have on the ground- and excited-state interactions between AN and CBr$_4$?

7. If one wanted to do a fluorescence quenching experiment that demonstrated (and thus maximized) the "transient effect" in diffusional processes, what qualities of solvent viscosity and fluorescence probe lifetime would be sought (i.e., high or low viscosity; long or short fluorescence lifetime)?

Further Readings

R. A. Alberty, "Physical Chemistry," 7th ed., pp. 752–755, Wiley (New York), 1987.

A. H. Alwattar, M. D. Lumb, and J. B. Birks, "Organic Molecular Photophysics," vol. 1, p. 403, Wiley (New York), 1973.

P. W. Atkins, "Physical Chemistry," 3rd ed., pp. 654–655, 674–682, 741–745, W. H. Freeman (New York), 1986.

J. B. Birks, "Photophysics of Aromatic Molecules," p. 518, Wiley (New York), 1970.

G. W. Castellan, "Physical Chemistry," 3rd ed., pp. 746–747, 757–758, Addison-Wesley (Reading, Mass.), 1983.

I. N. Levine, "Physical Chemistry," 2nd ed., pp. 464–472, 536–539, McGraw-Hill (New York), 1983.

R. M. Noyes, *Progress in Reaction Kinetics, 1,* 131 (1961).

N. J. Turro, "Modern Molecular Photochemistry," p. 311, Benjamin-Cummings (Menlo Park, Calif.), 1978.

W. R. Ware and J. S. Novros, *J. Phys. Chem., 70,* 3246 (1966).

Experiment 20

Diffusion-Controlled Rate Constant

NAME _____   DATE _____

Anthracene concentration: _____ g in _____ mL

CBr$_4$ stock solution: _____ g added

An/CBr$_4$ solutions percent stock	Fluorescence intensity, ca. 398 nm
0%	_____
_____	_____
_____	_____
_____	_____
_____	_____
_____	_____
_____	_____
_____	_____
_____	_____
_____	_____
_____	_____

Temperature _____

Diffusion-Controlled Rate Constant

NAME _____ DATE _____

Anthracene concentration: _____ g in _____ mL

CBr_4 stock solution: _____ g added

An/CBr_4 solutions percent stock	Fluorescence intensity, ca. 398 nm
0%	_____

Temperature _____

_____	_____
_____	_____
_____	_____
_____	_____
_____	_____
_____	_____
_____	_____
_____	_____
_____	_____
_____	_____

□ PART NINE

Polymers

□ Experiment 21

Molecular Weight and Monomer Linkage Properties of Polyvinyl Alcohol

Objective

To determine the viscosity-average molecular weight of polyvinyl alcohol and the fraction of "head-to-head" monomer linkages in the polymer.

Introduction

One of the fundamental molecular properties used to characterize polymers is molecular weight. Many of the physical characteristics of polymeric materials can be associated with the shape and weight distribution of the polymer. Some experimental techniques used to obtain this information are viscosity, osmotic pressure, and light-scattering measurements. These measurements are not made on the polymer itself but on solutions containing the dissolved polymer. Because the viscosity and osmotic pressure of these solutions depend systematically on the concentration of polymer, they are called colligative properties.

This experiment deals with viscosity measurements of solutions of polyvinyl alcohol in water. This technique is very straightforward and does not require specialized equipment. We begin by reviewing the fundamentals of fluid viscosity. If a liquid undergoes laminar or streamline flow* through a cylindrical tube, the differential equation that describes the mass transport in terms of the flow rate is known as the Poiseuille law [named after Poiseuille (1844), in whose honor the cgs unit of viscosity (the "poise," i.e., $1 \text{ g cm}^{-1} \text{ s}^{-1}$) is named]. This equation is

$$\frac{dV}{dt} = \frac{\pi r^4 (P^0 - P')}{8\eta \, (z' - z^0)}, \tag{9.21.1}$$

in which dV is the volume of fluid (liquid or gas) transported in a time dt through

---

*For more information about streamline flow, see the discussion in Experiments 15, 16, and 17.

a straight, cylindrical tube of radius r. The fluid is under a pressure difference ($P^0 - P'$) that extends over the length of the tube ($z^0 - z'$). If the pressure gradient is produced by the gravitational force acting on the fluid (the fluid flowing, for simplicity, in a downward vertical direction), $P^0 - P'$ is proportional to the *density* of the fluid, D. Hence for an arbitrary but constant volume of fluid, equation (9.21.1) yields (after integration):

$$V = \frac{t(\pi r^4 gD)}{8\eta},$$
(9.21.2)

where g is the gravitational constant and D is the bulk density of the fluid. Equation (9.21.2) can be rearranged to give an expression for the fluid viscosity:

$$\eta = \frac{\pi r^4 g}{8V} Dt = cDT \qquad (\text{g cm}^{-1}\text{ s}^{-1}),$$
(9.21.3)

where the constant c depends on the characteristics of the measuring device used, that is, the tube bore radius, r, and the total volume of fluid that flows in the time, t. The application of equation (9.21.3) to the measurement of liquid viscosities is realized with an apparatus called a *viscometer*. See Experiment 15. The constant c is determined by calibrating the viscometer with a liquid of a known viscosity, in this case water. The time, t, required for water to flow through the viscometer is measured. Knowledge of the water density (at the particular temperature) then allows c to be determined. It should be mentioned that for highly precise determinations of viscosity a kinetic energy correction term must be subtracted from equation (9.21.3). Because this correction is usually less than 1 percent of the quantity cDt, it can be ignored with some justification.

We seek a relationship between the viscosity of a solution containing a dissolved high polymer and some molecular weight property of the polymer itself. A significant contribution to this field was made by Einstein (1906), who showed that the *fractional change* in the viscosity of a solution—relative to the pure solvent—is related to the fraction of the total volume of solution occupied by the solute, in this case the polymer. It is assumed that the solute has simple *spherical* geometry. Thus the relevant equation is

$$\frac{\eta - \eta_0}{\eta_0} \equiv \eta_{sp} = \frac{Cv}{V},$$
(9.21.4)

where η and η_0 are the respective viscosities of the solution and the pure solvent. η_{sp} (which is dimensionless) is defined as the *specific viscosity* and is seen to be proportional to the ratio of the solute volume, v, to that of the solution, V. η_{sp} is thus a colligative property: its value depends on the amount of polymer in solution, i.e., its concentration. The constant C has a theoretical value of 5/2 (for spherical solutes). The specific viscosity is dimensionless and positive ($\eta > \eta_0$).

For a solution containing N spherical solute molecules each of radius R, equation (9.21.4) becomes (using $C = 5/2$ and $v = 4\pi R^3/3$)

$$\eta_{sp} = \frac{10\pi NR^3}{3V}.$$
(9.21.5)

This can be more conveniently expressed in terms of the mass concentration of the solute in the solution, c_m (defined as *grams per mL*), and the *molecular* mass of the solute, m, as (noting that $c_m = N_A m/V$)

$$\eta_{sp} = \frac{10\pi R^3 c_m}{3m}.$$
(9.21.6)

The colligative nature of η_{sp} is explicit in equation (9.21.6) because of the c_m dependence.

Another quantity that will prove to be very useful is the *intrinsic viscosity,* [η] (sometimes referred to as the *Staudinger index*):

$$[\eta] \equiv \lim_{c_m \to 0} \frac{\eta_{sp}}{c_m} = \frac{10\pi R^3}{3m}. \tag{9.21.7}$$

The limit of infinite dilution ($c_m \to 0$) is required to define a viscosity property that is intrinsic to the solute (in the given solvent), i.e., independent of the concentration. This approach essentially eliminates problems caused by the fact that the *bulk* properties of the polymer solution vary with concentration.

The limiting condition defined in equation (9.21.7) is needed to define [η] because the reduced specific viscosity (i.e., η_{sp}/c_m) depends on the polymer concentration due to shear forces between the dissolved macromolecule and the solvent medium. This concentration dependence is observed to follow the following relation:

$$\frac{\eta_{sp}}{c_m} = [\eta] + k[\eta]^2 c_m + k' c_m^2, \tag{9.21.8}$$

where k, known as the Huggins constant, has a value of about 2 for rigid, uncharged spheres, and near 0.35 for flexible polymers in a "good" solvent.[1] The higher order term in c_m in equation (9.21.8) can often be neglected. From equation (9.21.6) it can be seen that the ratio η_{sp}/c_m is not a colligative quantity; it is equal to an *intrinsic* property of the solute itself, namely, $10\pi R^3/3m$. The intrinsic viscosity has dimensions of $cm^3\ g^{-1}$ and is perhaps analogous to a molar volume. Because both η_{sp} and c_m are experimental quantities, equation (9.21.8) can be used to obtain [η] by extrapolation to zero concentration.

It is interesting to point out that if R is known, equation (9.21.7) can be used to determine m, the molecular mass of the solute, from [η]. Furthermore, the mass of the solute divided into its molecular weight provides Avogadro's number, and this suggests an experimental approach for obtaining this fundamental constant. Conversely, if m is known, a value of the solute radius, R, can be determined from [η]. Following this reasoning, Einstein was able to obtain a satisfactory value of Avogadro's number as well as the molecular radii of carbohydrates. Few other direct experimental methods can be used to measure Avogadro's number.

Intrinsic viscosity measurements can be applied to a high polymer to obtain its (spherical) radius, or its molecular weight. It must be recognized that the value of R in equations (9.21.5) to (9.21.7) denotes the *effective* radius of the solute because the solute may not be exactly spherical and, in addition, the solute may be associated with a (possibly large) number of solvent molecules. This effective radius is referred to as the *hydrodynamic radius* and can, in principle, vary for a given solute from solvent to solvent depending on what shape the solute adopts in a particular solvent, that is, the extent to which solvent molecules penetrate or stick to the solute.

Another complication that can be anticipated in applying (9.21.7) to molecules is that the assumption of spherical geometry may be invalid in some cases. Deviations from spherical geometry would manifest themselves theoretically in that C [in equation (9.21.4)] would have values other than 5/2. In principle, this can be accounted for if the correct *solvated* molecular shape (oblate or prolate spheroid, oblong, etc.) is known.

A more serious problem encountered in dealing with high polymers is the fact that these systems consist of polymers of various chain lengths. Thus, these are not homogeneous solutes (monodisperse) but have a distribution of molecular weights and are called *polydisperse*. The degree to which a polymer is polydisperse depends on the conditions under which it is synthesized. While in theory a polydisperse polymer can be fractionated into groups that have a more narrow molecular weight distribution (nearly monodisperse), this is usually a long and difficult process. Measurements on raw, polydisperse polymer solution provide information that is averaged over the molecular weights (and sizes) of the polymer. Such is the case in viscosity measurements.

In order to apply equation (9.21.7) to a system that is composed of a high molecular weight polydisperse polymer, a statistically averaged form of the molecular "radius" will have to be used. The polymer studied in this experiment—like many others—is a *linear* chain system, which means that the macromolecule is formed from a large number of monomer units in such a way that monomers add to the developing polymer chain without branching. Although the molecule is called a "linear" chain, its geometry surely does not resemble a straight line. Rather, it is coiled up and adopts an overall shape that is approximately spherical. It may be thought to resemble a loosely tangled ball of yarn. See Figure 9.21.1. This so-called random coil is not a static structure but continually undergoes contortional motion as the different segments go through various conformational transitions. On a time-averaged basis, however, one can define a *statistical* radius called the *radius of gyration, R_g*. This quantity is expressed quantitatively as

$$R_g = \left(\frac{I}{m}\right)^{1/2} \qquad (9.21.9)$$

where m is the molecular mass and I is the moment of inertia, a second-order moment defined as

$$I = \sum_i m_i r_i^2 \qquad \text{(g cm}^2\text{)}, \qquad (9.21.10)$$

in which the sum is over all of the point masses, m_i (atoms), of the molecule and r_i is the distance to the ith atom from the center of mass. I can be calculated only if some (static) structural model of the polymer is adopted. If the radius of gyration is substituted for the unique-valued radius R in equation (9.21.7):

$$[\eta] = \frac{N_A 10\pi R_g^3}{3M} \qquad \text{(cm}^3\text{ g}^{-1}\text{)}, \qquad (9.21.11)$$

FIGURE 9.21.1   Schematic diagram of a linear chain polymer in a random coil configuration.

where M, the molecular weight (along with N_A, Avogadro's number), replaces the molecular mass, m.

If one now assumes that the linear chain polymer is constructed of N monomer units linked together in such a way that *each* link is rotationally flexible, or unhindered (freely jointed chain), it can be demonstrated that the radius of gyration is proportional to the square root of the number of links in the polymer chain; thus $R_g \propto N^{1/2}$. This important conclusion is developed through the application of statistical mechanics to polymers.[2] Thus,

$$[\eta] = K'M^{1/2}, \qquad (9.21.12)$$

where K' is a proportionality constant that contains the conversion factor between R_g and $M^{1/2}$. Equation (9.21.12) is of particular importance in this experiment. It relates a measurable quantity, the intrinsic viscosity, to the desired molecular weight of the polymer. Although the square root dependence of $[\eta]$ on molecular weight is actually observed for some monodisperse polymer solutions, many others deviate from equation (9.21.12). The reasons for these discrepancies have to do with the nature of solvation and the effects brought about by solvent association on the structure of the polymer solute. Thus the molecules of a "good" solvent enter into the polymer coils to maximize solvolytic associations, and the polymer expands. Looked at another way, the polymer "swells" out into the solvent. In a "poor" solvent, on the other hand, the polymer knots up into itself avoiding the interactions with the solvent molecules. As you might expect, the solubility of a given polymer is larger in a good solvent than in a poor one. See Figure 9.21.2.

Another important consideration that has been ignored thus far in the discussion is called the "excluded volume." This is the effective space that is inaccessible to a random coil or freely jointed chain. This volume cannot be occupied because the segments of the polymer avoid each other as they become too close during the twisting and turning that the polymer undergoes.

A more general relationship between the intrinsic viscosity and molecular weight is provided by the Mark-Houwink equation

$$[\eta] = KM^a \qquad (\text{cm}^3 \text{ g}^{-1}), \qquad (9.21.13)$$

where K and a are parameters that depend on the particular polymer, the solvent medium, and the temperature. Typically, a ranges between 0.5 and 0.8 [cf. equation (9.21.12)]. The Mark-Houwink parameters are obtained from log-log plots of $[\eta]$ vs. M for a series of monodisperse polymers. Agreement between K and a

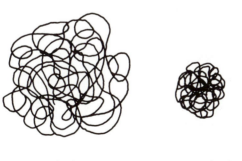

good solvent   poor solvent

FIGURE 9.21.2 Schematic diagrams of a random coil polymer in a "good" solvent (left) and a "poor" solvent (right).

TABLE 9.21.1   Mark-Houwink parameters for PVOH in water at 25°C.

K (cm³ g⁻¹)	a	Molecular Weight Range	Note*
0.020	0.76	$(0.6 - 2.1) \times 10^4$	4
0.30	0.50	$(0.9 - 17) \times 10^4$	5
0.14	0.60	$(1 - 7) \times 10^4$	6

*Notes appear at end of experiment.

values obtained by different workers is often apparently poor.[3] For example, for poly-vinyl alcohol (PVOH) in water at 25°C, the parameters shown in Table 9.21.1 have been reported.

The molecular weight can be obtained in explicit form from equation (9.21.13):

$$M_v = \left(\frac{1}{K}\right)^{1/a} [\eta]^{1/a}. \tag{9.21.14}$$

This expression (in which $[\eta]$ has dimensions of cm³ g⁻¹) is the computational basis of the experiment; it can be applied to *poly*disperse PVOH (which is used in this experiment), but the molecular weight obtained from equation (9.21.14) is a *viscosity average* molecular weight, M_v. Statistically this quantity is different from the *number average* molecular weight, M_n, which is obtained, for example, from osmotic pressure measurements. M_n is defined as

$$M_n = \sum_i f_i M_i, \tag{9.21.15}$$

where f_i is the fraction of polymers having molecular weight M_i; the sum in equation (9.21.15) extends (in theory) from 0 to ∞.

Flory and others[7] have shown that the relationship between M_v and M_n is

$$\frac{M_v}{M_n} = [(1 + a) \, \Gamma \, (1 + a)]^{1/a}, \tag{9.21.16}$$

where a is the same parameter used in equation (9.21.13), and Γ denotes the *gamma function* whose value (for a given a) can be obtained from mathematical tables. E.g., for $a = 0.76$ (for PVOH in water at 25°),[4] the ratio in equation (9.21.16) is

$$\frac{M_v}{M_n} \equiv S = 1.89. \tag{9.21.17}$$

For $a = 0.50$ and 0.60 see Table 9.21.1, $S = 1.77$ and 1.81, respectively.

Chemical Bonding in PVOH

Poly (vinyl alcohol) is obtained from the hydrolysis of poly (vinyl acetate). The latter is synthesized from vinyl acetate monomers. Let us consider the question of how the vinyl acetate monomers link up to form the polyvinyl acetate polymer. The structure of vinyl acetate in shown below.

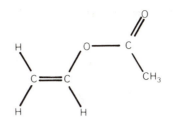

vinyl acetate

It is an unsymmetrically substituted ethylene molecule since only one end of the molecule is derivatized. Thus when a monomer is about to bond to the "growing" end of a polymer chain, it can do so in two ways with respect to the previously bonded monomer unit: either the carbon atom containing the functional group, X, (C_1) can bond to the terminal end of the polymer chain or the unfunctionalized carbon (C_2) can form the bond. These alternatives are illustrated below.

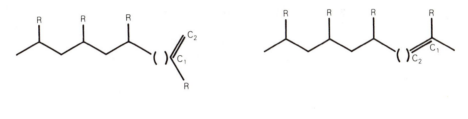

'head-to-head'   'head-to-tail'

An attachment in which the functional groups alternate is called "head-to-tail," while the one in which they are attached on adjacent carbon atoms is referred to as a "head-to-head" linkage. Because a head-to-head linkage involves considerable steric repulsion between the functional groups (they will have been added to *adjacent* carbon atoms on the polymer chain to form 1,2-substituents), it proceeds more slowly than a head-to-tail linkage. Therefore, a polymer will have a predominance of the latter arrangements and the functional groups will, for the most part, alternate in their positions on the polymer backbone (i.e., repetitive 1,3-substituents).

It should also be noted that if, after a head-to-head linkage has taken place, the next monomer attaches in a tail-to-tail fashion, the resulting arrangement between the substituents is a 1,4-disubstituted configuration. Considering the way this experiment is to be carried out (see below), it is only the presence of head-to-head attachments that is chemically significant. These result in the formation of 1,2-disubstituted structures on the polymer chain. The number of such events that occurs in polymerization relative to the more common head-to-tail attachments (and thus 1,3-disubstituted structures) is of interest because the physical and chemical properties of the polymer depend on the fraction of such head-to-head linkages. Thus some analytical method for establishing this information is desirable.

In the case of the polymerization of vinyl acetate and the subsequent hydrolysis of the polymer to form PVOH, the consequence of a head-to-head linkage is the formation of a 1,2-diol (or vicinal glycol). The more common head-to-tail link-ages result in the formation of 1,3-diols. The presence of 1,2-diol structures in the polymer can be conveniently distinguished from the 1,3-diols by a chemical means. The 1,2-diol is specifically cleaved (and oxidized) using periodic acid, HIO_4. Actually, the reagent used is the periodate anion (from KIO_4), which hydrolyzes in water to form HIO_4. The reaction is as follows:

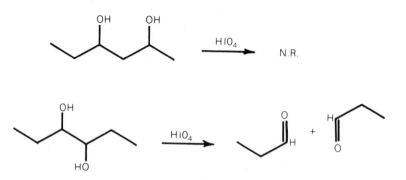

Hence treatment of a PVOH sample with KIO_4 will split the polymer wherever a 1,2-diol structure exists. This results in a decrease in the (average) molecular weight of the polymer, and this change can be detected via viscosity measurements.

   If we assume that the above reaction is 100 percent effective (so that all the 1,2-diol linkages are cleaved), the increase in the number of solute molecules in the solution after treatment with KIO_4, divided by the total number of monomer units represented in the polymer sample, is equal to the ratio of 1,2-diol structures to the total 1,2-diol *and* 1,3-diol arrangements in the system. This ratio, which is defined here as f, can be expressed as

$$f = \frac{\dfrac{1}{M'_n} - \dfrac{1}{M^0_n}}{\dfrac{1}{M^0}}, \tag{9.21.18}$$

where M^0_n and M'_n are the number average molecular weights of the polymer in the sample before and after the KIO_4 cleavage, respectively. M^0 is the molecular weight of the monomer unit (CH_2CHOH), which is 44 D. Using the relationship between the number average and viscosity average molecular weight for PVOH [see equation (9.21.17)], we get the following expression for the head-to-head fraction:

$$f = SM^0 \left\{ \frac{1}{M'_v} - \frac{1}{M^0_v} \right\}. \tag{9.21.19}$$

This equation provides the final result used in this experiment. It permits f to be determined from $[\eta]$ measurements of the PVOH sample before and after cleavage by KIO_4. The value of S depends on the choice of a used in equation (9.21.16) (see Table 9.21.1).

Safety Precautions

☐ Always wear safety glasses in the laboratory.
☐ Make sure you have been shown how to use proper pipeting techniques. *Never* pipet by mouth.
☐ Handle the viscometer carefully to avoid breakage.

Procedure

You will be provided with a clean viscometer. If you do not remember how to use it, review the material in Experiment 15, or ask your instructor for this information. A stock solution of PVOH in water (c_m about 0.016 g mL^{-1}) should be available; if it is not, you will be told how to prepare this solution. The viscosities of PVOH solutions of three or four different concentrations will be measured for both cleaved and uncleaved polymers. The viscometer will be calibrated using pure water. Temperature control for each of these measurements is important.

1. Prepare 100 mL of a 50 percent solution of the PVOH stock solution. This will be used to prepare the diluted samples of the uncleaved polymer for measurement. Use deionized water for the dilution in the volumetric flask. Cap and invert several times gently. Do not agitate; this causes foaming. Because the polymer tends to adhere to the surface of glassware, rinse the pipet with water, then acetone, immediately after use. Aspirate dry. Follow this procedure whenever glassware is used with PVOH solutions.

2. To prepare the *cleaved* polymer, pipet 50 mL of the stock solution into a 100-mL volumetric flask. Add 0.25 g of KIO$_4$ and about 20 mL of deionized water. Cap and swirl the liquid around a few times. Place the mixture in a hot water bath (ca. 70°C) for several minutes until the solid dissolves; periodically mix the solution if necessary. After the solid has dissolved, remove the flask and let it cool to room temperature; then fill to the mark with deionized water. Gently invert several times. Label the flask.

3. There are now two PVOH solutions (one cleaved and one uncleaved), each with c_m of about 0.0080 g mL^{-1}. From each of these prepare 50 mL of solutions corresponding to 80, 60, and 40 percent of that concentration. The 40 percent solution should be made up from the 80 percent one. Label these new samples *immediately* after preparation. Place them all in a 25°C bath.

4. Calibrate the viscometer with deionized water. Pipet 5 mL of water into the viscometer which is clamped and held in a 25°C bath. Using a pipet bulb, carefully "push" the water up through the capillary tube until it is above the upper fiducial mark. Remove the bulb and allow the water to drain. Start the timer or stopwatch exactly as the water meniscus passes the upper fiducial mark; stop the timer just as the meniscus passes the lower mark. Record this time. Repeat this measurement at *least* twice. The liquid can be "reset" by using the pipet bulb. Alternatively, a flexible tube connected to an aspirator can be used. The timings should be within 0.2 to 0.5 s. After the calibration is complete, remove the viscometer from the bath, rinse with acetone, and aspirate dry.

5. Sample measurements. Following the above procedure, measure the viscosities of the cleaved and uncleaved samples starting with the most dilute samples. Each measurement should be made in triplicate.

Calculations and Data Analysis

1. Calibrate the viscometer. Thus, determine the constant c in equation (9.21.3) from the mean flow time of water and its density and viscosity at 25°C. Three trials should be carried out.

2. Determine and tabulate the viscosities and specific viscosities of the PVOH samples. They can be assumed to have densities equal to that of pure water. In the same table list the bulk c_m values (g mL^{-1}) of the PVOH samples.

3. For each sample determine [η] for the uncleaved and cleaved polymer. Also determine the Huggins constant. See equation (9.21.8).

4. Choose a set of K and a values (see Table 9.21.1) and, using the appropriate relations, determine values of M_v for the uncleaved and cleaved PVOH, as well as the head-to-head ratio, f. Comment on the magnitudes of your results.

5. Determine M_v and f for another set of K and a values and compare with the results in step 4.

Questions and Further Thoughts

1. Consult a table of mathematical functions and, using the appropriate gamma functions and equation (9.21.16), verify the values of S cited in and after equation (9.21.17).

2. Using the definition of f (see the discussion before equation (9.21.18)), derive equation (9.21.18) and use the result in (9.21.17) to obtain equation (9.21.19).

3. Viscosity measurements are often performed on proteins and other biopolymers. What experiments could be carried out on such molecules to examine the effect of solution conditions (i.e., ionic strength, pH, etc.) on the extent of denaturation (breakdown of the biologically active structure)?

4. Specific and intrinsic viscosities provide structural information about macromolecules under the assumption that the solute is spherical [see equation (9.21.4)]. Is this a good assumption for a polymer such as PVOH? For what sort of macromolecule would this assumption be expected to be poor?

5. How would the structure of a polymer such as PVOH in the vapor phase differ from that in dilute aqueous solution?

6. The intrinsic viscosity of a sample of polydisperse PVOH in water was determined to be 86 $cm^3\ g^{-1}$ at 298 K. Using the information in Table 9.21.1 and the Mark-Houwink equation, what is the range of molecular weights that can be reported for the polymer?

Notes

1. C. Tanford, "Physical Chemistry of Macromolecules," pp. 390–392, Wiley (New York), 1961.
2. Loc. cit., pp. 138–168.
3. See J. Bandrup and E. H. Immergut, eds., "Polymer Handbook," vol. 2, IV-14, Wiley (New York), 1975.
4. P. J. Flory and F. S. Leutner, *J. Poly. Sci. 3,* 880 (1948); *5,* 267 (1950).
5. K. Dialer, K. Vogler, and F. Patat, *Helv. Chim. Acta, 35,* 869 (1952).
6. H. A. Dieu, *J. Poly. Sci., 12,* 417 (1955).
7. J. R. Schaefgen and P. J. Flory, *J. Am. Chem. Soc., 70,* 2709 (1948).

Further Readings

R. A. Alberty, "Physical Chemistry," 7th ed., pp. 813–816, Wiley (New York), 1987.

P. W. Atkins, "Physical Chemistry," 3rd ed., pp. 615–616, W. H. Freeman (New York), 1986.

G. W. Castellan, "Physical Chemistry," 3rd ed., pp. 940–941, Addison-Wesley (Reading, Mass.), 1983.

I. N. Levine, "Physical Chemistry," 2nd ed., pp. 463–464, McGraw-Hill (New York), 1983.

J. H. Noggle, "Physical Chemistry," pp. 453–455, Little, Brown (Boston), 1985.

D. P. Shoemaker, C. W. Garland, J. I. Steinfeld, and J. W. Nibler, "Experiments in Physical Chemistry," 4th ed., pp. 358–369, McGraw-Hill (New York), 1981.

J. F. Swindells, R. Ullman, and H. Mark, "Determination of Viscosity," in A. Weissberger, ed., "Techniques of Organic Chemistry," 3rd ed., vol. 1, part 1, chap. XII, Interscience (New York), 1959.

Experiment 21

Intrinsic Viscosity

NAME _____ DATE _____

I. *Viscometer calibration, bath temperature:* _____

   Stock solution concentration: _____

Flow times: _____   _____

   _____   _____

   _____   _____

II. *Cleaved samples, bath temperature:* _____

Sample 1 _____   Sample 2 _____   Sample 3 _____

_____   _____   _____

_____   _____   _____

_____   _____   _____

_____   _____   _____

III. *Uncleaved samples, bath temperature:* _____

Sample 1 _____   Sample 2 _____   Sample 3 _____

_____   _____   _____

_____   _____   _____

_____   _____   _____

_____   _____   _____

Experiment 21

Intrinsic Viscosity

NAME _____ DATE _____

I. *Viscometer calibration, bath temperature:* _____

   Stock solution concentration: _____

Flow times: _____   _____

   _____   _____

   _____   _____

II. *Cleaved samples, bath temperature:* _____

Sample 1 _____   Sample 2 _____   Sample 3 _____

_____   _____   _____

_____   _____   _____

_____   _____   _____

_____   _____   _____

III. *Uncleaved samples, bath temperature:* _____

Sample 1 _____   Sample 2 _____   Sample 3 _____

_____   _____   _____

_____   _____   _____

_____   _____   _____

_____   _____   _____

313

□ Experiment 22

The Thermodynamic Properties of Elastomers

Objective

To determine the equation of state and other properties of an elastomer.

Introduction

We are all familiar with the very useful properties of such objects as rubber bands, solid rubber balls, and tires. Materials such as these, which are capable of undergoing large reversible extensions and compressions, are called *elastomers*. An example of such a material is natural rubber, obtained from the plant, *Hevea brasiliensis*. An elastomer has rather unusual physical properties; for example, an ordinary elastic band can be stretched up to 15 times its original length and then be restored to its original size. While one might, at first thought, consider an elastomer to be a solid, it has an isothermal compressibility comparable with that of a liquid (e.g., toluene), about 10^{-4} atm^{-1} (cf. solids such as polystyrene or aluminum, which have values of ca. 10^{-6} atm^{-1}). Certain evidence suggests that an elastomer is a disordered ''solid,'' i.e., a glass, which cannot flow as a result of internal, structural restrictions.

The reversible deformability of an elastomer is reminiscent of a gas. In fact the term elastic was first used by Robert Boyle (1660) in describing a gas, ''There is a spring or elastical power in the air in which we live.'' In this experiment, certain formal thermodynamic similarities between an elastomer and a gas will be encountered.

One of the rather dramatic and anomalous properties of an elastomer is that once brought to an extended form, it *contracts* upon heating. This is in sharp contrast to the familiar response of expansion shown by other solids and liquids. This behavior, first reported by Gough (1805) and later studied in detail by Joule (1859), is called the Gough-Joule effect. Thermoelasticity is the basis of this experiment.

An understanding of the molecular properties and theoretical concepts that characterize elasticity was not fully developed until this century (1930s), when the foundations of polymer science were firmly established. The molecular requirements for elasticity are now rather well understood and can be described in terms of the most common (and historically most important) elastomer, natural rubber. This substance is a polymer having a molecular weight of about 350,000. The monomeric unit is isoprene, or 2-methylbutadiene.

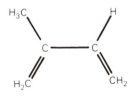

Isoprene

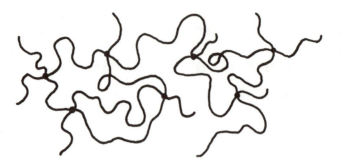

FIGURE 9.22.1   Schematic of a cross-linked polymer. The cross-links are joined at (or near) a common site.

When polymerized, isoprene can form the cis-configured chain which is *Hevea* rubber; in the trans configuration, the polymer, which is called gutta-percha, is crystalline at room temperature and thus not very elastic.

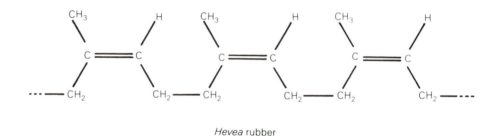

Hevea rubber

The elasticity of natural rubber is understood to be a consequence of three of its molecular properties: (1) the poly (isoprene) subunits can freely rotate, (2) forces between the polymer chains are weak (as in a liquid), and (3) the chains are linked together in a certain way at various points along the polymer. The latter property is especially important for reversible elasticity. It also eliminates the phenomenon called creep in which, once deformed, an elastomer will "relax" to the deformed shape. Property 3 is more specifically called *cross-linking,* and the phenomenon was first exploited by Charles Goodyear (1839) in a process he called vulcanization. Natural gum rubber is heated with sulfur (2 to 10 percent), and as a result, a number of cross-linking chains are imposed on the original polymer as shown in Figure 9.22.1 in which the black dots represent cross-linking sites.

In this type of cross-linking, which is essential to elastomers, the cross-links are bonded to a common site on the polymer backbone. A cross-linked elastomer thus resembles a fish net. The thermodynamic properties of an elastomer can be understood in terms of what happens to these cross-linked networks as the bulk material is stretched from its relaxed state.

Thermodynamics of Elasticity

Consider the consequence of subjecting an elastomer to an external force that causes it to undergo an extension (or a compression). The first law of thermodynamics may be written

$$dU = \dbar Q + \dbar w \tag{9.22.1}$$

where dU is the change in the elastomer's internal energy resulting from the absorption of heat, $\slashed{d}Q$, and the dissipation of work, $\slashed{d}w$, on it by the external force. (The symbol $\slashed{d}$ indicates that heat and work are inexact differentials; their integrated values depend on the path of the process carried out.) If we assume that the deformation process is *reversible* (i.e., carried out very slowly), the heat flow can be expressed as

$$\slashed{d}Q = T \, dS \tag{9.22.2}$$

where T is the absolute temperature and dS is the entropy change, and thus

$$dU = T \, dS + dw \quad \text{(reversible process)} \tag{9.22.3}$$

In considering the reversible deformation of an elastomer, it is convenient to use the Helmholtz free energy, $A = U - TS$. For the deformation process, the infinitesimal isothermal change in A is

$$dA = dU - T \, dS \tag{9.22.4}$$

hence, in this case, $dA = dw$. The work done on the elastomer by the external force in this isothermal, reversible process is simply equal to the change in the Helmholtz free energy.

We must now consider the kind of work being done on the elastomer. In addition to the work associated with the expansion of the system against the atmosphere ($-P \, dV$, where P is the external pressure and dV is the change in volume accompanying elongation), work also arises from the application of a force to an elastomer. If its original length, L_0, is changed by an amount dL as a result of a force, f, the work done on the elastomer is

$$dw = f \, dL \tag{9.22.5}$$

so if f is in the positive x direction and dL is positive, the elastomer stretches, and thus work is done *on* the polymer. The total work done on the solid is thus

$$dw = f \, dL - P \, dV \tag{9.22.6}$$

Because of the very small compressibility of elastomers (e.g., $\sim 10^{-4} \, \text{atm}^{-1}$), $P \, dV$ is much smaller than $f \, dL$ under most circumstances (typically by a factor of 10^{-4}), and we can approximate the work as $f \, dL$. In more detailed treatments of elasticity, the $P \, dV$ term cannot be neglected.

Since from equation (9.22.5), $f = (dw/dL)_T$, and $dA = dw$, it follows that

$$f = \left(\frac{\partial A}{\partial L}\right)_T \tag{9.22.7}$$

which means that the tensile force is equal to the (isothermal) change in the Helmholtz energy with respect to an infinitesimal change in elastomer length.

Differentiating equation (9.22.4) with respect to L (at constant T):

$$\left(\frac{\partial A}{\partial L}\right)_T = \left(\frac{\partial U}{\partial L}\right)_T - T\left(\frac{\partial S}{\partial L}\right)_T \tag{9.22.8}$$

and combining this with equation (9.22.7) gives

$$f = \left(\frac{\partial U}{\partial L}\right)_T - T\left(\frac{\partial S}{\partial L}\right)_T \tag{9.22.9}$$

It is desirable to express dA in terms of the infinitesimal (experimental) variables, dL and dT. This can be done by taking the total differential of A,

$$dA = dU - T \, dS - S \, dT \tag{9.22.10}$$

and inserting $dU = f \, dL + T \, dS$,

$$dA = f \, dL - S \, dT \tag{9.22.11}$$

We now make use of the Maxwell relation for dA (A is a thermodynamic state function, hence dA is an exact differential),

$$\left(\frac{\partial S}{\partial L}\right)_T = -\left(\frac{\partial f}{\partial T}\right)_L \tag{9.22.12}$$

This important result expresses the infinitesimal (isothermal) dependence of entropy on length in terms of an experimental quantity: the temperature dependence of the tension (at constant length). Now by substituting the left-hand side of (9.22.12) into equation (9.22.9), we can write the applied tensile force as

$$f = \left(\frac{\partial U}{\partial L}\right)_T + T\left(\frac{\partial f}{\partial T}\right)_L \tag{9.22.13}$$

Equations (9.22.9) and (9.22.13) are central to thermoelasticity. Equation (9.22.9) says that the force is composed of two components, one due to the change in the elastomer's internal energy as a result of the elongation (or compression), and the other due to the entropy change accompanying the deformation. Since f and T are dependent and independent experimental variables, respectively, equation (9.22.13) allows us to obtain $(\partial U / \partial L)_T$ from a plot of f vs. T.

Comparison With Gases

It is very enlightening to compare the force acting on an elastomer with the pressure of a gas. Extending the length of an elastomer is analogous to compressing a gas. If the work in equation (9.22.5) is replaced by $-P \, dV$, as is the case for a gas, the pressure can be expressed as

$$P = -\left(\frac{\partial U}{\partial V}\right)_T + T\left(\frac{\partial S}{\partial V}\right)_T \tag{9.22.14}$$

Again we see two components to the pressure, one due to internal energy and the other due to the entropy. The internal energy term is small (zero for the ideal gas); thus, if a gas is compressed, dV is negative, and the second term in equation (9.22.14) accounts for the pressure increase because dS is also negative for a compression.

The van der Waals equation of state, which expresses P as

$$P = \frac{-a}{V_m^2} + T\left[\frac{R}{V_m - b}\right] \tag{9.22.15}$$

attempts to take intermolecular attractions into account through the constant, a. Thus some of the thermal energy that is absorbed by a system undergoing an increase in temperature goes into overcoming these internal forces, and it is not all available to increase the kinetic energy. The significance of the two terms in the expressions for the force acting on the elastomer [see equations (9.22.9) and (9.22.13), can be better appreciated by comparing equations (9.22.9), (9.22.14), and (9.22.15) in light of the foregoing discussion. The first term in the van der Waals equation is equal to $(\partial U / \partial V)_T$.[3] In analogy to the ideal gas, an *ideal* elastomer is defined as one for which $(\partial U / \partial L)_T$ is zero.

Stress Versus Strain and Temperature

We will now consider the relationship between the applied force, f, and the elastomer length, L. While this can be studied directly from isothermal stress-strain measurements $\{f(L)_T\}$, we will obtain these data indirectly from measurements of the restoring force as a function of temperature carried out at various fixed elongations, L, i.e., $f(T)_L$. The analogous experiment, if carried out on a gas, would be to measure the pressure as a function of temperature at fixed volume. The gas law pertaining to this $P(T)_V$ isometric is called Rudberg's law (1842). It is customary to refer to the applied force as a *stress* (dimensions, force/area) and to the deformation as a *strain*. It is also convenient to use a dimensionless quantity for the strain, s, namely,

$$s = \frac{L - L_0}{L_0} \tag{9.22.16}$$

where L and L_0 are the stressed and unstressed lengths of the elastomer, respectively.

If Hooke's law were obeyed by an elastomer, the stress would be linearly dependent on the strain, i.e.,

$$f = Ys \tag{9.22.17}$$

where the force constant, Y, is commonly known as Young's modulus. Equation (9.22.17) is found to be valid only for small extensions (that is, small for an elastomer, but normal for other solids, i.e., ~ 1 percent or so). The fact that Hooke's law appears to be followed by many solids (including metals) but not by elastomers is a consequence of the very high deformation tolerated by an elastomer before its tensile limit, the point at which the material breaks, is reached, typically up to $s = 9$ (1000 percent).

The stress, which is the mass of the weight causing the elongation of the elastomer divided by the cross-sectional area of the elastomer, is usually determined on the basis of the *unstressed* elastomer's cross section, a_0 (when $L = L_0$). This is referred to as the *nominal* stress, as distinct from the actual stress. One of the objectives of this experiment is to determine Young's modulus for the elastomer. Not surprisingly, Y for an elastomer is much smaller (typically several hundred times) than that for other solids because it is, in fact, so "elastic."

It is possible to relate such measurables as Y to theoretical models of elastomer structure and behavior. For example, a statistical mechanical treatment of rubber elasticity provides an expression of strain vs. deformation. In this case, however, the deformation is expressed as the *fractional elongation*, σ, where $\sigma = L/L_0$. Assuming that the polymer chains are freely jointed (see the structure for *Hevea* rubber), i.e., there is no barrier to rotations about the CH_2—CH_2 single bonds, and that, moreover, the distance between the termini of a given polymer chain is characterized by a random, or Gaussian, distribution, the stress f can be related to the fractional elongation, σ

$$f = \frac{dRT}{zM} \left(\sigma - \frac{1}{\sigma^2} \right) = Y' \left(\sigma - \frac{1}{\sigma^2} \right) \tag{9.22.18}$$

where d is the density of the (unstrained) polymer having a *monomer* molecular weight M and z is the average number of monomer units between cross-links. R and T are the gas constant and absolute temperature. By comparing equations (9.22.17) and (9.22.18), it can be seen that the coefficient (dRT/zM) is an effective Young's modulus that can be obtained by plotting f vs. $(\sigma - 1/\sigma^2)$. In this way, z can be determined, and an important characteristic of the elastomer can be obtained. z can be related to the more immediate (and practical) concept of the fraction of monomers that are cross-linked in the polymer, F_{cl}. If a polymer chain consisting of N monomer units contains n cross-links, the average number of monomer units between cross-link nodes (z) is $z = N/n$. On the other hand, the *fraction* of cross-links is $F_{cl} \sim n/N$. Typically, F_{cl} is a few percent; its value depends on the particular vulcanization process used.

Thermal Effects

It is interesting to consider the thermal effects of elastomer deformation. The phenomenon can be experienced by suddenly (i.e., quasiadiabatically) stretching a rubber band or a balloon and then holding it to your lips (a sensitive temperature sensor). The rubber band will feel warmer relative to its initial state. Conversely, after the stretched elastomer is temperature-equilibrated and then rapidly returned to its initial length, it will feel cooler. The rapidity is necessary to minimize heat transfer from the rubber band to the surroundings and thus to keep the process nearly adiabatic. This process is thus analogous to the adiabatic compression and expansion of a gas. The formal expression for the temperature dependence of a reversible adiabatic (isentropic) deformation of an elastomer is $(\partial T/\partial L)_S$. This expression can be obtained from equation (9.22.12) by using the cyclic rule of partial derivatives, i.e.,

$$\left(\frac{\partial T}{\partial L}\right)_S = -\left(\frac{\partial T}{\partial S}\right)_L\left(\frac{\partial S}{\partial L}\right)_T = \frac{-(\partial S/\partial L)_T}{(\partial S/\partial T)_L} \qquad (9.22.19)$$

We can relate this expression to the tensile force, f, by using equations (9.22.12) and (9.22.13) and assuming that the rubber is "ideal" (i.e., that $(\partial U/\partial L)_T = 0$); thus,

$$\left(\frac{\partial S}{\partial L}\right)_T = \frac{-f}{T}$$

Further simplification results from identifying $(\partial S/\partial T)_L$ with the "constant-length" heat capacity, C_L:

$$C_L = T(\partial s/\partial T)_L$$

Equation (9.22.19) now becomes

$$\left(\frac{\partial T}{\partial L}\right)_S = \frac{f}{C_L} \qquad (9.22.20)$$

The temperature change is determined from the integration of equation (9.22.20):

$$\Delta T = \int_{L_0}^{L} \frac{1}{C_L} f \, dL$$

Assuming that C_L is independent of length and temperature (and that it is approximately equal to C_P of the elastomer—a more common physical constant), we have

$$\Delta T = \frac{1}{C_P} w \qquad (9.22.21)$$

in which w replaces $\int f \, dL$. w is the work done on the elastomer in stretching it from L_0 to L. Qualitatively, we see that in stretching an elastomer, $w > 0$ and, hence, $\Delta T > 0$, the temperature rises. Conversely, in relaxing from the stretched state, $w < 0$, $\Delta T < 0$, and the elastomer cools. To use equation (9.22.21) quantitatively, C_P is expressed as the specific heat, c_P. For many elastomers c_P is 1.8 to 2.0 J K^{-1} g^{-1}.

Safety Precautions

☐ Always wear safety glasses when in the laboratory.

☐ Compressed nitrogen or air is used in this experiment. The cylinder must be securely lashed to a firm support. A reducing valve provides an N_2 supply at a few psi above ambient pressure (<5 psig). Do not increase this pressure during the experiment.

☐ The gas heating chamber may be hot. Do not touch.

☐ If the elastomer sample is to be prestressed, it may give off an odor. This experiment should be performed in an open, well-ventilated laboratory.

Procedure

The elastomer sample is a commercial rubber band made from a synthetic poly(isoprene). While similar to the natural product, it is free of some of the impurities present in *Hevea* rubber. The instructor will tell you if the sample has been "prestressed" to eliminate the hysteresis that is usually observed in the $f(T)$ plots of new samples. (With hysteresis present, heating and cooling will produce different, although reproducible, results. This is obviously unsatisfactory for determinations of an equation of state of a substance.) The prestressing is accomplished by keeping an elongated elastomer at the maximum temperature for some time and then slowly reducing the temperature. If the sample requires prestressing, your instructor will give you the needed information.

The procedure is to measure the force as a function of temperature at various fixed elongations: $\{f(T)\}_L$. To be carried out successfully, this experiment (like so many others) requires much patience and attention to detail.

The apparatus is shown in Figure 9.22.2. The sample is suspended in a vacuum-jacketed condenser tube. The sample is heated by a stream of N_2 that is passed over a resistance coil contained in another insulated condenser. (The gas may be first passed through a metal coil immersed in an ice bath in order to speed up the cooling of the sample after it has been heated.) A thermocouple (or thermistor) senses and controls the sample temperature.

The elongation of the sample is set by adjusting the position of a small lab jack that is mounted on the platform of a digital balance. The readout of the balance indicates the tension of the elastomer. A marker wire attached to the elastomer mounting rod and suspended in front of a meter stick is used to determine the extent of elongation.

1. The balance and elongation indicator must first be initialized to zero elongation. Set the lab jack to the highest position and switch on the balance. The set screw coupling the sample rod to the lab jack should be loosened so that there is slack in the rod. "Tare" the balance and read the position of the wire marker on the meter stick. Gently tighten the set screw; be careful not to jar or misalign the sample rod. The balance should still read zero (or very close to zero). If it doesn't, check the alignment of the sample rod; retare the balance if necessary (leaving the set screw tight).

2. Open the control valve for the N_2 gas and allow the gas to flow through the apparatus. The instructor will tell you what an appropriate flow rate is. Obtain a value of L_0 by measuring the length of the unstressed elastomer using a ruler held behind the sample. Gently lower the lab jack so that the sample is extended between 20 and 30 percent. This will usually correspond to an elongation of about 4 mm. You will notice that the balance indicates a *negative* weight.

3. Set the controller to a temperature between 70 and 80°C. Wait for a few minutes after the set point is reached and read both the sample temperature and the balance. After these readings have stabilized, record the values.

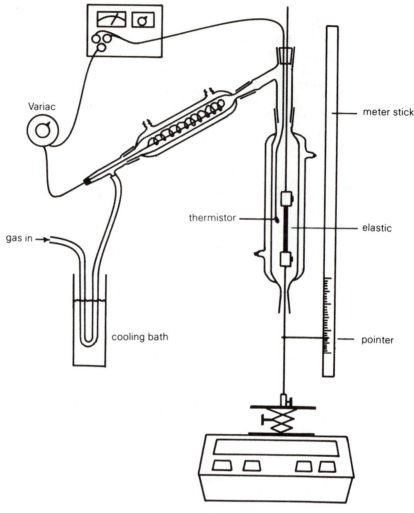

temperature controller and readout

Variac

gas in →

cooling bath

meter stick

thermistor

elastic

pointer

electronic balance

FIGURE 9.22.2 The apparatus used to vary stress, strain, and temperature of an elastomer sample.

4. Lower the temperature set point by 10 to 15° and wait for the temperature and sample tension to stabilize. Record the values. To speed up cooling, the N_2 flow rate can be temporarily increased slightly.

5. Repeat step 4 until 30° or ambient temperature is reached.

6. Lower the lab jack to increase the elongation by another 20 to 30 percent (4 mm). Now perform a set of measurements in which the temperature is *increased* in increments of 10 to 15°C until 70 to 80°C is reached. The temperatures used need not correspond exactly to those previously recorded. In each case, wait for equilibration and read the sample temperature and tension.

7. Repeat steps 4 through 6 until the maximum elongation is reached (usually between 150 and 200 percent) and the data acquisition is complete. Thus a series of decreasing and increasing temperature scans will be performed.

8. Shut off the temperature controller, balance, and gas flow.

9. A sample of the elastomer will be provided. Using a micrometer, measure the length and width of the sample. Be careful not to compress the elastomer.

Data Analysis

1. Convert all the tension readings to strain values, f, using the cross section of the sample. Convert the elongations to stress values, s, using equation (9.22.16).

2. Plot f vs. T (in K) for each elongation used. You should assume that these plots are linear (if they are clearly nonlinear, reexamine your data). Using linear regression, determine the intercept $(f, 0)$ and the slope for each plot. Because of the very large extrapolation between the studied temperatures and 0 K, a considerable error in $(f, 0)$ is inevitable. Determine these errors as well as those of the slopes.

3. From equation (9.22.13) determine the nonideal contribution, $(\partial U/\partial L)_T$, to the total force referenced to 20°C for each elongation. Tabulate these values.

4. From the $f(T)$ plots, and the least squares fit to these data, construct $f(s)_T$ isotherms for about five temperatures that fall in your data range. Thus draw vertical lines and tabulate f values for the different elongations. Convert the s values into $\sigma(L/L_0)$. From these $f(\sigma)_T$ isotherms, plot f_T vs. $(\sigma - 1/\sigma^2)$ and obtain Y'; see equation (9.22.18). If the plots are curved, use the limiting values of the slopes.

5. Plot Y' vs. T (in K) and obtain an estimate of the fraction of cross-links, $1/z$, in the elastomer. Use a density of 0.970 g/cm^3. See equation (9.22.18) and the discussion thereafter.

Note: If very small initial elongations are used (<7 percent), you might notice that these $f(T)$ plots will have a *negative* slope, contrary to the indication of equation (9.22.9); $(\partial S/\partial L)_T$ is a negative quantity. This effect, which is referred to as thermoelastic inversion, is the result of the thermal expansion of the elastomer. Thus as the sample is heated, the increase in tension is counterbalanced (for small extensions) by its slackening. This apparent anomaly can be "corrected" by first transforming the $f(T)_s$ data to the $f(s)_T$ isotherms and then recalculating the elongations as follows:

$$s' = \frac{s - \alpha(T - T_0)}{1 + \alpha(T - T_0)}$$

where s' and s are the corrected and uncorrected stress values, respectively; α is the linear coefficient of thermal expansion for the rubber (ca. 2.2×10^{-4} deg^{-1}); T is the sample temperature; and T_0 is the temperature at which L_0 is determined. After these corrections are made and the $f(s')_T$ data are used to construct $f(T)_{s'}$ plots, the latter will show positive slopes for all elongations.

Questions and Further Thoughts

1. Using the Maxwell relations and the thermodynamic equation of state, show that each of the terms in equations (9.22.14) and (9.22.15) are equivalent; i.e.,

$$-\left(\frac{\partial U}{\partial V}\right)_T = \frac{-a}{(V_m)^2} \quad \text{and} \quad T\left(\frac{\partial S}{\partial U}\right)_T = \frac{RT}{V_m - b}$$

for a van der Waals gas.

2. Young's modulus, defined by equation (9.22.17), is several hundred times smaller for an elastomer as compared with other solids. Why is this to be expected? Thus, what would be the result of trying this experiment with a steel rod rather than a rubber band?

3. Show that equation (9.22.14) follows from $dA = dU - T\,dS = -P\,dV - S\,dT$.

Notes

1. Note that $(\partial U/\partial V)_T = T(\partial P/\partial T)_V$.

Further Readings

R. A. Alberty, "Physical Chemistry," 7th ed., pp. 124–129, Wiley (New York), 1987.

R. L. Anthony, R. H. Caston, and E. Guth, *J. Phys. Chem.*, *46*, 826 (1942).

P. W. Atkins, "Physical Chemistry," 3rd ed., pp. 635–636, W. H. Freeman (New York), 1986.

J. E. Mark, *J. Chem. Educ.*, *58*, 898 (1981).

P. Meares, "Polymers: Structure and Bulk Properties," chap. 6–8, D. Van Nostrand (London), 1965.

J. H. Noggle, "Physical Chemistry," pp. 154–161, Little, Brown (Boston), 1985.

M. Shen, W. F. Hall, and R. E. DeWames, "Molecular Theories of Rubber-Like Elasticity and Polymer Viscoelasticity," in G. B. Butler and K. F. O'Driscoll, eds., "Reviews of Macromolecular Chemistry," vol. 3, Marcel Dekker, Inc., New York, 1968.

L. R. G. Treloar, "The Physics of Rubber Elasticity," Clarendon Press (Oxford), 1975.

L. R. G. Treloar, "Introduction to Polymer Science," Wykeham Publications (London), 1974.

F. T. Wall, "Chemical Thermodynamics," pp. 335–350, W. H. Freeman (San Francisco), 1974.

Experiment 22

Thermoelasticity

NAME _____ DATE _____

Elastomer data: width, _____; thickness, _____; L_0 _____

Marker setting at zero strain, _____. Tension

Marker: _____

Temp.	Tension
_____	_____
_____	_____
_____	_____
_____	_____
_____	_____
_____	_____
_____	_____

Marker: _____

Temp.	Tension
_____	_____
_____	_____
_____	_____
_____	_____
_____	_____
_____	_____
_____	_____

Marker: _____

Temp.	Tension
_____	_____
_____	_____
_____	_____
_____	_____
_____	_____
_____	_____
_____	_____

Marker: _____

Temp.	Tension
_____	_____
_____	_____
_____	_____
_____	_____
_____	_____
_____	_____
_____	_____

Marker: _____

Temp.	Tension
_____	_____
_____	_____
_____	_____
_____	_____
_____	_____
_____	_____
_____	_____

Marker: _____

Temp.	Tension
_____	_____
_____	_____
_____	_____
_____	_____
_____	_____
_____	_____
_____	_____

(continues)

Marker: _____   Marker: _____   Marker: _____

Temp.	Tension	Temp.	Tension	Temp.	Tension
_____	_____	_____	_____	_____	_____
_____	_____	_____	_____	_____	_____
_____	_____	_____	_____	_____	_____
_____	_____	_____	_____	_____	_____
_____	_____	_____	_____	_____	_____
_____	_____	_____	_____	_____	_____
_____	_____	_____	_____	_____	_____

Experiment 22

Thermoelasticity

NAME _____ DATE _____

Elastomer data: width, _____; thickness, _____; L_0 _____

Marker setting at zero strain, _____. Tension

Marker: _____

Temp.	Tension
_____	_____
_____	_____
_____	_____
_____	_____
_____	_____
_____	_____
_____	_____

Marker: _____

Temp.	Tension
_____	_____
_____	_____
_____	_____
_____	_____
_____	_____
_____	_____
_____	_____

Marker: _____

Temp.	Tension
_____	_____
_____	_____
_____	_____
_____	_____
_____	_____
_____	_____
_____	_____

Marker: _____

Temp.	Tension
_____	_____
_____	_____
_____	_____
_____	_____
_____	_____
_____	_____
_____	_____

Marker: _____

Temp.	Tension
_____	_____
_____	_____
_____	_____
_____	_____
_____	_____
_____	_____
_____	_____

Marker: _____

Temp.	Tension
_____	_____
_____	_____
_____	_____
_____	_____
_____	_____
_____	_____
_____	_____

(*continues*)

Marker: _____   Marker: _____   Marker: _____

Temp.   *Tension*   *Temp.*   *Tension*   *Temp.*   *Tension*

_____   _____   _____   _____   _____   _____
_____   _____   _____   _____   _____   _____
_____   _____   _____   _____   _____   _____
_____   _____   _____   _____   _____   _____
_____   _____   _____   _____   _____   _____
_____   _____   _____   _____   _____   _____
_____   _____   _____   _____   _____   _____

□ PART TEN

Photophysics and Molecular Spectroscopy

□ Experiment 23

Excited State Properties of 2-Naphthol

PART I: ACIDITY CONSTANT

Objective

To determine the acidity constants of the ground and lowest electronically excited states of 2-naphthol in aqueous solution.

Introduction

The electronic structure of a molecule determines such physical and chemical properties as its charge distribution, geometry (therefore dipole moment), ionization potential, electron affinity, and, of course, chemical reactivity. If the electronic structure of a molecule were to be changed, therefore, one would expect its physical and chemical properties to be altered. Such a rearrangement in electronic structure can, in fact, be brought about (and *very* rapidly, $\sim 10^{-13}$ s) if the molecule is raised to an electronically excited state via the absorption of a quantum of light (photon) whose energy matches the gap between the molecular ground and excited state energy levels.

For most organic molecules that contain an even number of electrons, the ground state is characterized by having all electron spins paired; the net spin angular momentum is zero, and such an arrangement is called a *singlet* state. When considered in terms of molecular orbitals (mo), electronic excitation involves the promotion of an electron from a filled mo to a higher, vacant mo. This new orbital configuration, which characterizes the electronically excited state, may be one in which the two electrons in the singly occupied mo's have opposite spins. Accordingly, this electronically excited state is also a singlet. The ground, and lowest electronically excited, singlet states are often denoted as S_0 and S_1, respectively. Higher excited singlet states are referred to as $S_2, S_3, \cdots, S_n$. This experiment deals with excited *singlet* states.

Although measurements of the physical and chemical properties of a molecule in its ground state can be carried out, more or less, at leisure (assuming that the molecule is thermally stable), the examination of these properties in its excited states is severely hampered by the fact that these states are very short-lived. For most molecules, S_1 states have lifetimes ranging from $10^{-6} - 10^{-11}$ s. Excited states are metastable; they undergo decay processes that dissipate the energy they possess relative to more stable products. For example, the excited state of a molecule may, in general: spontaneously return to the ground state via photon emission (fluorescence), convert electronic excitation into ground state vibrational energy (heat), undergo bond dissociation or rearrangement or a change in electron spin multiplicity. Because spontaneous emission from an excited state (i.e., fluorescence) often takes place very rapidly, fluorescence can be used as a probe, or measurement, of excited state concentration (e.g., fluorescence assay). In addition, fluorescence studies can provide information about the physical and chemical properties of these short-lived singlet states. This field of experimentation is called photophysics.

In this experiment, some ground and excited state properties of the organic molecule, 2-naphthol (ArOH) will be determined.

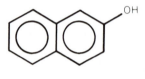

2-naphthol (ArOH)

In aqueous solution, ArOH behaves as a weak acid, forming the hydronium ion and its conjugate base, the naphthoxy ion, ArO^-.

$$ArOH + H_2O \rightleftarrows H_3O^+ + ArO^-$$

It is instructive to measure the acidity constant of ArOH in its lowest excited electronic state, denoted as K_a^*, and to compare this value with that of the ground state, K_a. This information indicates how the change in electronic structure alters the charge density at the oxygen atom. The experimental method is best introduced in terms of the energy-level diagram shown in Figure 10.23.1.

The relative energies of the free acid and its conjugate base (the naphthoxy ion) are indicated for both the electronic ground (S_0) and (lowest) excited (S_1) states in aqueous solution. Each anion is elevated with respect to its free acid by an energy, ΔH and ΔH^*, respectively. These are the enthalpies of deprotonation. Both the ground state acid and its conjugate base can be transformed to their respective excited states via the absorption of photons of energy $h\nu_{ArOH}$ and $h\nu_{ArO^-}$. For simplicity, these absorptive transitions are shown to be equal to the fluorescence from the excited to the ground states of the acid and conjugate base. (The ground and excited state vibrational levels that are involved in the transitions are not indicated.)

The free energy of deprotonation of ArOH can be expressed in terms of the enthalpy and entropy of deprotonation and the equilibrium (ionization) constants:

$$\Delta G = \Delta H - T \Delta S = -RT \ln K_a, \qquad (10.23.1)$$

and

$$\Delta G^* = \Delta H^* - T \Delta S^* = -RT \ln K_a^*, \qquad (10.23.2)$$

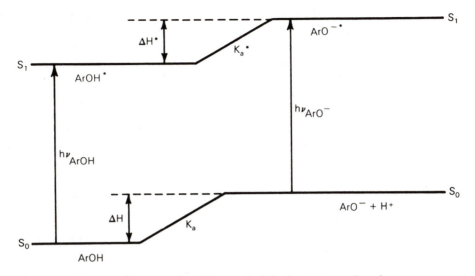

FIGURE 10.23.1 Schematic diagram of the ground and first excited singlet state energies of free naphthol and its conjugate base, the naphthoxy ion, in aqueous solution.

for the S_0 and S_1 states, respectively. If we make the assumption that the entropies of dissociation of ArOH and ArOH* are equal, it follows that

$$\Delta H - \Delta H^* = -RT \ln\left(\frac{K_a}{K_a^*}\right), \qquad (10.23.3)$$

and thus from Figure 10.23.1, it can be deduced that

$$\Delta H + N_A h\nu_{\text{ArO}^-} - \Delta H^* = N_A h\nu_{\text{ArOH}}, \qquad (10.23.4)$$

where h is Planck's constant. Avogadro's number, N_A, has been included to put each energy term on a molar basis. Combining equations (10.23.4) and (10.23.3) and rearranging gives

$$\ln\left(\frac{K_a^*}{K_a}\right) = \frac{N_A h\{\nu_{\text{ArOH}} - \nu_{\text{ArO}^-}\}}{RT}. \qquad (10.23.5)$$

Thus knowledge of the energy gap between the ground and first excited states for both the free acid and its conjugate base leads to an estimate of K_a^*, if K_a is known. The analysis presented above, which accounts for the observed thermodynamic and spectroscopic energy differences, was first developed by Th. Förster (1949, 1950). This approach is thus often referred to as a *Förster cycle*. Equation (10.23.5) can be recast into a more convenient form

$$pK_a^* = pK_a - \frac{N_A hc}{2.303 RT}[\bar{\nu}_{\text{ArO}^-} - \bar{\nu}_{\text{ArOH}}], \qquad (10.23.6)$$

where c, the speed of light, has been incorporated in the expression to convert the transition energies of ArOH and ArO$^-$ into *wavenumbers* (cm^{-1}), a common spectroscopic energy unit. The acidity constants are expressed as pK values.

The question now is how to obtain the spectroscopic energy difference $(\bar{\nu}_{\text{ArO}^-} - \bar{\nu}_{\text{ArOH}})$, or $\Delta\bar{\nu}$, pertinent to the Förster cycle in 2-naphthol. Three approaches can be considered. The first is to base the measurement on the absorption maxima of the free acid and conjugate base. The second is to use the fluorescence maxima of the two species, and the third is to use what is called the 0-0 energies of ArOH and ArO$^-$. The first two methods are somewhat more straightforward. Obtaining $\Delta\tilde{\nu}$ from absorption data has the advantage that the energy difference that is obtained does not require instrumental correction. Although obtaining $\Delta\tilde{\nu}$ from the fluorescence maxima seems simple enough, many fluorimeters produce fluorescence spectra that are distorted by the wavelength response of the monochromator/photomultiplier combination. Thus unless these spectra are corrected for this sensitivity distortion, the recorded fluorescence maxima will depend (although perhaps subtly) on the particular instrument used and thus will not represent the "true" or absolute spectroscopic properties of the ArOH/ArO$^-$ system.

The third approach provides the value of $\Delta\tilde{\nu}$ that best represents the energy difference between S_1 and S_0 implied in the Förster cycle, $\tilde{\nu}_{0-0}$. Unfortunately $\tilde{\nu}_{0-0}$ cannot be determined directly in all cases (such as ArOH), but it can be estimated from an analysis of *both* the absorption and fluorescence spectra. This is shown schematically in Figure 10.23.2 for a single species (ArOH or ArO$^-$).

The absorption and (preferably instrument-corrected) fluorescence spectra are both plotted on a common energy axis. Furthermore, the spectra are presented so that they are normalized to have identical maxima. The point of intersection of these spectra can be approximated as the 0-0 energy gap between S_0 and S_1. This approach does not apply to those cases in which the S_0-S_1 transition is distorted by a nearby (and especially more intensely absorbing) S_0 to S_2 transition.

Whichever method is used to determine $\Delta\tilde{\nu}$, it should be used consistently with both species, ArOH and ArO$^-$.

Safety Precautions

☐ Always wear safety goggles in the laboratory. Ultraviolet light–absorbing eye protection is required. Some plastic safety goggles do not completely absorb ultraviolet radiation.

☐ 2-naphthol is an irritant. If you are to prepare the solutions from solid material, you *must* wear gloves; if possible, work in a fume hood.

☐ Be sure you have been instructed to use the proper pipeting techniques when handling 2-naphthol solutions. *Never* pipet using suction by mouth.

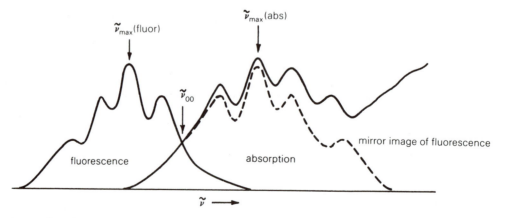

FIGURE 10.23.2   Schematic diagram of the absorption and fluorescence spectra of a molecule. It is assumed that these transitions are between the same two electronic states, i.e., $S_0 \leftrightarrow S_1$.

☐ If you are to obtain fluorescence spectra, be sure that any ozone produced by the ultraviolet source is vented. Ozone is a noxious, dangerous gas that has an acrid odor. If you detect this gas, leave the vicinity, inform your instructor, and increase air circulation at once.

Procedure

In this experiment, K_a will be determined from the pH dependence of the absorption spectrum of aqueous 2-naphthol.

1. Obtain absorption spectra of two solutions of ArOH (approximately 2×10^{-4} M; the actual concentration must be accurately known). In one solution, the free acid must predominate (low pH), and in the other, the conjugate base must be the major naphthol component (high pH). This can be achieved by using concentrated stock solutions of HCl and NaOH (e.g., $0.10M$) to create $[H^+]$ and $[OH^-]$ of $0.02M$, respectively. It is important that the ArOH concentration be the same in each case. Label and save these solutions. These spectra should be recorded on the same chart paper having a common wavelength axis so that the spectra overlap.

2. Now obtain absorption spectra of ArOH (also, 2×10^{-4} M) at intermediate pH values. Appropriate pH levels can be obtained by using ammonium chloride buffer solutions (NH_4OH/NH_4Cl), for example, $0.1M/0.1M$, $0.1/0.2$, or $0.2/0.1$. These solutions can be conveniently prepared from $1.00M$ stock solutions of NH_4OH and NH_4Cl. At least three spectra, run over the entire wavelength range as in step 1, illustrating both free acid and conjugate base absorption, should be obtained. If necessary, appropriate ratios of the NH_4OH and NH_4Cl stock solutions can be chosen in order to produce a satisfactory series of ArOH/ArO$^-$ spectra. It is *essential* that the 2-naphthol concentrations be identical (and accurately known) in each case. Immediately after each spectrum is recorded, measure the actual pH of the solution using a properly calibrated pH meter. Label and save these solutions. It is instructive to overlap each spectrum on the same strip chart paper, making sure that the wavelength is synchronized in all cases. Make sure that the temperature of the samples is constant (to within 1°C) or otherwise controlled throughout the experiment. If the *bulk* ArOH concentration, e.g., [ArOH] + [ArO$^-$], is invariant, the spectra, when properly overlapped, should intersect at a common wavelength called the *isosbestic point* (equal extinction). The presence of an isosbestic point indicates that there is a closed system (as a function of the variable, pH) consisting of two species in equilibrium.

3. Using the same solutions in step 1, obtain the fluorescence spectra of naphthol and its conjugate base. If possible, these should be corrected for the monochromator/photomultiplier response of the fluorimeter.

4. (Optional) Record the fluorescence spectra of the intermediate pH solutions on the same recorder paper so the spectra overlap. There should, again, be a common wavelength where the spectra overlap. This is sometimes referred to as an *isostilbic point* (equal brightness). Likewise, the presence of an isostilbic point indicates that there are two excited state species that, aside from emitting light, *only* interconvert with each other.

Calculations

Since the bulk naphthol concentrations, $[ArOH]_0$ are identical in each of the solutions studied, the following material balance applies:

$$[ArOH]_0 = [ArOH] + [ArO^-]. \qquad (10.23.7)$$

Under the condition that both the free acid and conjugate base absorb at λ_{max}ArOH, the wavelength of maximum absorption for ArOH, and that Beer's law holds,

$$A(\lambda_{max}\text{ArOH}) = \epsilon_{\text{ArOH}}[\text{ArOH}] + \epsilon_{\text{ArO}^-}[\text{ArO}^-], \qquad (10.23.8)$$

where A is the absorbance (for a one cm path length), and ϵ_{ArOH} and ϵ_{ArO^-} are the extinction coefficients of the free acid and conjugate base at λ_{max}(ArOH), respectively. Combining equations (10.23.7) and (10.23.8) produces

$$A(\lambda_{max}\text{ArOH}) = \{\epsilon_{\text{ArOH}} - \epsilon_{\text{ArO}^-}\}[\text{ArOH}] + \{\epsilon_{\text{ArO}^-}\}[\text{ArOH}]_0. \quad (10.23.9)$$

This relation allows [ArOH] to be determined under equilibrium conditions. Once this is known, the value of [ArO$^-$] at the same pH can be obtained from equation (10.23.7). Since the acidity constant is

$$K_a = \frac{[\text{H}_3\text{O}^+][\text{ArO}^-]}{[\text{ArOH}]}, \qquad (10.23.10)$$

(assuming that the activity coefficient ratio is unity), the pK_a value for ArOH can be obtained from a graphical analysis of

$$\text{pH} = \text{p}K_a + \log\frac{[\text{ArO}^-]}{[\text{ArOH}]}, \qquad (10.23.11)$$

using linear (first order) regression. The least-squares value of pK_a (and its confidence limits) should be reported. Compare your results with the literature values.

Determination of K_a^*

From the high and low pH absorption spectra, determine λ_{max}ArOH and the extinction coefficients of ArOH and ArO$^-$ at this wavelength. Using these values, along with the absorbance (pH) data, determine pK_a for ArOH using equations (10.23.7), (10.23.9), and (10.23.11).

Calculate pK_a^* for (ArOH)* from the pK_a value obtained above and whatever method(s) can be used to determine $(\bar{\nu}_{\text{ArOH}} - \bar{\nu}_{\text{ArO}^-})$. Compare these values and perform an error analysis. Discuss the errors that the primary measurements have on the derived value of pK_a^*.

Other molecules that can be readily studied using this procedure are 2-naphthoic acid, acridine, and quinoline.

2-naphthoic acid   Acridine   Quinoline

Questions and Further Thoughts

1. In comparing the values of K_a and K_a^*, what can you deduce about the change in electron density at the 0 atom in 2-naphthol in the electronically excited state relative to the ground state?

2. If $pK_a^* < pK_a$ (i.e., the excited state species is a stronger acid), can you comment on the relative (absolute) magnitudes of the enthalpies of deprotonation? See Figure 10.23.1.

3. On what basis can one justify the assumption that the entropies of deprotonation of the ground and excited state 2-naphthol molecules are equal [see equations (10.23.1)–(10.23.3)]. Can you think of another possibility in which this assumption is a poor one?

4. Other molecules that can be studied using this technique are 2-naphthoic acid, acridine, and quinoline (see above). Indicate the protolytic reactions for these molecules, i.e., write the aqueous acid-base reactions.

5. Can you predict before doing an experiment whether the exited state of a molecule is a stronger or weaker acid relative to its respective ground state? What information would you need to perform such an assessment?

Further Readings

Förster Cycle and Excited State Acidity

Th. Förster, *Naturwiss., 36,* 186 (1949).

Th. Förster, *Z. Electrochem., 54,* 531 (1950).

C. Parker, ''Photoluminescence of Solutions,'' pp. 328–341, Elsevier (Amsterdam), 1968.

J. Van Stam and J. E. Loefroth, *J. Chem. Educ., 63,* 181 (1986).

A. Weller, in G. Porter, ed., ''Progress in Reaction Kinetics,'' vol. 1, pp. 189–214, Pergamon (New York), 1961.

Fluorescence and Photophysics

R. A. Alberty, ''Physical Chemistry,'' 7th ed., pp. 543–547, Wiley (New York), 1987.

P. W. Atkins, ''Physical Chemistry,'' 3rd ed., pp. 471–475, W. H. Freeman (New York), 1986.

G. N. Castellan, ''Physical Chemistry,'' 3rd ed., pp. 889–896, Addison-Wesley (Reading, Mass.), 1983.

I. N. Levine, ''Physical Chemistry,'' 2nd ed., pp. 725–729, McGraw-Hill (New York), 1983.

J. N. Noggle, ''Physical Chemistry,'' pp. 754–760, Little, Brown (Boston), 1983.

N. J. Turro, ''Modern Molecular Photochemistry,'' Benjamin/Cummings (Menlo Park, Calif.), 1978.

Excited State Properties of 2-Naphthol (Part I)

NAME _____   DATE _____

2-naphthol concentration _____ M (_____ mg in _____ mL)

Solution #   pH

_____   _____

_____   _____

_____   _____

_____   _____

_____   _____

_____   _____

_____   _____

_____   _____

_____   _____

_____   _____

Temperature _____

Excited State Properties of 2-Naphthol (Part I)

NAME _____ DATE _____

2-naphthol concentration _____ M (_____ mg in _____ mL)

Solution #   pH

_____   _____

_____   _____

_____   _____

_____   _____

_____   _____

_____   _____

_____   _____

_____   _____

_____   _____

_____   _____

Temperature _____

□ Experiment 24

Excited State Properties of 2-Naphthol

PART II: DEPROTONATION/PROTONATION RATE CONSTANTS

Objective

To determine the deprotonation and protonation rate constants of 2-naphthol in its lowest excited singlet state in aqueous solution.

Introduction

In the previous experiment, acidity constants for aqueous 2-naphthol (ArOH) were determined for both the ground and lowest excited (singlet) states. These constants pertain to the equilibrium

$$\text{ArOH} + \text{H}_2\text{O} \underset{k_p'}{\overset{k_d'}{\rightleftarrows}} \text{ArO}^- + \text{H}_3\text{O}^+, \qquad (10.24.1)$$

where the rate constants for the forward (deprotonation) and reverse (protonation) reactions are indicated as k_d' and k_p', respectively. A similar equilibrium can be written for the excited state species:

$$\text{ArOH}^* + \text{H}_2\text{O} \underset{k_p}{\overset{k_d}{\rightleftarrows}} \text{ArO}^-{}^* + \text{H}_3\text{O}^+, \qquad (10.24.2)$$

in which the values of the forward and reverse rate constants may be different from those in the ground state because of differences in the properties of the 2-naphthol in these two states (e.g., different K_a values).

The ratio of the concentrations of free acid and the conjugate base can be expressed as a function of the pH, viz.,

$$\log\left(\frac{[\text{ArOH}]}{[\text{ArO}^-]}\right) = \text{p}K_a - \text{pH}, \qquad (10.24.3)$$

where molar concentrations are used to approximate activities. An analogous equation,

$$\log\left(\frac{[\text{ArOH}^*]}{[\text{ArO}^-{}^*]}\right) = \text{p}K_a^* - \text{pH} \qquad (10.24.3a)$$

applies to excited state species. Equations (10.24.3) and (10.24.3a) show that if the pH of the solution is less than $\text{p}K_a$ of naphthol (in either electronic state), the free acid form will predominate over that of conjugate base, i.e., $[\text{ArOH}] \gg [\text{ArO}^-]$. Likewise, if $\text{pH} > \text{p}K_a$, then $[\text{ArO}^-] > [\text{ArOH}]$.

Suppose that by using a suitable buffer, the pH of the medium is established to be less than $\text{p}K_a$ but greater than $\text{p}K_a^*$. The ground state of the system will then consist primarily of ArOH. Electronic excitation via light absorption will "instantaneously" ($\sim 10^{-13}$ s) transform ArOH into ArOH*. We may assume that in this experiment, the buffer holds the pH of the medium constant during and after electronic excitation. This is a valid assumption because the number of photons absorbed per unit volume is much less than the ground-state concentration of ArOH. Thus [ArOH*] ≪ [ArOH]; moreover, ArOH* will spontaneously dissociate to form ArO⁻* in order to establish new equilibrium conditions. Under these circumstances, [ArO⁻*] must be greater than [ArOH*] because pH > pK_a^* [see equation (10.24.3a)]. In fact, most of the fluorescence observed from ArO⁻* takes place from species that were formed via ArOH* deprotonation *after* electronic excitation. As excited state equilibrium is approached, the ArOH* concentration decreases, while that of ArO⁻* increases. The strategy of this experiment in determining k_d and k_p is to measure the dependence of [ArOH*] on the pH of the solution. [ArOH*] is monitored through its fluorescence intensity, I_f, assuming that it is proportional to [ArOH*] (see below). The pH is varied (and established) using an ammonium acetate buffer.

Kinetic Analysis

Because in this experiment pH < pK_a, only light absorption by the protonated form, ArOH, is considered:

$$ArOh + h\nu_{abs} \rightarrow ArOH^* \qquad \text{absorption.}$$

The ArOH* thus produced is (like any excited state) metastable and in relaxing, can undergo a number of different decay processes, e.g.:

1: $\quad ArOH^* \xrightarrow{k_f} ArOH + h\nu_{fluor} \qquad$ fluorescence $\qquad$ rate $= k_f[ArOH^*]$.

2: $\quad ArOH^* \xrightarrow{k_{nr}} ArOH + $ heat $\qquad$ nonradiative decay
$\qquad\qquad\qquad$ possible photoproduct

$\qquad\qquad\qquad\qquad\qquad\qquad\qquad\qquad\qquad$ rate $= k_{nr}[ArOH^*]$.

3: $\quad ArOH^* \xrightarrow{k_d} ArO^-{}^* + H^+(aq) \qquad$ deprotonation $\qquad$ rate $= k_d[ArOH^*]$.

4: $\quad ArOH^* + Ac^- \xrightarrow{k_{Ac^-}} ArO^-{}^* + HAc \qquad$ deprotonation via Ac⁻
$\qquad\qquad\qquad\qquad\qquad\qquad\qquad\qquad$ rate $= k_{Ac^-}[ArOH^*][Ac^-]$.

In addition to radiative (fluorescence) decay (1) and nonradiative relaxation (2), ArOH* can undergo "unassisted" (3) and "acetate-assisted" (4) deprotonation. This distinction is significant because the rate of deprotonation will be *enhanced* in the presence of Ac⁻ in the bimolecular step indicated above in (4). Undoubtedly, solvent plays a role in the unassisted deprotonation (3), but this step can be considered pseudo-first-order in ArOH* because the concentration of "solvent" is so much larger than [ArOH*]. The reverse steps of the deprotonation processes, which are bimolecular and proportional to [H₃O⁺] and [HAc] [see steps (3) and (4), respectively], are ignored because under these experimental conditions, [H₃O⁺] and [HAc] are very small.

If the pH of the solution is much lower than pK_a^* (e.g., in the presence of sulfuric acid), deprotonation by either process will be suppressed, and fluorescence from ArOH* predominates. In this case, the *fluorescence intensity, I_f^0,* which is proportional to the ratio of the rate of radiative decay to the *total* ArOH* decay rate, is

$$I_f^0 = \frac{Ck_f[ArOH^*]}{k_f[ArOH^*] + k_{nr}[ArOH^*]}, \qquad (10.24.4)$$

or, canceling [ArOH*],

$$I_{\mathrm{f}}^0 = \frac{Ck_f}{k_f + k_{nr}}, \tag{10.24.5}$$

where C is an instrumental constant.

On the other hand, when $\mathrm{pH} > \mathrm{pK}_a^*$ (but less than pK_a), i.e., under $\mathrm{NH_4Ac}$ buffer conditions, the deprotonation steps become kinetically important, and thus the denominator of equation (10.24.4) will contain the additional terms, $k_d[\mathrm{ArOH^*}]$ and $k_{\mathrm{Ac}^-}[\mathrm{ArOH^*}][\mathrm{Ac}^-]$. Therefore the ArOH* fluorescence intensity, now denoted as I_f, becomes [see equation (10.24.5)]

$$I_f = \frac{Ck_f}{k_f + k_{nr} + k_d + k_{\mathrm{Ac}^-}[\mathrm{Ac}^-]}. \tag{10.24.6}$$

The deprotonation of ArOH* causes a diminution in its fluorescence intensity; thus $I_f < I_f^0$. Assuming that [ArOH*] is identical in all the solutions studied, the ratio of ArOH* fluorescence intensity in a solution containing sulfuric acid (low pH) to that containing ammonium acetate buffer (high pH) is obtained by dividing equation (10.24.5) by (10.24.6); thus,

$$\frac{I_f^0}{I_f} = \frac{k_f + k_{nr} + k_d + k_{\mathrm{Ac}^-}[\mathrm{Ac}^-]}{k_f + k_{nr}}. \tag{10.24.7}$$

Equation (10.24.7) can be rearranged to the more convenient form

$$\left(\frac{I_f^0}{I_f} - 1\right) = \frac{k_d}{k_f + k_{nr}} + \frac{k_{\mathrm{Ac}^-}[\mathrm{Ac}^-]}{k_r + k_{nr}},$$

or
$$\tag{10.24.8}$$

$$\left(\frac{I_f^0}{I_f} - 1\right) = \tau_0 k_d + \tau_0 k_{\mathrm{Ac}^-}[\mathrm{Ac}^-],$$

where $\tau_0 = 1/(k_f + k_{nr})$ and is called the "lifetime" of ArOH* in the absence of significant deprotonation. A plot of $(I_f^0/I_f) - 1$ vs. [Ac$^-$] (called a *Stern-Volmer* plot) should be linear with a slope of $\tau_0 k_{\mathrm{Ac}^-}$ and an intercept of $\tau_0 k_d$. In order to determine k_d, τ_0 must be determined separately. Stern-Volmer quenching kinetics is also discussed in Experiment 20.

The information provided by the Stern-Volmer plot (a time-independent, or steady-state, method) could be obtained directly using a transient, or kinetic approach by monitoring the fluorescence *decay* of ArOH*. After instantaneous photoexcitation ($<$ ca. 10^{-9} s), the ArOH* fluorescence intensity follows the decay law:

$$I_f(t) \propto [\mathrm{ArOH^*}]_0 \exp\{-(k_f + k_{nr} + k_d + k_{\mathrm{Ac}^-}[\mathrm{Ac}^-])t\}, \tag{10.24.9}$$

where $[\mathrm{ArOH^*}]_0$ is the concentration of photoexcited ArOH produced immediately after excitation (at $t = 0$), and t is the time after excitation. It should again be noted that equation (10.24.9) represents the proportionality between fluorescence intensity and excited state concentration. Equation (10.24.9) also indicates that ArOH* decays via a net first-order process; the coefficient of t in equation (10.24.9) is the reciprocal of the *lifetime* of ArOH*, $(1/\tau_{\mathrm{Ac}^-})$, *in the presence* of the $\mathrm{NH_4Ac}$ buffer. By plotting $1/\tau_{\mathrm{Ac}^-}$ vs [Ac$^-$], the information provided by the Stern-Volmer plot, i.e., equation (10.24.8), could be directly obtained. This approach can be carried out using a nanosecond (10^{-9} s) fluorescence spectrometer, a sophisticated and expensive apparatus.

In this experiment, k_d will be obtained using the steady-state approach previously described. The value of τ_0 will be provided from data for the time dependence of ArOH* fluorescence intensity determined from a fluorescence kinetic experiment (see the appendix). It should be emphasized again that in deriving equations (10.24.4) to (10.24.8), it is assumed that throughout the series of fluorescence intensity measurements, first in sulfuric acid, and then in different NH_4Ac buffer solutions, the concentration of ArOH* is *invariant*. This condition requires that the amount of light absorbed by ArOH per unit time be constant; thus, not only must the formal concentration of ArOH be identical in each of the samples, but also the excitation source must not fluctuate. Satisfying these conditions is crucial for the success of the experiment.

Once a value of k_d is obtained from the analysis of the data as discussed above, the value of k_p (the ArO^-* protonation rate constant) can be determined from K_a^*, since

$$K_a^* = \frac{k_d}{k_p}, \tag{10.24.10}$$

for this set of elementary reactions.

Safety Precautions

☐ Always wear safety goggles; these glasses should block ultraviolet light. Ordinary plastic safety goggles or glasses may not be effective in absorbing all of the ultraviolet radiation. Check with your instructor.

☐ 2-naphthol is an irritant. If you are to prepare the solutions from solid material, you *must* wear gloves; if possible, work in a fume hood.

☐ Be sure you have been instructed to use proper pipeting techniques when handling 2-naphthol. *Never* pipet by mouth.

☐ When using the fluorimeter, be sure that any ozone produced by the ultraviolet source is vented. Ozone is an irritating, dangerous gas that has a sharp, pungent odor. If you notice this during the experiment, inform your instructor *at once*. The laboratory must be immediately ventilated and, if necessary, evacuated.

Procedure

1. Prepare a series of aqueous solutions, each having the *same* ArOH concentration (about $4.0 \times 10^{-4} M$). One of these will be $0.1M$ in H_2SO_4 (for I_f^0), and another ca. $0.1M$ in NaOH (in order to observe predominantly ArO^-* fluorescence). The remainder of the solutions will have varying NH_4Ac concentrations, between ca. $0.01M$ and ca. $0.10M$. At least five NH_4Ac-containing ArOH solutions should be studied. 25-mL volumetric flasks are appropriate. Stock solutions of H_2SO_4, NaOH, ArOH, and NH_4Ac will be provided. Label each of the solutions.

Procedure Using Scanning Fluorimeter

2. You will be shown how to operate the fluorimeter. Using an excitation wavelength of 320 nm, obtain the fluorescence spectra of the ArOH solutions starting with the solution containing H_2SO_4. Next obtain the fluorescence spectrum of the NaOH-containing solution; then proceed with the NH_4Ac-containing solutions in order of increasing NH_4Ac concentration. It is desirable to record these spectra on the same chart paper. If the excitation source remains steady and the solutions have the same bulk ArOH concentrations, a distinct *isostilbic point*

(equal brightness) should be produced. This is the common wavelength point of all the ArOH fluorescence spectra. See Experiment 25 for a further discussion of the isostilbic point.

3. If a fixed-wavelength fluorimeter is used, or if time permits, a second determination of data should be performed at the emission wavelength of the fluorimeter corresponding to the maximum for the protonated ArOH* (ca. 360 nm). First, place the H_2SO_4-containing sample in the cavity and establish the "maximum" intensity setting on the chart paper by moving the pen near to its full displacement (or just read and record the signal strength). Carefully and systematically obtain values of the fluorescence intensity of the buffered solutions in order of increasing NH_4Ac concentrations. If there is any doubt about the constancy of the instrument, remeasure I_f^0 and compare with the original value. If the agreement is unsatisfactory, the series of measurements will have to be repeated.

Data Analysis

1. Using the time dependent fluorescence intensity data provided in the appendix for ArOH in $0.10M$ H_2SO_4, determine τ_0. The fluorescence quantum efficiency of ArOH under these conditions has been determined to be 0.18; this is equal to $k_f\tau_0$. Report values of both k_f and k_{nr} for ArOH.

2. Tabulate I_f^0 and I_f values for the samples and construct a Stern-Volmer plot indicated by equation (10.24.8). Determine the values of k_d and k_{Ac^-} using linear (first-order) regression. Consider the error in τ_0 [obtained in (10.24.1)] in your error analysis of the rate constants.

3. Determine k_p from your previously obtained value of K_a^* [see equation (10.24.10)].

4. Tabulate all the rate constants determined and include the respective error limits.

5. Using the Stokes-Einstein-Smoluchowski (SES) equation for a diffusion-controlled rate constant (see also Experiment 20):

$$k_{diff} = \frac{8RT}{3000\eta} \qquad (dm^3\ mol^{-1}\ s^{-1}),$$

where η is the solvent viscosity, compare your values of the second-order rate constants, k_d and k_{Ac^-}, with k_{diff}. You can use the viscosity of water at the appropriate temperature for this comparison. The SES equation applies to the reaction between two neutral species (or a neutral and a charged reaction pair). For anion-cation pairs (each of single charge), the rate constant will be about an order of magnitude larger than k_{diff}.

Questions and Further Thoughts

1. The pH of the aqueous medium is assumed to be unchanged as a result of electronic excitation of 2-naphthol. Why is this justified?
2. Can you predict how the value of the deprotonation rate constant of ArOD* (ArOD) would differ from that of ArOH* (ArOH)? What would you need to know in order to answer this question?
3. Can you suggest an experiment in which the deprotonation rate constant of ArOH (ground state) could be determined?
4. Ozone is sometimes produced by ultraviolet light sources. What is the mechanism by which ozone is produced? Can you suggest a way in which the ozone *production* by a given ultraviolet source can be eliminated?

Further Readings

2-Naphthol Protolysis

R. Boyer, G. Deckey, C. Marzzacco, M. Mulvaney, C. Schwab, and A. M. Halpern, *J. Chem. Educ., 62,* 630, 1985.

J. Van Stam and J. E. Loefroth, *J. Chem. Educ., 63,* 181 (1986).

Fluorescence and Photophysics

R. A. Alberty, ''Physical Chemistry,'' 7th ed., pp. 543–547, Wiley (New York), 1987.

P. W. Atkins, ''Physical Chemistry,'' 3rd ed., pp. 471–475, W. H. Freeman (New York), 1986.

G. N. Castellan, ''Physical Chemistry,'' 3rd ed., pp. 889–896, Addison-Wesley (Reading, Mass.), 1983.

I. N. Levine, ''Physical Chemistry,'' 2nd ed., pp. 725–729, McGraw-Hill (New York), 1983.

J. N. Noggle, ''Physical Chemistry,'' pp. 754–760, Little, Brown (Boston), 1983.

N. J. Turro, ''Modern Molecular Photochemistry,'' Benjamin/Cummings (Menlo Park, Calif.), 1978.

Appendix

Fluorescence Decay Data for 2-Naphthol in $0.10M$ H_2SO_4 at 25°C

Time (ns)	Intensity (photons emitted per unit time)
0.00	21753
1.00	18907
2.00	16380
3.00	14171
4.00	12432
5.00	10757
6.00	9288
7.00	8138
8.00	7083
9.00	6014
10.00	5350

1 ns = 10^{-9} s.

Excited State Properties of 2-Naphthol (Part II)

NAME _____   DATE _____

2-naphthol concentration _____ M (_____ mg in _____ mL)

Solution #   [NH₄Ac]

_____   _____

_____   _____

_____   _____

_____   _____

_____   _____

_____   _____

_____   _____

_____   _____

_____   _____

_____   _____

_____   _____

Temperature _____

Excited State Properties of 2-Naphthol (Part II)

NAME _____ DATE _____

2-naphthol concentration _____ M (_____ mg in _____ mL)

Solution #	[NH₄Ac]
_____	_____
_____	_____
_____	_____
_____	_____
_____	_____
_____	_____
_____	_____
_____	_____
_____	_____
_____	_____
_____	_____

Temperature _____

□ **Experiment 25**

The Enthalpy and Entropy of Excimer Formation

Objective

To determine the enthalpy and entropy of formation of the dimer between ground state and electronically excited pyrene molecules in solution.

Introduction

The physical and chemical properties of a molecule can, in principle, be understood in terms of its electronic structure. It is not surprising, therefore, that such properties as acidity, dipole moment, and chemical reactivity may change considerably upon electronic excitation. In this experiment, the pairwise interaction between aromatic pyrene molecules will be investigated. The properties studied are sensitive to the difference in the characteristics of the interaction between two ground state molecules on the one hand, and a ground state and an electronically excited molecule.

Consider the approach of two pyrene molecules along an axis perpendicular to their molecular planes. If both molecules are in the ground electronic state, there will be a very weak van der Waals attraction at moderate distances. At much closer approach, significant intermolecular repulsion occurs. If, on the other hand, one of the molecules is in its electronically *excited* state (as a result of light absorption) and is allowed to approach a ground state species with the appropriate orientation, a stable dimer will form. This dimer, whose structure is proposed to have a sandwich configuration (see below), is formed reversibly; that is, it can thermally dissociate back into an electronically excited pyrene molecule (excited monomer) and a ground state species.[1] This unusual dimeric species is stable *as long as it possesses electronic excitation* but becomes dissociative as soon as this electronic excitation is dissipated and the system returns to the ground state. Such a species is called an *excimer* (from *exci*ted di*mer*).

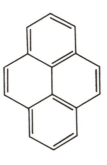

Pyrene

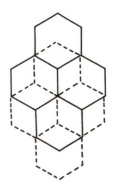

Pyrene crystal dimer

A bound state arising from the interaction between two dissimilar species when one is electronically excited is sometimes called an *exciplex* (*exci*ted com*plex*) or heteroexcimer. Like an excimer, an exciplex is dissociative in the electronic ground state. Because the excimer forms only when an electronically excited molecule and a ground state molecule come into specific and close contact with each other, excimer fluorescence is often used as a probe of molecular interactions. For example, pyrene-labeled polymers are used to study the conformational structure and dynamics of linear chains. Excimer (or exciplex) emission is also used as a probe of the intercalation or binding of aromatic molecules into the crevices of certain macromolecules.

Figure 10.25.1 illustrates the potential energy diagram for electronic transitions and excimer formation. At large intermolecular separation (which occurs at low concentration), electronic transitions involving the excited monomer (absorption and fluorescence) are shown. The involvement of molecular vibrations in these transitions is also indicated. Below the potential energy diagram in Figure 10.25.1 is a schematic illustration of monomer and excimer fluorescence spectra. Although the monomer absorption and fluorescence spectra show vibrational structure (usually involving a skeletal mode), excimer emission is structureless. This is characteristic of electronic transitions (absorption or emission) between bound and unbound states.

The maximum in the excimer emission spectrum corresponds to transitions between the minimum of the excimer potential well (which exists at the equilibrium separation between the excimer components) and the unbound, ground state dimer having the *same* intermolecular separation as the excimer. Hence, excimer formation involves the creation of a bound species subsequent to photon absorption. Conceptually, this is the reverse of photo*dissociation* in which light absorption results in bond breakage. Moreover, photon emission in the excimer (fluores-

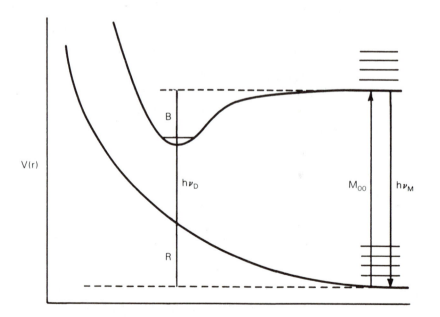

FIGURE 10.25.1   Potential energy of two pyrene molecules as a function of intermolecular separation. The coordinate *r* describes the approach of the molecules having the equilibrium orientation of the excimer. The right-hand transitions represent the isolated monomer.

cence) brings about molecular dissociation since the repulsive ground state pair rapidly dissociates.

The potential energy diagram in Figure 10.25.1 can also be used to compare the emission energies of monomer and excimer in terms of the excimer binding energy, B, and the ground state repulsion energy, R. Thus,

$$hv_M = hv_D + B + R, \tag{10.25.1}$$

where hv_M represents the monomer fluorescence energy (called the 0-0 energy because the transition takes place between excited and ground state monomers possessing zero quanta of vibrational energy). hv_D is the energy of the excimer fluorescence maximum. The binding energy of the excimer is just the negative of its enthalpy of formation, i.e., $B = -\Delta H$.

Kinetic Scheme

Excimer formation (called photoassociation) and reversible dissociation can be represented by the following set of elementary processes, or mechanism:

Process	Equation	Rate Constant
1. Electronic excitation:	$M + hv_{abs} \rightarrow M^*$	Rate $= I_{abs}$
2. Photoassociation	$M^* + M \rightarrow D^*$	k_{DM}
3. Excimer dissociation	$D^* \rightarrow M + M^*$	k_{MD}
4. Excited monomer decay	$M^* \rightarrow M + hv_M$	k_{fM}
	$M^* \rightarrow M + \cdots$	k_{iM}
5. Excimer decay	$D^* \rightarrow (M_2)' + hv_D$	k_{fD}
	$D^* \rightarrow 2M + \cdots$	k_{iD}

$$M + hv_{abs} \longrightarrow M^* \underset{k_{MD} \, -M}{\overset{+M \, k_{DM}}{\rightleftharpoons}} D^*$$

with branches: $M^* \xrightarrow{k_{fM}} M + hv_M$, $M^* \xrightarrow{k_{iM}} M$, $D^* \xrightarrow{k_{iD}} 2M$, $D^* \xrightarrow{k_{fD}} 2M + hv_D$

In the above scheme, I_{abs} is the number of moles of photons (einsteins) absorbed by ground state monomer per dm^3 per second. For convenience, k_M and k_D are defined as the total *intrinsic* decay rate constants of monomer and excimer, respectively, independent of photoassociation or dissociation, i.e.:

$$k_M = k_{fM} + k_{iM} \qquad k_D = k_{fD} + k_{iD}. \tag{10.25.2}$$

The formation rate of excited monomer (in moles dm^{-3} s^{-1}) is

$$\frac{d[M^*]}{dt} = I_{abs} - (k_M + k_{DM}[M])[M^*] + k_{MD}[D^*], \tag{10.25.3}$$

while that for the excimer is

$$\frac{d[D^*]}{dt} = k_{DM}[M][M^*] - (k_D + k_{MD})[D^*]. \tag{10.25.4}$$

It should be noted here that the term $k_{DM}[M]$ is the pseudo-first-order rate of formation of excimer from excited and ground state monomers; it is implied that $[M^*] \ll [M]$; i.e., a very small fraction of monomer is electronically excited. In experiments using conventional light sources (arc lamps, not lasers) this is indeed the case.

Under conditions of steady-state illumination, the rates in both equations

(10.25.3) and (10.25.4) are equal to zero (photostationary conditions). The ratio of [D*] to [M*] can be thus obtained:

$$\frac{[D^*]}{[M^*]} = \frac{k_{DM}[M]}{k_D + k_{MD}}.$$ (10.25.5)

It stands to reason that since the objective of this experiment is the determination of thermodynamic quantities, the temperature dependence of the excimer formation equilibrium constant must be determined. Therefore, the first methodological imperative is to obtain the *equilibrium concentrations* of M, M^*, and D^*. First, it will be assumed that the equilibrium ground state concentration $[M]$ is equal to the bulk pyrene concentration. This is consistent with the inequality $[M^*] \ll [M]$ discussed above. The second crucial, and fundamental, assumption is that the fluor concentration (i.e., $[M^*]$ or $[D^*]$) is proportional to its respective *fluorescence intensity*. This is a basic tenet of spectrofluorimetry and is analogous to Beer's law in absorption spectrophotometry.

This experiment is based on the fact that it is possible to obtain the ratio of excimer to excited monomer concentrations from the measured fluorescence intensities, each measured at the appropriate wavelength. Both of the fluor concentrations are assumed to be proportional to the respective *integrated fluorescence intensities* (areas under the monomer and excimer fluorescence spectra):

$$I_{fM} = k_{fM}[M^*] \qquad \text{and} \qquad I_{fD} = k_{fD}[D^*],$$ (10.25.6)

where the proportionality constants are the *radiative rate constants*. The dimensions of I_f are einsteins $\mathrm{dm}^{-3}\,\mathrm{s}^{-1}$. Combining equations (10.25.5) and (10.25.6), we have

$$\frac{I_{fD}}{I_{fM}} = \frac{k_{fD}k_{DM}[M]}{k_{fM}(k_D + k_{MD})}.$$ (10.25.7)

Equation (10.25.7) is central to the application of photostationary techniques to the study of photoassociation. (The distinction between the integrated and "instantaneous" fluorescence intensities will be discussed below.)

Because we seek thermodynamic data, the temperature dependence of equation (10.25.7) will be examined. First, we will make the general observation that both the formation and dissociation rate constants are temperature-dependent and can be expressed in Arrhenius form:

$$k_{DM} = A_{DM}\exp\left(\frac{-W_{DM}}{RT}\right) \qquad \text{and} \qquad k_{MD} = A_{MD}\exp\left(\frac{-W_{MD}}{RT}\right),$$ (10.25.8)

where the A's and W's are the preexponential factors and activation energies, respectively. Because intermolecular excimer formation is restricted only by molecular transport, or diffusion, the activation energy, W_{DM}, is related to the activation to viscous flow of the solvent medium. W_{MD}, however, reflects the intrinsic strength of the "excimer bond" as well as the energetics of molecular diffusion. This is because W_{MD} represents the activation energy for the *dissociation* of the bound excimer into *separated* and individually solvated excited and ground state constituents.

If we assume that the temperature dependence of k_{MD}, the rate constant for thermally activated excimer dissociation, is much larger than that for k_D, the intrinsic decay rate of excimer (indeed, for many systems, k_D is nearly temperature-independent), we have $k_{MD} \gg k_D$ in the limit of high temperature. Applying this result to equation (10.25.7):

$$\frac{I_{fD}}{I_{fM}} \simeq \frac{k_{fD}k_{DM}[M]}{k_{fM}k_{MD}} \qquad \text{(high temperature).} \qquad (10.25.9)$$

If we furthermore make the general assumption (which has been confirmed by measurements of several monomer/excimer systems) that both of the radiative rate constants (k_{fM} and k_{fD}) are temperature-independent, an Arrhenius plot of I_{fD}/I_{fM} yields, in the limit of high temperature, a slope

$$\frac{d \ln\left(\frac{I_{fD}}{I_{fM}}\right)}{d\left(\frac{1}{T}\right)} = \frac{-(W_{DM} - W_{MD})}{R} \qquad \text{(high temperature).} \qquad (10.25.10)$$

See equations (10.25.7) and (10.25.8). Because $W_{MD} > W_{DM}$, the desired binding energy (the negative of the enthalpy of formation) of the excimer can be determined as the slope of such a plot.

At low temperatures, where excimer dissociation is slow compared with its intrinsic decay rate (i.e., $k_{MD} << k_D$), (10.25.7) can be written

$$\frac{I_{fD}}{I_{fM}} \simeq \frac{k_{fD}k_{DM}[M]}{k_{fM}k_D} \qquad \text{(low temperature).} \qquad (10.25.11)$$

An Arrhenius plot of the left-hand side of equation (10.25.7) will have a slope approaching

$$\frac{d \ln\left(\frac{I_{fD}}{I_{fM}}\right)}{d\left(\frac{1}{T}\right)} = \frac{-W_{DM}}{R} \qquad \text{(low temperature).} \qquad (10.25.12)$$

W_{DM} has been found to be closely related to the activation energy associated with molecular diffusion (hence bulk viscous flow) in the solvent medium. This is one of the reasons that excimer formation is interpreted as a diffusion-controlled process.

An Arrhenius plot of equation (10.25.7) illustrates the behavior of the monomer and excimer fluorescence spectra over a wide temperature range (Figure 10.25.2). At low temperature, the ratio of excimer to monomer emission is small because excimer formation is impeded by the high viscosity of the solvent; indeed, in a glass or rigid medium, excimer formation does not take place because of the inability of excited monomer and ground state species to diffuse. (It is possible, however, that weak van der Waals dimers may be stable at low temperatures and that emission from these directly photoexcited dimers may be observed. Strictly speaking, this is not excimer emission.) At very high temperature, excimer emission is also suppressed because of the high dissociation rate of the excimer. Thus there is an intermediate temperature at which there is a maximum excimer emission intensity relative to the excited monomer. This temperature depends not only on the nature of the excimer system (its binding energy and entropy of formation) but also on the solvent characteristics such as its viscosity and activation to viscous flow.

Another very important condition, which is pertinent to the high-temperature limit and crucial to this experiment, concerns the fact that if interconversion between photoexcited monomer and excimer is rapid relative to the intrinsic decay rates of M^* and D^*, these species are in *dynamic equilibrium*. Thus with

$$k_{DM}[M] \gg k_M \qquad \text{and} \qquad k_{MD} \gg k_D,$$

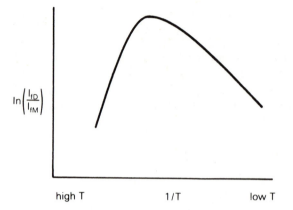

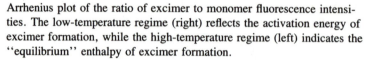

FIGURE 10.25.2 Arrhenius plot of the ratio of excimer to monomer fluorescence intensities. The low-temperature regime (right) reflects the activation energy of excimer formation, while the high-temperature regime (left) indicates the "equilibrium" enthalpy of excimer formation.

$K_{eq} = [D^*]_{eq}/[M^*]_{eq}[M]_{eq}$, the true *equilibrium constant* (we assume unit activity coefficients) for photoassociation can be expressed as the ratio of the formation and dissociation rate constants. Using this result and rearranging equation (10.25.9), we have

$$K_{eq} = \frac{k_{DM}}{k_{MD}} = \frac{k_{fM}I_{fD}}{k_{fD}I_{fM}[M]}.$$ (10.25.13)

The high-temperature or dynamic equilibrium regime can be depicted by an analogy to two leaky containers connected to each other by two tubes, each containing a pump. Refer to Figure 10.25.3. The containers are filled with a liquid. If the leak rate in container *A* is small compared with the rate of transport of liquid from container *A* to container *B*, *and*, likewise, the leak rate of container *B* is small

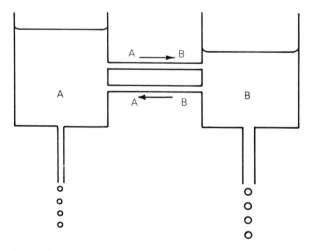

FIGURE 10.25.3 Hydrodynamic model of reversible reactions between two metastable (electronically excited) species.

relative to the flow rate from B to A, the amounts of liquid present in each container depend only on the $A \to B$ and $B \to A$ flow rates. This situation thus represents the dynamic equilibrium state; the volumes of liquid in A and B are intrinsic to the plumbing between them. On the other hand, if one of the leak rates is too large, the volume of liquid in that container will be depleted and would not reflect the "equilibrium" condition.

The link between K_{eq} and the enthalpy and entropy of excimer formation is

$$-RT \ln K_{eq} = \Delta G^\circ = \Delta H^\circ - T\,\Delta S^\circ, \qquad (10.25.14)$$

and if the expressions in (10.25.13) and (10.25.14) are combined, the temperature dependence of I_{fD} and I_{fM} can be represented as

$$\ln\left(\frac{I_{fD}}{I_{fM}}\right) = -\ln\left(\frac{k_{fM}}{k_{fD}[M]}\right) + \frac{\Delta S^\circ}{R} - \frac{\Delta H^\circ}{RT}. \qquad (10.25.15)$$

Assuming that k_{fM} and k_{fD} are temperature-independent, a plot of $\ln(I_{fD}/I_{fM})$ vs. $1/T$ should be linear, having a slope equal to $-\Delta H^\circ/R$. The intercept can also provide a value of ΔS° *if* the ratio of the radiative rate constants of the excimer and monomer is known. While this information can be obtained from fluorescence lifetime and efficiency studies, Stevens and Ban[2] have described a photo-stationary fluorimetric technique for obtaining ΔS° and ΔH°; this will be discussed below.

There is another interesting consequence of high-temperature conditions. Although the *distribution* between D^* and M^* changes with temperature in this regime, the *total* concentration of excited states ($D^* + M^*$) remains constant. In other words, M^* and D^* form a "closed system." Hence,

$$[M^*]_T + [D^*]_T = \text{constant} = [M^*]_0, \qquad (10.25.16)$$

where the temperature dependence of the excited monomer and excimer concentrations is explicit. The spectrometric significance is that there will be a specific wavelength between the maxima of the monomer and excimer emission spectra at which the emission intensity is independent of temperature. This position, called the *isostilbic* (equal brightness) point, is analogous to the isosbestic point (equal extinction) seen in absorption spectra when there is linear relationship between the concentrations of two absorbing species.

In order to proceed with the fluorimetric analysis, we must consider the important distinction between the total molecular fluorescence intensity, I_{fM} (area under the spectrum), and the fluorescence intensity at a specific wavelength, f_M. The latter is a function of energy or wavenumbers [$f_M(\bar\nu)$ represents the monomer emission spectrum], while the former is a constant at a given temperature and monomer concentration. This distinction is important, because in this experiment, one measures fluorescence intensities of the monomer and excimer emission spectra at *specific* wavelengths (e.g., the respective maxima), rather than the total areas under the spectra. The relationships between the total intensity and the instantaneous intensity are

$$I_{fM} = C\int f_M(\bar\nu)d\bar\nu \qquad \text{and} \qquad I_{fD} = C\int f_D(\bar\nu)d\bar\nu. \qquad (10.25.17)$$

C is an instrumental constant that links the integrals of instantaneous intensities with the molecular fluorescence strengths.

One problem with the analysis presented thus far is that the ratio of the excimer and monomer radiative rate constants, k_{fD}/k_{fM}, is needed [see equations (10.25.7), (10.25.9), and (10.25.15)]. While this information can be obtained independently from fluorescence lifetime and (absolute) quantum efficiency measurements, Stevens and Ban have described an approach that gets around this problem. It should be observed in (10.25.15) that the value of k_{fD}/k_{fM} is needed only in order to determine the *entropy* of photoassociation (i.e., an absolute intercept).

Although the method for obtaining this information from fluorimetric measurements is straightforward, the development of the result is presented in the appendix. The approach is as follows.

Let R_D° and R_M° be the *recorded* intensities of excimer and monomer at their respective maxima. In the high-temperature regime, a plot of R_D° vs. R_M° (for different temperatures) is expected to be linear with a negative slope ($-a/b$). Thus, as the temperature increases, the excimer intensity decreases with a concomitant increase in the monomer intensity. The relation used to obtain the entropy and enthalpy of excimer formation is

$$\ln\left(\frac{R_D^\circ}{R_M^\circ}\right) = \ln\left\{[M]\left(\frac{b}{a}\right)\right\} + \frac{\Delta S^\circ}{R} - \frac{\Delta H^\circ}{RT}. \qquad (10.25.18)$$

Safety Precautions

☐ Safety glasses that block ultraviolet light must be worn during this experiment.

☐ Wear gloves when handling pyrene.

☐ Make sure you are instructed how to use proper pipeting techniques. Never pipet by mouth.

☐ If solid pyrene or its solutions come in contact with the skin, immediately wash the affected area with soap and water.

☐ A cylinder containing N_2 at high pressure is used. Be sure it is securely attached to a firm foundation. A reducing valve is used to deliver the N_2 at a pressure slightly above ambient (< 5 psig). Do not change this pressure.

☐ The fluorimeter may produce ozone. Make sure the ultraviolet source is vented. If you notice a sharp, pungent odor, inform your instructor immediately.

☐ The experiment must be performed in an open, well-ventilated laboratory.

Procedure

Prepare 25 mL of a 5.0×10^{-3} M solution of pyrene in methylcyclohexane (ca. 25 mg/25 mL). The pyrene should be of high purity (and should be sublimed or recrystallized before use if necessary). The solvent must have a negligible fluorescence background and preferably should be of spectrometric quality.

Add sufficient solution to a fluorescence cell that is equipped with an inert, gastight stopper. Deaerate by bubbling a fine stream of pure dry N_2 through the solution for 4 to 6 min. Control the gas flow so that solution is not expelled from the cell. [Deaerating with N_2 is essential because it displaces dissolved oxygen which significantly quenches the pyrene fluorescence. This procedure is an application of Henry's law; the air above the liquid is replaced by an atmosphere of nitrogen and thus the solubility of oxygen approaches a very low value commensurate with the residual oxygen composition in the nitrogen used. Bubbling of the nitrogen through the liquid hastens gas-liquid equilibrium.]

Immediately after deaerating, firmly stopper the fluorescence cell. Wipe the four windows (cleaning if necessary with ethanol) using lens paper or the equivalent; be sure not to scratch the cell. Place the cell in the fluorimeter cavity.

Your instructor will show you how to obtain emission spectra using the particular fluorimeter and how to vary and measure the temperature of the sample. An excitation wavelength of ca. 370 to 380 nm should be used. When the sample is exposed to the ultraviolet radiation, you should be able to see the bright blue-violet fluorescence. Emission should be scanned between 370 and about 500 nm. The number of spectra to be obtained depends on the available time, but *at least* five spectra, taken between 60 and 100°C, should be acquired. The temperature should remain constant (to within 0.5°C) during a given scan.

If possible, overlap these spectra on a common wavelength axis. Preferably, and if time permits, spectra should be run over a wider temperature range, e.g., 0 to 20 to 100°C. It should be noted that even if the excitation lamp fluctuates or drifts during the experiment, the data are still valid as long as this change does not occur *during a particular scan*.

Data Analysis

Tabulate the recorded fluorescence intensities of the monomer and excimer (R_M° and R_D°) at each temperature; choose convenient wavelengths (e.g., at or near the respective maxima).

Plot R_M° vs. R_D° and determine b/a from the linear portion (corresponding to the high-temperature regime).

From equation (10.25.18) determine ΔS° and ΔH° for the pyrene excimer formation in methylcyclohexane using the values of b/a and $[M]$.

Determine the binding energy of the pyrene excimer as well as the repulsion energy of the ground state pyrene pair having the excimer geometry [equation (10.25.1)].

If low enough temperatures were achieved in the experiment, estimate the activation energy to excimer formation and compare with the activation energy of viscosity of the solvent (or similar hydrocarbon).

Estimate the errors in ΔS°, ΔH°, B, and R.

Questions and Further Thoughts

1. What effect would solvent polarity (e.g., acetonitrile vs. cyclohexane) have on the transition energies of the monomer and excimer and the binding energy of the excimer? Consider the same effect(s) on an exciplex, e.g., pyrene (or anthracene) and *N,N*-dimethylaniline.

2. The "excimer" laser is based on transitions between the bound state formed with rare gas and halogen atoms [e.g., $(ArF)^*$, $(XeCl)^*$] and the dissociative ground state of these species. One of the advantages of an "excimer" laser is that, after emission takes place from the excited complex, the ground state is rapidly removed because it is dissociative. This allows a significant population inversion to be achieved, and this enhances lasing efficiency.

3. Does this statement make sense: the ground state of an excimer is an excited state?

4. The probability that each end of a linear chain oligomer (or polymer) comes into close proximity can be probed by attaching pyrene probes to the ends of the molecule (end-capped polymers). How does such a technique work? What type of experiment is needed to gather information about the end-to-end interactions?

5. The fluorescence of some excimers can only be readily observed at moderately low temperatures (and high concentrations), yet at very low temperatures where the solvent becomes glassy or crystalline, excimer fluorescence nearly vanishes. How can you explain these observations?

6. Suppose the pyrene excimer were formed in the gas phase. Each ground state pyrene molecule formed as a result of excimer fluorescence has a recoil energy. Estimate the speed of the recoiling molecules. Describe the trajectory.

Notes

1. J. B. Birks, "Photophysics of Aromatic Molecules," pp. 301–371, Wiley-Interscience (London), 1970.
2. B. Stevens and M. I. Ban, *Trans. Faraday Soc., 60,* 1515 (1964).

Further Readings

J. B. Birks, D. J. Dyson, and I. H. Munro, *Proc. Roy. Soc. A., 275,* 575 (1963).
B. Stevens, "Photoassociation in Aromatic Systems," in J. N. Pitts, Jr., G. S. Hammond, and W. A. Noyes, Jr., eds., "Advances in Photochemistry," vol. 8, pp. 161–226, Wiley-Interscience (New York), 1971.

Appendix

The fluorescence intensity at the isostilbic point (wavenumber i) is composed of the separate contributions of monomer and excimer emission

$$f_i = f_{iM} + f_{iD}. \tag{10.25.19}$$

Each emission is, in turn, proportional to its respective integrated fluorescence spectrum, i.e.,

$$f_{iM} = m \int f_M(\bar{\nu})d\bar{\nu} \quad \text{and} \quad f_{iD} = d \int f_D(\bar{\nu})d\bar{\nu}. \tag{10.25.20}$$

After combining equations (10.25.17), (10.25.19), and (10.25.20), we have for the isostilbic intensity

$$f_i = \frac{m}{C}I_{fM} + \frac{d}{C}I_{fD}. \tag{10.25.21}$$

Now using equation (10.25.6), which expresses the absolute fluorescence intensities in terms of radiative rate constants and excited state concentrations, and equation (10.25.16), which states the "closed system" constraint of the high-temperature limit, equation (10.25.21) becomes

$$f_i = \frac{m}{C}k_{fM}[M^*] + \frac{d}{C}k_{fD}\{[M^*]_0 - [M^*]\},$$

which can be factored to give

$$f_i = \frac{d}{C}k_{fD}[M^*]_0 + [M^*]\left[\frac{m}{C}k_{fM} - \frac{d}{C}k_{fD}\right]. \tag{10.25.22}$$

Because both f_i and the first term of the right-hand side of (10.25.22) are independent of temperature (in the high-temperature regime), while $[M^*]$ is temperature-dependent, the bracketed term in (10.25.22) must be zero. Thus, $d/m = k_{fM}/k_{fD}$, and combining this result with equations (10.25.20) and (10.25.17) furnishes the relation

$$\frac{f_{iD}}{f_{iM}} = \frac{dI_{fD}}{mI_{fM}} = \frac{k_{fM}I_{fD}}{k_{fD}I_{fM}},$$

or

$$\frac{I_{fD}}{I_{fM}} = \left(\frac{k_{fD}}{k_{fM}}\right)\left(\frac{f_{iD}}{f_{iM}}\right). \tag{10.25.23}$$

The importance of equation (10.25.23) is that we can now express the enthalpy and entropy of photoassociation in terms of the experimentally determined contributions of monomer and excimer fluorescence to the isostilbic intensity. Combining equations (10.25.23) and (10.25.15) we have

$$\ln\left(\frac{f_{iD}}{f_{iM}[M]}\right) = \frac{\Delta S^{\circ}}{R} - \frac{\Delta H^{\circ}}{RT}. \tag{10.25.24}$$

The problem *now* is to find a way of decomposing f_i into f_{iM} and f_{iD}. We again follow the Stevens and Ban procedure. To understand this approach, the distinction must be made between the *absolute* fluorescence intensity (f_{iM}, referred to above) and the *observed instrumental* response, or signal strength (R_{iM}). Because the combination of the analyzing monochromator and photomultiplier tube in the particular fluorimeter used has a unique, nonuniform wavelength dependence, a given emission spectrum will be distorted by an instrumental sensitivity function, $S(\bar{\nu})$. This can be expressed as

$$R(\lambda) = [S(\lambda)][I(\lambda)]. \tag{10.25.25}$$

There are various experimental techniques by which the sensitivity function $S(\lambda)$ can be determined. For example, a standard emission source having a known output spectrum [absolute power density as a function of wavelength, i.e., $I(\lambda)$] may be used to calibrate the analyzing system. These sources are often a quartz halogen tungsten lamp—a black body radiator—for the near infrared and visible region, and a deuterium discharge lamp for the ultraviolet range. The observed emission profile of the lamp is obtained using the particular fluorimeter, $R(\lambda)$, and the sensitivity function is then obtained by performing a point-by-point division of $R(\lambda)$ and $I(\lambda)$.

At the isostilbic point, the observed emission intensity, R_i, is composed of contributions from excited monomer and excimer at that wavelength:

$$R_i = R_{iD} + R_{iM}. \tag{10.25.26}$$

Since the sensitivity function, S_i, is the same for both species at a common wavelength (e.g., the isostilbic point), the ratio of the response contributions will be equal to that of absolute intensities:

$$\frac{f_{iD}}{f_{iM}} = \frac{R_{iD}}{R_{iM}}. \tag{10.25.27}$$

Now we relate these response values to those at different reference wavelengths at which it is assumed *only* excimer and excited monomer, respectively, emit (these reference wavelengths may be taken to be the maxima of the monomer and excimer emission spectra—see Figure 10.25.1). Referring to the observed instrumental responses at these reference wavelengths as R_D° and R_M°, respectively, we can assume the following proportion between R_{iD} and R_D°, etc.:

$$R_{iD} = aR_D^{\circ} \quad \text{and} \quad R_{iM} = bR_M^{\circ}, \tag{10.25.28}$$

where the constants a and b are temperature-independent. Combining equations (10.25.26) and (10.25.27) gives

$$\frac{f_{iD}}{f_{iM}} = \frac{aR_D^{\circ}}{bR_M^{\circ}}. \tag{10.25.29}$$

The ratio a/b can be obtained from a plot of R_M° vs. R_D° in which the set of $\{R_M^\circ, R_D^\circ\}$ is obtained over the temperature range for which $k_{MD} \gg k_D$ and $k_{DM}[M] \gg k_M$ (i.e., the high-temperature regime). This is justified by using equations (10.25.26) and (10.25.28):

$$R_i = \text{constant} = aR_D^\circ + bR_M^\circ.$$

Differentiation of R_M° with respect to R_D° gives

$$\frac{dR_M^\circ}{dR_D^\circ} = \frac{-a}{b} \qquad \text{(high temperature)}. \qquad (10.25.30)$$

Hence, a/b can be obtained by plotting R_M° vs. R_D°, and in this way the needed fluorescence intensity ratio can be obtained from the observed instrumental response data, see equation (10.25.29).

The final working result in the Stevens-Ban method, obtained by combining equations (10.25.24) and (10.25.29), is equation (10.25.31):

$$\ln\left(\frac{R_D^\circ}{R_M^\circ}\right) = ln\left\{[M]\frac{b}{a}\right\} + \frac{\Delta S^\circ}{R} - \frac{\Delta H^\circ}{RT}. \qquad (10.25.31)$$

Excimer Thermodynamics

NAME _____ DATE _____

Pyrene concentration _____ M (_____ mg in _____ mL)

Solvent _____   Excitation wavelength _____

Spectrum #	Temperature	Comments
_____	_____	_____
_____	_____	_____
_____	_____	_____
_____	_____	_____
_____	_____	_____
_____	_____	_____
_____	_____	_____
_____	_____	_____
_____	_____	_____
_____	_____	_____
_____	_____	_____
_____	_____	_____

Experiment 25

Excimer Thermodynamics

NAME _____   DATE _____

Pyrene concentration _____ M (_____ mg in _____ mL)

Solvent _____   Excitation wavelength _____

Spectrum #	Temperature	Comments
_____	_____	_____
_____	_____	_____
_____	_____	_____
_____	_____	_____
_____	_____	_____
_____	_____	_____
_____	_____	_____
_____	_____	_____
_____	_____	_____
_____	_____	_____
_____	_____	_____
_____	_____	_____
_____	_____	_____

□ Experiment 26

The Rotation-Vibration Spectrum of HCl

Objective

To obtain the rotationally resolved vibrational absorption spectrum of HCl; to determine the rotational constant and rotational-vibrational coupling constant; to compare the rotational transition intensities with theoretical predictions.

Introduction

In this experiment, we will examine the energetics of vibrational and rotational motion in the diatomic molecule, HCl. The experiment involves detecting transitions between different molecular vibrational *and* rotational levels brought about by the absorption of quanta of electromagnetic radiation (photons) in the infrared region of the spectrum. The introductory discussion consists of three parts: (1) vibrational and rotational motion and energy quantization, (2) the influence of molecular vibration on rotational energy levels, and (3) the intensities of rotational transitions.

Vibrational Motion

Consider how the potential energy of a diatomic molecule *AB* changes as a function of internuclear distance. A simple way of expressing this mathematically is

$$V(x) = 1/2 \, kx^2, \tag{10.26.1}$$

where the coordinate x represents the *change* in internuclear separation relative to the equilibrium position; thus, $x = 0$ represents the equilibrium configuration (i.e., the position with the lowest potential energy). For $x < 0$ the two atoms are closer to each other, while $x > 0$ represents the two atoms being farther apart. k is called the *force constant* and its value reflects the strength of the force holding the two masses to each other. Equation (10.26.1) is a consequence of Hooke's law, which states that the force between two linked masses is proportional to the displacement of the two masses from their equilibrium position: $F(x) = -kx$. The minus sign indicates that the restoring force is opposite in direction to the displacement.

Equation (10.26.1) is referred to as the *harmonic potential,* and the application of Newtonian mechanics to this problem results in a description of the behavior of the masses called simple harmonic (or periodic) motion. The system is commonly referred to as the *harmonic oscillator*. The harmonic potential (10.26.1), which is parabolic, is shown in Figure 10.26.1. According to this model, the steepness of the potential is determined by the force constant, k. This simple picture of molecular energetics predicts that the potential energy of a diatomic molecule increases symmetrically with respect to bond elongation or compression, and does so without limit. Chemical intuition suggests that this description cannot be entirely true. It is correct that as the nuclei are brought very close to each other, strong repulsive forces are brought to bear and these forces increase sharply with decreasing x. On the other hand, as the nuclei are pulled apart, the chemical bond is weakened and eventually broken, and the potential energy therefore becomes constant as the two "free" (i.e., unassociated) atoms are produced. Thus the harmonic potential is unrealistic for large internuclear displacements. This in-

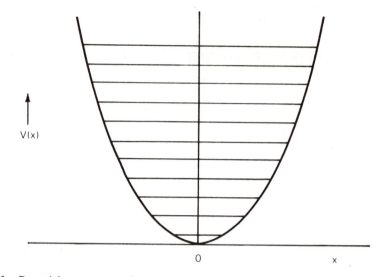

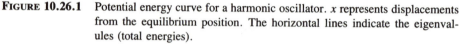

FIGURE 10.26.1   Potential energy curve for a harmonic oscillator. *x* represents displacements from the equilibrium position. The horizontal lines indicate the eigenvalues (total energies).

adequacy of the harmonic potential is referred to as *anharmonicity* and will be discussed further below.

The application of quantum mechanics to the harmonic oscillator reveals that the total vibrational energy of the system is constrained to certain values; it is quantized. The expression for these energy levels (or eigenvalues) is

$$E_v = (v + 1/2)h\nu_e, \qquad \text{(ergs)} \qquad (10.26.2)$$

where $v = 0,1,2,3, \ldots$ is the vibrational quantum number, h is Planck's constant, and ν_e is called the *harmonic frequency* (s^{-1}), which is equal to $(1/2\pi)$ $(k/\mu)^{1/2}$. k is the force constant [see equation (10.26.1)], and μ is the reduced mass of the oscillator: $\mu = m_A m_B/(m_A + m_B)$. Notice that the lowest eigenvalue (i.e., $v = 0$), is nonzero; $E_0 = 1/2\ h\nu_e$. This is the *zero point energy* of the oscillator. It represents the residual vibrational energy possessed by a harmonic oscillator at 0 K; it is a "quantum mechanical effect." The eigenvalues, E_v, which represent the levels of total energy (kinetic and potential) of the harmonic oscillator are superimposed on the potential energy function in Figure 10.26.1. As indicated by equation (10.26.2), these levels are equally spaced with a separation of $h\nu_e$.

It is convenient to express the harmonic frequency in a unit called the *wavenumber, $\bar{\nu}$,* or *reciprocal centimeter,* cm^{-1}. The wavenumber, which is the product of frequency, ν, and the speed of light, c, is very widely used in atomic and molecular spectroscopy. Expressed in this unit, the vibrational energy in equation (10.26.2) becomes

$$E_v = (v + 1/2)hc\bar{\nu}_e \qquad \text{(cm}^{-1}\text{)}. \qquad (10.26.3)$$

Even though $\bar{\nu}_e$ has the dimensions of cm^{-1}, it is often referred to as the harmonic frequency (cm^{-1}). For diatomic molecules, $\bar{\nu}_e$ is typically on the order of hundreds to thousands of wavenumbers. For example, for I_2 and H_2, $\bar{\nu}_e$ values (which represent, roughly, the extremes of the vibrational energy spectrum for diatomic molecules) are 215 and 4403 cm^{-1}, respectively.

Next let us introduce an important and necessary complication to the harmonic

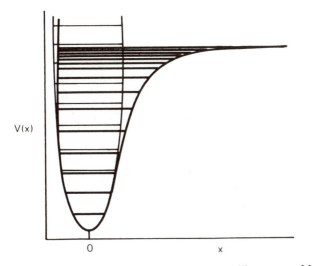

$V(x)$

0   x

FIGURE 10.26.2   Potential energy curve for an anharmonic oscillator, e.g., Morse function.

oscillator: anharmonicity. Rather than describing the potential energy function as parabolic [equation (10.26.1)], we will use a more realistic potential, the Morse function, which takes into account molecular dissociation at large values of x.

$$V(x) = D_e[1 - \exp(-ax)]^2 \qquad (10.26.4)$$

A plot of this function is shown in Figure 10.26.2. D_e is the potential energy (relative to the *bottom* of the well) at infinite A-B separation ($x = \infty$), and a is a constant that, like k in equation (10.26.1), determines the shape of the potential well and hence reflects the vibrational frequency; in fact $a = k/(2D_e)^{1/2}$. The use of the Morse potential instead of harmonic potential results in the following expression for the vibrational eigenvalues:

$$E_v = (v + 1/2)hc\bar{\nu}_e - (v + 1/2)^2 hc\bar{\nu}_e\chi_e \qquad (10.26.5)$$

where $\bar{\nu}_e$ is the harmonic frequency (as above) and $\bar{\nu}_e\chi_e$ is called the *anharmonicity constant*. It is a molecular constant that, for the Morse oscillator, is equal to $ha^2/(4\pi\mu)$. Because the anharmonicity term in the eigenvalue expression (10.26.5) is multiplied by $-(v + 1/2)^2$, the spacing between eigenvalues rapidly becomes smaller for higher v. As the total energy approaches the dissociation limit, the levels become very closely bunched together. The harmonic potential and its energy levels are also shown in Figure 10.26.2 so that these characteristics can be compared with the Morse oscillator. For relatively low excitations (only a few vibrational quanta), the Morse potential provides a good description of molecular vibrations.

Rotational Motion

Consider the rotation of the molecule AB. In the absence of external electric or magnetic fields, the potential energy is invariant with respect to the rotational coordinates; rotational motion is *isotropic* (independent of orientation). If the AB bond length is assumed to be constant, i.e., independent of rotational energy, the AB molecule is called a *rigid rotator*. A quantum mechanical treatment of this simple (but not unreasonable) model gives the eigenvalue expression

$$E_J = J(J + 1)hcB_e, \qquad (10.26.6)$$

where J is the rotational quantum number ($J = 0,1,2, \ldots$) and B_e is the *rotational constant* (cm^{-1}), which is equal to

$$B_e = \frac{h}{8\pi^2 c I_e}. \tag{10.26.7}$$

I_e is a molecular constant, the *moment of inertia*, which for a diatomic molecule is $I_e = \mu r_e^2$. μ *and* r_e are, in turn, the reduced mass, $m_A m_B/(m_A + m_B)$, and the equilibrium bond length of the molecule AB. r_e, it should be noted, is the internuclear separation for which $x = 0$ in equations (10.26.1) and (10.26.4) (i.e., the bottom of the potential well). For the general case in which r is considered a variable, $I = \mu r^2$ and thus the rotational constant is a function of r:

$$B = \frac{h}{8\pi^2 c \mu r^2}. \tag{10.26.8}$$

Notice that the rotational constant contains structural information about the molecule. In fact, the determination of molecular structure can be achieved with high precision from studies of rotational transitions, usually in the gas phase (microwave spectroscopy).

One possible complication that must be considered is the fact that molecules are not perfectly rigid. If a chemical bond is weak enough so that the nuclei respond to increasing rotational energy by moving farther apart, the molecule experiences centrifugal elongation: r_e increases with increasing J. The rotational eigenvalues for the nonrigid rotator are approximated by

$$E_J \simeq J(J + 1)hcB_e - J^2(J + 1)^2 hcD \qquad \text{(ergs)}, \tag{10.26.9}$$

where D is called the *centrifugal distortion constant*. The effect of nonrigidity is to bring the rotational levels closer together for higher-energy states. While, in principle, this complication is expected to alter the energy levels implied by the rigid rotator model, the efffect is often very small and can be ignored. For example, D is typically several orders of magnitude smaller than B_e, and therefore centrifugal distortion is significant only for very highly rotationally excited states (large J). D can be expressed in terms B_e and $\bar{\nu}_e$:

$$D = \frac{4B_e^3}{\bar{\nu}_e^2} \qquad \text{(cm}^{-1}\text{)}. \tag{10.26.10}$$

Rotational and Vibrational Motion

If rotational and vibrational motion were completely separable, that is, if molecular vibrations had no effect on rotational states and vice versa, the total energy of a rotating, vibrating diatomic molecule (i.e., a Morse oscillator) would be expressed as the sum of equations (10.26.5) and (10.26.9), i.e.,

$$E_{v,J} = (v + 1/2)hc\bar{\nu}_e - (v + 1/2)^2 hc\bar{\nu}_e\chi_e +$$

$$J(J + 1)hcB_e - J^2(J + 1)^2 hcD \qquad \text{(ergs)}. \tag{10.26.11}$$

In this experiment, we are justified in neglecting centrifugal distortion, and thus the last term in equation (10.26.11) will be ignored (see Questions and Further Thoughts).

However, the complete separation of rotational and vibrational motion is not realistic. Because the mean internuclear separation depends on the vibrational energy (i.e., v) and the rotational constant is a function of $1/r^2$ [see equation (10.26.8)], the "effective" rotational constant for a vibrating molecule will vary with the mean value of $1/r^2$ for the vth eigenstate, i.e., $<1/r^2>_v$. The rotational constant can thus be approximated by

$$B_v = B_e - \alpha_e(v + 1/2) \qquad (\text{cm}^{-1}), \qquad (10.26.12)$$

where B_v is the rotational constant taking vibrational excitation into account and α_e is defined as the *rotational-vibrational coupling constant*. It turns out that for an anharmonic potential (e.g., the Morse potential), $\alpha_e > 0$, while for the harmonic potential, $\alpha_e < 0$. This is evident in the expression for α_e

$$\alpha_e = \frac{6B_e^2}{\bar{\nu}_e}\left[\left(\frac{\bar{\nu}_e\chi_e}{B_e}\right)^{1/2} - 1\right]. \qquad (10.26.13)$$

Notice that for an anharmonic oscillator, B_v decreases as v increases. Also observe that according to equation (10.26.12), $B_v < B_e$ even for the case of *no* vibrational excitation ($v = 0$). Thus B_e, as well as r_e and $\bar{\nu}_e$, are intrinsic molecular parameters that are referenced to the *bottom* of the potential well (see Figures 10.26.1 and 10.26.2); they are independent of vibrational energy. Finally, in applying this treatment of coupled rotational-vibrational motion, we replace B_e in equation (10.26.11) by B_v [equation (10.26.12)] and write

$$E_{v,J} = (v + 1/2)hc\bar{\nu}_e - (v + 1/2)^2 hc\bar{\nu}_e\chi_e +$$
$$J(J + 1)hcB_e - (v + 1/2)J(J + 1)hc\alpha_e \qquad (\text{ergs}). \qquad (10.26.14)$$

Equation (10.26.14) shows that there is a "pure" vibrational contribution (containing only v), a "pure" rotational term (containing only J), and the vibrational-rotational coupling containing both v and J. It should be emphasized again that equation (10.26.14) represents the *total* energy of a vibrating, rotating molecule; furthermore, the molecular constants $\bar{\nu}_e$, $\bar{\nu}_e\chi_e$, B_e, and α_e are all expressed in cm^{-1} while the total energy is in ergs. To make things consistent, we will divide equation (10.26.14) by hc and replace $E_{v,J}/hc$ by $T_{v,J}$

$$T_{v,J} = \frac{E_{v,J}}{hc} = (v + 1/2)\bar{\nu}_e - (v + 1/2)^2\bar{\nu}_e\chi_e +$$
$$J(J + 1)B_e - (v + 1/2)J(J + 1)\,\alpha_e \qquad (\text{cm}^{-1}). \qquad (10.26.15)$$

$T_{v,J}$ is called the *term value;* it is the rotational-vibrational state energy expressed in wavenumbers, $\bar{\nu}$. Equation (10.26.15) is the key expression that will be used to explain and analyze the experimental data.

Experimental Method

Rotational energy levels are much more closely spaced than vibrational levels. For example, for HCl the spacing between the lowest two levels ($J = 0$ and $J = 1$) is about 20 cm^{-1}, whereas the gap between the lowest vibrational level ($v = 0$, ground state) and the next highest ($v = 1$, first vibrational excited state) is about 2900 cm^{-1}. The arrangement of rotational and vibrational levels is shown schematically in Figure 10.26.3. Because we wish to examine transitions between different vibrational-rotational levels, the spectrometer must be able to resolve the rotational "fine structure" in the vibrational spectrum. The absorption spectrum involving the lowest-energy vibrational transition, $v = 0$ to $v = 1$, called the *fundamental,* is observed in the infrared near 2900 cm^{-1}, or 3400 nm. [The wavelength of a transition is the reciprocal of its energy in cm^{-1}; the nm (10^{-9} m) is a common wavelength unit.] Many ordinary infrared spectrophotometers do not possess the required resolution for this experiment.

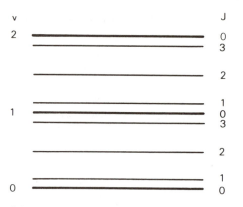

FIGURE 10.26.3   Schematic representation of rotational and vibrational (rovibronic) eigenstates and eigenvalues denoted by *J* and *v*, respectively.

A satisfactory alternative is to examine not the fundamental transition but the next higher one, $v = 0$ (ground state) to $v = 2$ (second excited vibrational state), called the *first overtone* transition. This nomenclature is in analogy with sound waves, or a vibrating string, in which the first overtone occurs at half the wavelength (twice the frequency) of the fundamental, or longest, wavelength vibration. The energy range in which this transition occurs lies just below the visible region of the spectrum from about 12,000 cm^{-1} (800 nm) to 4000 cm^{-1} (2500 nm); this is called the near infrared region. Many spectrophotometers operating in this range possess the needed resolution to perform this experiment satisfactorily (1 to 2 cm^{-1}).

One complication that arises in examining the overtone transition is that it is much weaker than the fundamental. It has a much smaller extinction coefficient; in fact, the overtone transition is formally forbidden and is made weakly allowed because of anharmonicity. For HCl, the absorption intensity of the overtone transition is less than 2 percent that of the fundamental. For this reason, a high concentration (and long optical path length) of HCl vapor is needed to make the experiment practical. In this case, HCl at 1 atm pressure in a 10-cm pathlength cell is used.

The discussion that follows treats both the overtone and fundamental transitions. Equations that pertain to the latter will be identified by the letter a.

The fundamental relation in spectroscopic transitions is derived by equating the photon energy with the difference in state energies involved in the transition. For the case where radiation of frequency *v* is absorbed, the energy of the transition is

$$E = hv = hc\tilde{v} = E_{v',J'} - E_{v'',J''},$$

where $\tilde{v}$ is the photon frequency in cm^{-1} and v',J' and v'',J'' are, respectively, the vibrational/rotational quantum numbers of the upper and lower energy states involved in the transition. Expressed in wavenumber units, the above equation is

$$\tilde{v} = T_{v',J'} - T_{v'',J''} = T_{2,J'} - T_{0,J''} \quad \text{(overtone)}, \quad (10.26.16)$$

$$\tilde{v} = T_{1,J'} - T_{0,J''} \quad \text{(fundamental)}, \quad (10.26.16a)$$

in which the overtone and fundamental transitions are made explicit ($v' = 2$; $v'' = 0$) and ($v' = 1$; $v'' = 0$).

Equations (10.26.16) and (10.26.16a) imply that the frequencies of the overtone and fundamental transitions are not precisely equal to the gap, $v' - v''$, but are instead altered by the simultaneous changes in rotational energy. Logically, we can anticipate three general possibilities: an increase, decrease, or no change in rotational quantum number. In proceeding further, we introduce an important (and simplifying) constraint in the way in which J changes during photon absorption. This restriction is called a *selection rule,* and for rotational transitions it states that

$$\Delta J = 0 \qquad \text{is forbidden,}$$

$$\Delta J = \pm 1 \qquad \text{is allowed,}$$

all other combinations of J values are forbidden. For vibrational transitions, the selection rule requires that $\Delta v = \pm 1$; this rule is rigorous only for the *harmonic* oscillator. Anharmonicity causes the vibrational selection rule to break down and thus the $v' = 2 \leftarrow v'' = 0$ transition is weakly allowed.

Returning to the rotational-vibrational spectrum of HCl, we can group the transitions into two classes. In the first, there is a *decrease* in rotational quantum number ($\Delta J = -1$, and $J' = J'' - 1$), and in the second case, there is an *increase* in rotational energy ($\Delta J = +1$, and $J' = J'' + 1$). The allowed transitions are depicted schematically in Figure 10.26.4. The $\Delta J = 0$ transitions are shown by dotted lines. Each vertical line in Figure 10.26.4 is expected to give rise to an observable rotational band in the high-resolution vibrational absorption spectrum of the molecule. The lines for which $\Delta J = -1$ are all positioned at lower energy, and these are called the *P branch.* The lines for which $\Delta J = +1$ are situated at higher energy and are referred to as the *R branch.* If $\Delta J = 0$ transitions were allowed (they are under some circumstances and are called the Q branch), they would ''pile up'' in a narrow grouping of lines between the *P* and *R* branches. The *Q* branch would appear as a single line only if the rotational constants for the upper and lower vibrational states were identical.

The rotationally resolved overtone and fundamental transitions can be expressed as a function of rotational quantum numbers, J'' (ground state, $v = 0$) and

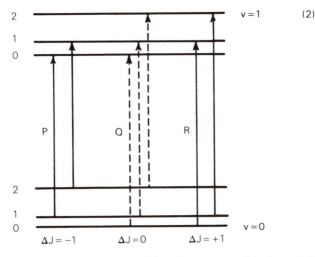

FIGURE 10.26.4 Rotational-vibrational transitions between $v = 0$ and $v = 1$ (fundamental). Dipole allowed transitions are for $\Delta J = +1$ (R) and $\Delta J = -1$ (P). $\Delta J = 0$ forbidden transitions (Q) are indicated by dashed lines.

J' (upper state, $v = 2$ or $v = 1$). This is done by combining equations (10.26.16), (10.26.16a), and (10.26.15). After grouping terms containing B_e and α_e, we have

$$T_{2,J'} - T_{0,J''} = 2\tilde{\nu}_e - 6\tilde{\nu}_e\chi_e + B_e[J'(J' + 1) - J''(J'' + 1)]$$

$$- \left(\frac{\alpha_e}{2}\right)[5J'(J' + 1) - J''(J'' + 1)], \tag{10.26.17}$$

$$T_{1,J'} - T_{0,J''} = \tilde{\nu}_e - 2\tilde{\nu}_e\chi_e + B_e[J'(J' + 1) - J''(J'' + 1)]$$

$$- \left(\frac{\alpha_e}{2}\right)[3J'(J' + 1) - J''(J'' + 1)]. \tag{10.26.17a}$$

These relations are simplified under the constraint of the selection rule for rotational transitions $\Delta J = \pm 1$; therefore, we can write the transition energy in terms of a common rotational quantum number which we choose to be J''. For the P branch, $J' = J'' - 1$ where $J'' = 1, 2, 3, \ldots$ and equations (10.26.17) and (10.26.17a) become

$$\tilde{\nu}(J'') = 2\tilde{\nu}_e - 6\tilde{\nu}_e\chi_e - (2B_e - 3\alpha_e)J'' - 2\alpha_e(J'')^2, \tag{10.26.18}$$

$$\tilde{\nu}(J'') = \tilde{\nu}_e - 2\tilde{\nu}_e\chi_e - (2B_e - 2\alpha_e)J'' - \alpha_e(J'')^2, \tag{10.26.18a}$$

these results show that the energies of the P-branch lines decrease nonlinearly with increasing J''.

For the R branch, $J' = J'' + 1$, where $J'' = 0, 1, 2, 3, \ldots$ and equations (10.26.17) and (10.26.17a) read

$$\tilde{\nu}(J'') = 2\tilde{\nu}_e - 6\tilde{\nu}_e\chi_e + 2B_e - 5\alpha_e + (2B_e - 7\alpha_e)J'' - 2\alpha_e(J'')^2, \tag{10.26.19}$$

$$\tilde{\nu}(J'') = \tilde{\nu}_e - 2\tilde{\nu}_e\chi_e + 2B_e - 3\alpha_e + (2B_e - 4\alpha_e)J'' - \alpha_e(J'')^2. \tag{10.26.19a}$$

Note that the transition energies of the R-branch members increase (also nonlinearly) with increasing J''. Above a certain J'' value the R lines will begin to *decrease* in energy because of the negative coefficient of the $(J'')^2$ term. For HCl, this does not occur until $J'' \geq 27$.

If the Q branch $(J' = J'')$ were observable in this experiment, its components would appear at

$$\tilde{\nu}(J'') = 2\tilde{\nu}_e - 6\tilde{\nu}_e\chi_e - 2\alpha_e[(J'')^2 + J''] \qquad \text{for the overtone,}$$

and

$$\tilde{\nu}(J'') = \tilde{\nu}_e - 2\tilde{\nu}_e\chi_e - \alpha_e[(J'')^2 + J''] \qquad \text{for the fundamental,}$$

and would be displaced to lower energies. In the absence of rotational vibrational coupling ($\alpha_e = 0$), the Q branch would appear as a single line at an energy equal to the gap in the vibrational levels, $v = 0$, $v = 2$ (or $v = 0$, $v = 1$). For convenience, this gap is defined as

$$\tilde{\nu}_0 \equiv 2\tilde{\nu}_e - 6\tilde{\nu}_e\chi_e, \tag{10.26.20}$$

$$\tilde{\nu}_0 \equiv \tilde{\nu}_e - 2\tilde{\nu}_e\chi_e. \tag{10.26.20a}$$

Determination of the Molecular Constants

We see from equations (10.26.18), (10.26.18a), (10.26.19), and (10.26.19a) that the P- and R-branch lines (abbreviated $P1$, $P2$, $P3$, . . . and $R0$, $R1$, $R2$, . . . , respectively) are quadratic in J''. Thus if the observed lines were to be fit to these equations, three constants could be obtained. These correspond to the coefficients a, b, and c of the equation

$$y = a + bx + cx^2.$$

The three constants obtainable from the $\bar{\nu}(J'')$ data are $\bar{\nu}_0$, B_e, and α_e; thus, the anharmonicity constant $\bar{\nu}_e \chi_e$ cannot be directly determined in this experiment.

The P- and R-branch data could be directly fit to the second-order equations (10.26.18), (10.26.18a), (10.26.19), and (10.26.19a), or alternatively (perhaps more simply), these equations could be linearized. One approach is to examine the *successive differences* in energies of the lines in both branches. For example, for the P branch, we develop the expression for $P1 - P2$; $P2 - P3$; $P3 - P4$; . . . and have, in general,

$$P(J'') - P(J'' + 1) = \Delta\bar{\nu}_P = 2B_e - \alpha_e + 4_e J'' \quad J'' = 1,2,3, ..., \quad (10.26.21)$$

$$\Delta\bar{\nu}_P = 2B_e - \alpha_e + 2\alpha_e J'' \quad J'' = 1,2,3, ... \quad (10.26.21a)$$

Notice that $\bar{\nu}(P1) > \bar{\nu}(P2)$, etc.

Likewise, for the successive differences in the R-branch lines, i.e., $R1 - R0$; $R2 - R1$; etc., the result is

$$R(J'' + 1) - R(J'') = \Delta\bar{\nu}_R = 2B_e - 9\alpha_e - 4\alpha_e J'' \quad J'' = 0,1,2, ..., \quad (10.26.22)$$

$$\Delta\bar{\nu}_R = 2B_e - 5\alpha_e - 2\alpha_e J'' \quad J'' = 0,1,2, ... \quad (10.26.22a)$$

Inspection of equations (10.26.21), (10.26.21a), (10.26.22) and (10.26.22a) reveals that the P-branch lines become more widely spaced as one goes out farther into the tail of the spectrum; likewise, the R-branch lines should become more closely spaced. Using these equations quantitatively, values of B_e and α_e can be obtained from plots of $\Delta\bar{\nu}_P$ and $\Delta\bar{\nu}_R$ vs. J''.

Equations (10.26.21), (10.26.21a), (10.26.22) and (10.26.22a) are redundant because each provides the same information; however, the examination of the P- and R-branch data separately indicates how well each set fits the vibrationally coupled rigid rotator model.

To obtain a value of $\bar{\nu}_0$, one plots the respective term values of the P branch vs. $(2B_e - 3\alpha_e)J'' + 2\alpha_e(J'')^2$ [or $(2B_e - 2\alpha_e)J'' + \alpha_e(J'')^2$], once B_e and α_e are known [see equations (10.26.18) and (10.26.18a)]. The intercept is equal to $\bar{\nu}_0$. A similar approach could be used with the R-branch data and equations (10.26.19) and (10.26.19a).

Transition Intensities

We now examine the intensity dependence of the P and R branches. The integrated absorption strength, $I_{0,J'';2,J'}$, of a particular rotational-vibrational transition can be written as

$$I_{J'',J'} \propto \bar{\nu} N_{J''} |M_{J'',J'}|^2, \quad (10.26.23)$$

where $\bar{\nu}$ is the frequency (photon energy) of the transition; $N_{J''}$ is the (relative) population of molecules in the J''th rotational eigenstate; and $M_{J'',J'}$ is called the transition moment integral. In equation (10.26.23), only the initial and final rotational quantum numbers, J'' and J', are indicated because we are interested in the intensity of rotational fine structure *within* a given vibrational transition, in this case the overtone. Throughout this transition $\bar{\nu}$ varies very slightly (e.g., from 6000 to 6500 cm^{-1} for the overtone transition). One should notice that, all things being equal, the transition intensity depends directly on the frequency of the transition. Thus as a rule of thumb, pure rotational transitional transitions are intrinsically much weaker than vibrational transitions, which in turn are weaker than electronic (visible and ultraviolet) transitions. The population factor in equation (10.26.23) indicates the fraction of molecules that exists in the absorbing rotational energy level denoted by J'' at thermal equilibrium. Specifically,

$$N_{J''} \propto g_{J''}\exp\left(\frac{-E_{J''}}{kT}\right), \tag{10.26.24}$$

where $g_{J''}$, called the *degeneracy factor*, is equal to $(2J'' + 1)$; $E_{J''}$ is the energy of the J''th rotational quantum level, $E_{J''} = J''(J'' + 1)hcB_e$; and k is the Boltzmann constant. The exponential term in equation (10.26.24) is called the Boltzmann factor.

The transition moment integral in equation (10.26.23) reflects the degree to which the electromagnetic radiation field interacts with the initial and final rotational quantum states, J'' and J'. There is a slight but important J dependence to the square of this integral:

$$|M_{J'',J'}|^2 \propto \frac{J'' + J' + 1}{2J'' + 1}. \tag{10.26.25}$$

The above information can be used to predict the shapes of the P- and R-branch structure in the rotational-vibrational spectrum of HCl. Combining equations (10.26.24), (10.26.25), and (10.26.23), we have

$$I_{J''} \propto \bar{\nu}J''\exp\left[\frac{-J''(J'' + 1)hcB_e}{kT}\right] \quad \textit{for the P branch,} \tag{10.26.26}$$

and

$$I_{J''} \propto \bar{\nu}(J'' + 1)\exp\left[\frac{-J''(J'' + 1)hcB_e}{kT}\right] \quad \textit{for the R branch.} \tag{10.26.27}$$

Note that in equation (10.26.26) $J'' = 1,2,3, \ldots$ while in (10.26.27) $J'' = 0,1,2,3$. These equations predict that the P and R branches have somewhat different line profiles.

Safety Precautions

☐ Safety glasses must be worn during this experiment.
☐ Fill the cell in a fume hood. If you are using a pressurized HCl cylinder and a gas-handling vacuum rack, make sure you understand the filling procedure.
☐ If the HCl supply develops a leak, or the vacuum system breaks or malfunctions, notify your instructor immediately.
☐ Do not use or handle liquid nitrogen until your instructor has explained its hazards and proper use.
☐ If the cell filled with HCl breaks outside of the fume hood, notify your instructor immediately.

Procedure

1. Your instructor will describe which procedure you will follow to fill the absorption cell. In addition, the type of spectrum to be obtained will be made known to you beforehand (i.e., fundamental or first overtone). This will determine the nature of the cell (Infrasil or NaCl windows) and the type of spectrophotometer (near infrared or infrared).

If you are to use the vacuum rack, the instructor will review its design and use. Make sure there is sufficient liquid nitrogen in the trap (dry ice should *not* be used). A 10-cm absorption cell will be filled with about 1 atm of HCl vapor. Lightly grease the cell ball joint and attach to the rack. Evacuate the cell and pump for several minutes to remove adsorbed water; *if necessary,* flame-dry the cell (use caution!). Evacuate the gas line connecting the HCl cylinder to the manifold. After isolating the manifold, fill the system with HCl vapor until the pressure gauge reads about 1 atm. Carefully, but firmly, close the cell stopcock. Close the HCl cylinder valve(s). Pump out the manifold, gas filling line, and the station to which the cell is attached. Shut off the gas inlet and cell station stopcocks and remove the gas inlet tubing and the absorption cell; wipe off the stopcock grease from the joints. Let the rack pump down for several minutes before shutting it down.

2. Obtain an absorption spectrum of HCl that reveals the isotope separation and shows *at least* six members in both the P and R branches. If possible record the spectrum so that it is linear in cm^{-1}; this simplifies data reduction.

A Cary 14 spectrophotometer, for example, operating in the near infrared is adequate for obtaining the overtone spectrum (the resolution of the spectrophotometer should be less than ca. 2 to 3 cm^{-1}). [The HCl^{35} and HCl^{37} features have an intrinsic line width of < 0.1 cm^{-1}. Hence the observed broadness is a direct manifestation of the relatively large bandpass (low resolution) of the spectrophotometer.] Scan from about 1680 to about 1530 nm. An expanded wavelength scale (ca. 1.5 nm/in.) is convenient for resolving the structure observed. If possible, expand the pen sensitivity to enhance the features at the extremes of the P and R branches.

Data Analysis

1. You will notice that each rotational "line" shows two components. This results from the presence of two isotopes of Cl; thus, the sample is a mixture of HCl^{35} (the major component) and HCl^{37}, presumably representing the natural abundance of these isotopes.

2. Tabulate the P- and R-branch components corresponding to HCl^{35}. Assign these features as $P1$, $P2$, $P3$, . . . and $R0$, $R1$, $R2$, ... If the spectrum is linear in wavelength, convert each wavelength position to wavenumbers, cm^{-1}.

3. From these data, determine and tabulate the *consecutive* differences in the P- and R-branch "lines"; see equations (10.26.21) and (10.26.22). From this information, determine B_e and α_e by plotting $\Delta \bar{\nu}$ vs. J''. Once B_e and α_e are obtained, determine $\bar{\nu}_0$, and I, the moment of inertia of HCl^{35}; see equations (10.26.18) to (10.26.20). From the reduced mass, obtain a value of r_e.

4. If the two isotopic HCl lines are reasonably well resolved, determine (from the mean line heights for several features) the relative abundance of the two Cl isotopes. Estimate the atomic weight of Cl assuming that Cl^{35} and Cl^{37} have masses of 34.968 and 36.956 amu, respectively. Compare with the reported atomic weight of Cl.

5. Using the values of B_e and α_e, calculate D, the centrifugal distortion constant [equation (10.26.10)]. What must the value of J be in order that the centrifugal distortion term will be 1 percent of that of the rigid rotator term in equation (10.26.9)? Does this result justify ignoring D in your experimental results?

6. (Optional) Compare the intensities of the P- and R-branch lines of your spectrum with those predicted from equations (10.26.26) and (10.26.27). To do this, calculate *relative* intensities for the P- and R-branch lines using the rigid rotator model [equations (10.26.26) and (10.26.27)]. The *integrated* line intensities should be used in this calculation; as an alternative, however, the *peak* intensities can be taken. Scale these intensities to your data using the observed band

heights of the R0 and P1 features and compare the predicted "sticks" with the observed bands. In other words, divide the peak heights of all the tabulated R-branch bands by the height of R0 and do likewise the P-branch bands vis-à-vis P1. Comment on the quality of this comparison.

Questions and Further Thoughts

1. Why *must* the cryogenic medium used with the trap on the vacuum line be liquid nitrogen rather than dry ice?
2. The force constant for HCl^{35} is 5.1669×10^5 dynes/cm (ergs/cm^2). What are the expected values of $\bar{\nu}_e$ and the anharmonicity constant?
3. Use the above result with your value of B_e in equation (10.26.13) to calculate the value of α_e. Compare with the value determined in this experiment.
4. What additional data would be required to determine the anharmonicity constant? Outline the equations and calculations required.
5. A high power infrared laser, a very narrow band light source that can deliver substantial energy in a very small time, can be used to achieve separation of the Cl^{35} and Cl^{37} isotopes from a sample of gas phase, natural abundance HCl. At what (single) frequency should such a laser be "tuned?" How do you think this process works? How would the chlorine isotopes be collected?

Further Readings

R. A. Alberty, "Physical Chemistry," 7th ed., pp. 500–509, Wiley (New York), 1987.

P. W. Atkins, "Physical Chemistry," 3rd ed., pp. 441–454, W. H. Freeman (New York), 1986.

P. W. Atkins, "Molecular Quantum Mechanics" 2nd ed., pp. 298–303, Oxford (New York), 1983.

G. W. Castellan, "Physical Chemistry," 3rd ed., pp. 625–632, Addison-Wesley (Reading, Mass.), 1983.

G. Herzberg, "Molecular Spectra and Molecular Structure I. Spectra of Diatomic Molecules," 2nd ed., chap. III, pp. 66–82, 90–97, 103–115, 121–130, Van Nostrand Reinhold (New York), 1950.

M. Karplus and R. N. Porter, "Atoms and Molecules," pp. 458–484, W. A. Benjamin (New York), 1970.

I. N. Levine, "Physical Chemistry," 2nd ed., pp. 683–694, McGraw-Hill (New York), 1983.

J. H. Noggle, "Physical Chemistry," pp. 707–724, Little-Brown (Boston), 1985.

C. H. Townes and A. L. Schawlow, "Microwave Spectroscopy," pp. 18–24, Dover (New York), 1975.

Rotational-Vibrational Absorption Spectrum of HCl

Name _____   Date _____

Notes

Rotational-Vibrational Absorption Spectrum of HCl

Name _____   Date _____

Notes

INDEX

PROSTAT

Charles Ward and James Reeves
University of North Carolina at Wilmington

Prostat is a set of statistical tools developed to meet many computational needs of students in physical chemistry. **Prostat** provides linear regression, numerical analysis, curve fitting, and a number of other data manipulation and analysis routines for scientific applications.

With **Prostat** you can carry out the statistical analyses needed for the lab experiments in this course:

• Menu driven. Eliminates the need to constantly refer to Reference Manual.
• Extensive use of graphics. Allows you to experiment and quickly see the results of changes in parameters.
• Shows P-values for all statistics. No need to refer to standard/statistical tables for this information.
• Accepts data from the keyboard, common "spreadsheet" programs, or directly from laboratory instruments.
• All programs on a single diskette. You avoid the troublesome disk swapping that is common to many statistics programs.

Through special arrangements with the authors and publishers of this laboratory textbook, you can purchase **Prostat** - not a student version, but the full program diskette and user manual – for **$19.95.**

Send for your copy today and start using **Prostat** in all your courses that require statistical analysis of data. Your satisfaction is guaranteed; you'll receive a full refund or credit if **Prostat** does not meet your needs.

CONTENTS
•Descriptive Statistics
•Linear Regression and Correlation
•Curvefitting and Nonlinear Regression
•t-Tests
•Multiple Linear Regression
•Analysis of Variance
•Numerical Analysis
•Prostat Utilities

HARDWARE REQUIRED: IBM PC, XT, AT or compatible with 256K, Color Graphics Adapter and Graphics monitor.

(See reverse side for order coupon)

To Order Your Copy: Complete this coupon and return it today with your payment or credit card information.

Please send me **Prostat** at the special student price of only **$19.95**.

[ ] My payment is enclosed. (Make checks payable to COMPress.)

[ ] Charge my VISA ____, MasterCard ____.

Account # _____ Exp Date _____

Signature _____
(signature required)

Ship Prostat to:

Name:
Address:
City/State/ZIP:

School:
Name of course:
Professor:

Mail this coupon to: COMPress, P.O. Box 102, Wentworth, NH 03282. Or call (603) 764-5831.